Springer Water

Series Editor

Andrey Kostianoy, Russian Academy of Sciences, P. P. Shirshov Institute of Oceanology, Moscow, Russia

The book series Springer Water comprises a broad portfolio of multi- and interdisciplinary scientific books, aiming at researchers, students, and everyone interested in water-related science. The series includes peer-reviewed monographs, edited volumes, textbooks, and conference proceedings. Its volumes combine all kinds of water-related research areas, such as: the movement, distribution and quality of freshwater; water resources; the quality and pollution of water and its influence on health; the water industry including drinking water, wastewater, and desalination services and technologies; water history; as well as water management and the governmental, political, developmental, and ethical aspects of water.

More information about this series at http://www.springer.com/series/13419

Omid Bozorg-Haddad
Editor

Essential Tools for Water Resources Analysis, Planning, and Management

Springer

Editor
Omid Bozorg-Haddad
Department of Irrigation and Reclamation
Engineering, Faculty of Agriculture
Engineering and Technology
College of Agriculture
and Natural Resources
University of Tehran
Karaj, Iran

ISSN 2364-6934 ISSN 2364-8198 (electronic)
Springer Water
ISBN 978-981-33-4294-1 ISBN 978-981-33-4295-8 (eBook)
https://doi.org/10.1007/978-981-33-4295-8

This Springer imprint is published by the registered company Springer Nature Singapore Pte Ltd.
The registered company address is: 152 Beach Road, #21-01/04 Gateway East, Singapore 189721, Singapore

Contents

Chapter 1
Modeling and Simulation in Water Resources Management

Masoumeh Zeinalie, Omid Bozorg-Haddad, and Barkha Chaplot

Abstract One of the most crucial and influential factors in the formation and development of a civilization is the available water resources. The most common and practical way to manage water resources is to model and simulate water systems. In general, modeling is the mapping of a natural phenomenon in the form of physical components or mathematical relationships and involves both physical modeling and mathematical modeling, which are used in mathematical modeling in water resources management issues. Simulation is one of the methods for solving mathematical programming models in situations where the use of algebraic analysis methods is not possible or cannot be tested in the real world. Simulation-based on trial and error examines the effect of different conditions on the system and evaluates the results. With the above-mentioned explanations, through modeling and simulation, in the least time and cost, researcher investigate different options for achieving the goal. This chapter includes details of modeling and simulation in water resources management.

Keywords Mathematical modeling · Simulation · Water resources · Trial and error · Experiment · Natural phenomenon

---

The original version of this chapter was revised: For detailed information, please see Correction. The correction to this chapter is available at https://doi.org/10.1007/978-981-33-4295-8_14

M. Zeinalie
Department of Irrigation & Reclamation Engineering, Faculty of Agricultural Engineering & Technology, College of Agriculture & Natural Resources, University of Tehran, Karaj, Iran
e-mail: masoumeh.zeinalie@ut.ac.ir

O. Bozorg-Haddad (✉)
Distinguished Professor, Department of Irrigation & Reclamation Engineering, Faculty of Agricultural Engineering & Technology, College of Agriculture & Natural Resources, University of Tehran, Karaj, Tehran, Iran
e-mail: OBHaddad@ut.ac.ir

B. Chaplot
Assistant Professor, Department of Geography, M.J.K. College Bettiah, Babasaheb Bhimrao Ambedkar Bihar University, Muzaffarpur, Bihar, India
e-mail: barkhachaplotjain@gmail.com

O. Bozorg-Haddad (ed.), *Essential Tools for Water Resources Analysis, Planning, and Management*, Springer Water, https://doi.org/10.1007/978-981-33-4295-8_1

1.1 Introduction

Nowadays, modeling, simulation and optimization are widely used in water resources management. To this end, numerous articles have been published on water resource modeling and simulation. Belaineh et al. (1999) have investigated the simulation and optimization of a hypothetical zone model with more focus on water use management and reservoir management under different scenarios. Their results indicated that the more details used in the model, the better the shared water management would be so that the model with the most detail was 13% more water. Khoshfatrat et al. (2006) model the optimal use of multi-barrier systems, simulation method with the limited operation and trial and error to determine the optimal water transfer pattern from the upstream dam to the downstream dam in a three-dimensional system (including two consecutive two-way systems) used monthly intervals. In this study, a simulation model of water resources analysis has been used and its unique features for modeling optimal water transfer in two-dimensional systems have been shown. The dams studied include Zayandehrood, Koohrang (1) and Sardar dams, which both located in Iran. Khayami et al. (2007) in their study, using Simulation Reservoir Dynamic Model and available statistics, evaluated water quality conditions of the Torogh dam reservoir located in Iran, in terms of temporal changes in temperature, salinity and dissolved oxygen parameters. The results showed that in high water years such as 1998, when the inflow discharge exceeds its long-term average discharge, due to the high water level (more than 50 m), the thermal stratification is complete from mid-spring. It occurs within the reservoir until late summer, which results in changes in the chemical, biological and physical properties of the water at different levels, but in years such as 2002 due to the low volume of inlet water and the relatively high average annual temperature, the water level in the reservoir decreases by about 16–20 m. In these conditions, the thermal layering either does not form or, if formed, starts earlier in time and shortens the installation period. Hooshmandzadeh (2007), after presenting the principles and foundations of simulation of water resources systems, has briefly reviewed the simulation of the water resources system of the Sante reservoir dam in Kurdistan province of Iran. Hitchcock and Collins (2007) reviewed several criteria for water resource management and planning analysis. Samadyar and Samadyar (2008) examined the process of modeling and simulation in water resources and what models are eligible for simulation. They have defined the steps of modeling and simulation in such a way that first the preparation of the model is appropriate to the goal, then the initial simulation and completion of information, model calibration, model approval, uncertainty analysis, sensitivity analysis and finally providing management solutions. Azari et al. (2009) simulated and warned of floods by combining hydrological models in GIS and estimating rainfall by remote sensing in Golestan province in Iran, especially the Mother river basin, due to its flooding. Since the precipitation on August 11, 2005 has led to floods in this area, images of NOAA satellite measuring AVHRR have been selected for this study. In this study, first, all the required layers in GIS were prepared and made based on the factors influencing the flood, then the local database of the required parameters including river route, cross-sections, coasts, for runoff flow and MODClark network precipitation model was formed and it has entered the hydrological model. NOAA/AVHRR satellite data, object-oriented classification and cloud index method has been used to detect and classify clouds and estimate rainfall, respectively. Thus, the amount of precipitation was estimated by each of the classified clouds. Also, through soil maps, land use and other inputs of the hydrological model, the amount of runoff due to rainfall is estimated and finally, due to the topographic situation of the region, the depth and speed of water flow from this runoff from the hydraulic model GIS and re-sent to the environment and the flood map was obtained. In this study, by combining remote sensing data in the form of satellite images and precipitation-runoff model and hydraulic model the extent of flood spread in the region was determined. The results of this study revealed that with the help of object-oriented cloud classification, acceptable results can be achieved, and the overall accuracy of the classification is estimated to be about 0.905 and the cap coefficient is estimated to be about 0.887. According to the flood hydrograph obtained between the upstream and downstream of the basin, and assuming the onset of calculations from the onset of precipitation in the study area, the flood zone, and its occurrence was predicted to be between 20 and 33 h earlier than the flood peak. Meyer and Monzehrnands (2009) reviewed the optimization of accumulated models of water resources. Abdo Kolahchi et al. (2010) simulated the performance of the underground dam using metal sheets with a thickness of 0.5 m in the Hosseinabad Strait area of Isfahan in Iran and used PLAXIS software for modeling. Salavi Tabar et al. (2011) simulated the combined surface and groundwater resources of the Haraz River catchment area. For this purpose, after evaluating and estimating the surface water resources in the basin and accurately determining the amounts of surface and underground uses in different parts of the Haraz River catchment area in Iran, to model the surface and underground water resources separately by mountain and plain with interest. The system dynamics method is taken with Vensim software. Finally, the performance of the Haraz river catchment area has been combined and analyzed according to the water cycle in surface and groundwater resources and return water from consumption, losses in the basin level and the amount of aquifer storage has been calculated as the output of the model. Finally, it can be said that the results of this model can provide a solution for evaluating and examining the components in the water cycle to conduct more strategic management to use water resources in the basin. Komasi et al. (2015) have modeled the pheromenon of dam failure. In this regard, the output hydrograph was calculated from the dam failure and then, due to the downstream morphology, flooding was carried out. This study evaluates the comparison of the results of Mike11 software with the results of the analytical solution for flood hydrograph due to the failure of Dez Dam in Iran and the results of experimental equations. The results of this study indicate that the effect of flood peak due to the failure of Dez Dam is only 30 km below it and in the area of Dezolia Dam in Dezful city in Iran and is not essential for other areas of the river. The results also showed that the Dez Dam, if broken, floods about 60,000 cubic meters per second to the city of Dezful and 11,000 cubic meters to the city of Ahvaz. Moradi et al. (2015) simulated the resources and expenditures of the Tajan Irrigation and Drainage Network of Mazandaran Province, Iran. They used the dynamic model of the Vensim system and software for this operation. According to their results, on average, over 214 million cubic meters per year at the diversion dam or 286 million cubic meters per year on the coastline, the excess water of the Tajan River basin system, located in North of Iran, can be transferred to the adjacent river basin.

Zeinali et al. (2020a, b), in a study of the interaction of surface and groundwater through their systematic simulation and creating a dynamic atmospheric connection between surface water and groundwater resources in the Loor-Andimeshk plain, located in southwestern Iran, by relevant mathematical models. They have reviewed it. Based on this, the hydrological soil moisture method and MODFLOW model were used to simulate unsaturated and unsaturated areas, respectively and the study revealed the interaction between surface and groundwater in any spatial and temporal period, in the form of a coupled model. Radmanesh et al. (2020) used the Madflu conceptual model and intelligent simulator models to model the hydrograph representing the aquifer. Finally, the results showed that the MODFLOW and GEP models, among the similar intelligent simulator models, performed almost the same in aquifer hydrographic modeling.

Zeinali et al. (2020a, b) examined the link between genetic algorithm II and a coupled surface and groundwater model in southwestern Iran. The advantage of this structure is the achievement and maintenance of the equilibrium balance between surface and groundwater, taking into account various limitations. Therefore, the structure of the proposed model can provide decision-makers to simulate the interaction of outputs between surface and groundwater and to calculate the simultaneous decrease in river flow and groundwater, especially in dry years.

1.2 Definitions and Terms

In the last decades, the use of various models, especially mathematical models, has become widespread in various sciences. The model is an abstract representation of the components and relationships of a phenomenon that exhibits the relationships between the various entities and/variables of that phenomenon. Since it is not possible to experience all the facts and phenomena practically, models are used to depict events, facts, or situations. Managers, for example, can measure the impact of different advertising tools (newspapers, television, and billboards) on sales using statistical models.

In general, modeling is the art of judging how to summarize the components of the real world that are important in decision-making and can be described in a few ways. Therefore, modeling requires judging the expression of these components and the relationships between them in the mathematical language (Bozorg-Haddad and Seifollahi Aghmioni 2013). Scientific modeling is a scientific activity to make it easier for a part of the phenomenon or the world in general to understand, define, see, determine quality or simulate by referring them to available and accepted knowledge. This requires identifying and selecting different aspects of the phenomenon in the real world to build different models for different purposes, such as conceptual models to understand, practical models to perform operations, mathematical models to determine quality and graphical models for illustrating the subject (Cartwright 1983; Hacking 1983).

Simulation is the implementation of a model in which the fundamental features of the system appear. Using simulation, it is possible to study the impact of different conditions on the system and analyze the results. The basis of the study of different conditions in system simulation is based on trial and error (Bozorg-Haddad and Seifollahi Agmuni 2013). Simulation is a kind of revitalization of the model and shows how an instance or phenomenon behaves. Simulation is used to test, analyze, and train real-world systems that can be modeled. For example, in a water resource system (a system is a set of related components that converts multiple inputs into multiple outputs), if the system outputs are unknown for its inputs, the system is called simulation. The outputs of the system can be determined and can be determined according to different input conditions (Bozorg-Haddad and Seifollahi Agmuni 2013).

There are two types of simulation: static and dynamic. Static simulation provides system information at a specific and constant time, and dynamic simulation provides system information over time.

Water resources management involves identifying and developing water resource projects to maximize net profit or minimize costs, including useful non-commercial items such as potential ecosystems for destruction and negative social impacts (Mir and Mons Hernandez 2009). In this regard, water resource models are used to make decisions about water supply, ecological restoration, and water management in complex systems (Loux 2008). Simulation and optimization are the two main approaches to the river basin model. In water resource simulation, its behavior is simulated based on the set of rules governing water allocation and infrastructure operations (McKinney et al. 1999).

1.3 Fundamentals and Logic of Modeling and Simulation

Models are often used when it is impossible to construct laboratory conditions and samples that directly measure results. Direct measurement of results under controlled conditions has always been more accurate and reliable than models constructed from results (Tolk 2015).

One of the essential aspects of modeling is cognition. That is, in similar modeling models mentioned above, the purpose of modeling is only to understand the model environment. Another aspect of modeling is an explanation. That is, sometimes, a model is presented to introduce and present the properties of a real entity. Geography is an excellent example of this aspect of modeling.

So it can be said that the purpose of modeling is two things:

(A) Cognition (B) explanation

In general, models can be segmented from comprehensibility, nature and structure.

Table 1.1 The three divided classes of the models

Equations are known Physical knowledge Information under model	Primitive sciences Model information	Model information Displaying inputs and outputs
White	Gray	Black

From an understandable perspective, the model is divided into three classes: gray box, white box and a black box. In white-box models, the process that goes into the model is clear. In the gray box models, what happens in the model is not clear. These models have an interface between white box and black box models. In black-box models, what happens in the model is unclear (Table 1.1).

In terms of nature, models are divided into four categories: physical, deductive, mathematical, and computer and laboratory models. In the physical model, the model is physically similar to the actual model, but on a different scale. For example, the physical model of a dam or river provided by hydraulic laboratories are examples of physical models. Physical models have a special place in the world right now. These types of models are usually expensive. In scaling models, the rational basis is that there are various phenomena in nature that are similar in physics and mathematical equations. For example, the movement of electric current or heat movement is in many cases similar to the physics of water movement.

Accordingly, it can be assumed that the pressure equivalent voltage, the electrical conductivity equivalent to the permeability coefficient, the electric potential gradient equivalent to the hydraulic gradient, and the electric current intensity vector are equivalent to the water velocity vector. In this case, the formula of ohm is used instead of the Darcy formula. Mathematical models are meant to solve the basic equations of the subject, which in water engineering are the equations governing soil and water physics. The steps that go into creating and implementing a mathematical model generally include:

(A) Understanding the physical behavior of the system: At this stage the relationships and interactions of the factors within a system are determined.
(B) Mathematical Equations Definition: At this stage, the physical relationships that have been identified in the previous step are interpreted as mathematical expressions. In this section one should consider simple assumptions and obtain the equations governing the flow. The results of this step generally include one or more differential equations with appropriate initial and boundary conditions.
(C) Solving Mathematical Equations: At this stage, a proper solution to the formula presented in the previous step must be obtained.

There are generally two basic methods for solving mathematical equations at this stage, which are the analytical method and the numerical method. On this basis, two types of mathematical models can be defined: analytical mathematical models and numerical mathematical models.

Many computer programs are commonly used to analyze mathematical models today. Computer models are mathematical models, but they are called computer models because they are analyzed by computers. The use of these types of models is nowadays widespread. This is because of the simplicity and low cost of these models.

In modeling and simulation, the task-oriented, concise, and purposeful model is an abstraction of a real phenomenon and consisting of physical, legal, and identified constraints (Tolk 2015). The model is task-oriented because it comes with a question in mind or a task. It is brief because it eliminates all observed and identified entities and their interactions with each other if they are not relevant to the purpose. It is digest because it collects all the information it needs. Both the brief and the abstract are purposefully done. However, they are all built on a real phenomenon. The phenomenon, if accompanied by physical limitations, is itself a model in itself. In the meantime, there are limits to what we can see using existing methods and tools. There are also cognitive limitations that limit us to what we can explain using existing theories. Such a model incorporates concepts, behaviors and relationships in a formal format and is known as a conceptual model. To run it, the model must be run by simulation software. This requires more than quantitative estimates or the use of discoveries (Oberkampf et al. 2002). Despite all the cognitive and computational constraints, simulation is considered to be the third pillar of scientific methods, because there are three options for examining a real system: conducting experiments, forming theories, and simulating, which is shown in Fig. 1.1.

In the process of modeling water resources, firstly, one must understand the nature of the problem. To this end, the purpose of modeling should be clearly defined for the modeler. Also, prediction of system behavior based on simulation methods, if not using physical mechanisms, can lead to incorrect results. Even when the model can fully represent the original design, incomplete results are obtained if sufficient

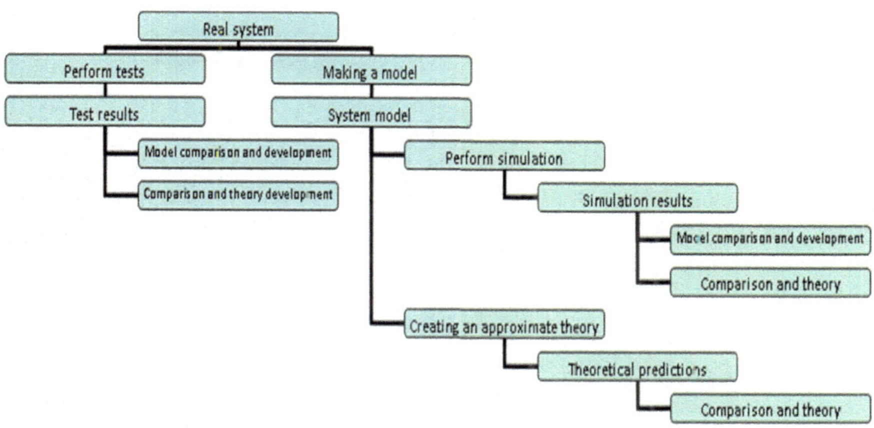

Fig. 1.1 Flowchart showing the relationship among experiment, simulation and theory

data and information are not available for estimation and rate constants (Torabian and Ashami 2002).

In the area of water resources, the issue of groundwater is one of the important topics. In general, any system that can show the response of groundwater storage to the stresses inflicted is called the groundwater model. Overall, groundwater modeling is the simulation of water movement in porous environment. An underground water the model has been simplified form of an underground water system that shows the correlation between hydrodynamic reaction and reaction. Groundwater models are also generally divided into three categories: physical, analog, and mathematical.

1.3.1 Physical Models

These models have attempted to create simplified conditions of nature in the laboratory. The aquifer elements are used to construct these models, namely, the water and the type of soil-forming the aquifer. These models are made on a smaller scale, depending on the type of aquifer, their principles, and their natural physical properties. Examples of such models are models created in the laboratory for checking dam construction, channel overflow, etc. Aquifer modeling is difficult and costly in most cases due to natural conditions and often unknown to them.

1.3.2 Analog Models

Such models are built on the similarity of mathematical equations that express physical phenomena in aquifers and other systems.

1.3.3 Mathematical Models

Mathematical models are either deterministic or probabilistic or a combination of both. The probabilistic model provides the whole range of solutions based on the probability of incidence and is often used to predict flood and rainfall. Certain models are based on the cause and effect relationships of recognized systems and processes and are generally used to solve regional groundwater problems. These models are also classified into numerical and analytical models.

1.3.4 Analytical Models

The purpose of analytical models is to obtain a precise mathematical solution from the description of physical processes. However, the groundwater flow equation is subject to analytical methods and requires simplification of system assumptions involving both initial conditions and boundary conditions. These models are used for pumping tests to estimate aquifer parameters, excavation calculations, inverse solutions to interpret water flow experiments and to evaluate numerical models.

1.3.5 Numerical Models

In numerical methods, the aquifer is subdivided into small elements and an equation is obtained for each element whose overall form is approximately the same for all internal elements. This equation can be approximated by one of the numerical methods. The result of this approximation for each equation is the production of an algebraic equation that yields a total n × n matrix for the aquifer in question. (n is the number of elements).

1.3.6 Definitive Simulation Models

Definitive simulation models of water resources systems do not consider uncertainties in hydrological parameters and variables. As a result, these models present limited planning and management issues. These models are generally used for initial decision making and general comparison between different options, and for more accurate decisions, uncertain models should be used. For the preliminary analysis of designs, before studying more closely the random simulation optimization, definitive models can be useful using selected values of inputs, parameters, and variables. The basis of most models for simulating water resources systems is the law of mass conservation, often called quantitative or qualitative balances. For example, in simulating the volume of water reservoirs in a dam reservoir, based on the law of mass conservation or continuity equation, the volume of the reservoir at the beginning of period t, the inflow rate, the volume of the reservoir at the end of the previous period, the amount of outflow during a similar period. It depends on the buildup, evaporation, infiltration and other available water. Other models of quantitative simulation of water resources systems include the runoff rainfall model which is widely used in studies of flood management and control. Various deterministic simulation models have also been developed to study the temporal and spatial variations of water quality in river systems and reservoirs. Following is one of the definitive simulation methods.

1.3.6.1 Artificial Neural Networks

The Basics of Neural Networks

Artificial neural networks are based on the biological model of the animal brain. These networks are essentially a data-processing system that is derived from the generalization of their mathematical models. Artificial neural networks are systems that are capable of performing functions similar to the human brain. Neurons are divided into three categories of sensory neurons, stimulus neurons, and communication neurons based on the structures between which messages are directed. Artificial neural networks (ANNs) or, more simply, neural networks are new computational systems and machine learning techniques, knowledge representation and, ultimately, applied knowledge to predict the output responses of intricate systems. The idea behind such networks is partly inspired by the way the biological nervous system works for processing data and information, for learning and knowledge creation. The main element of this idea is the creation of new structures for the information processing system. The system consists many of highly interconnected processing elements called neurons that work together to solve a problem. Using computer programming knowledge, one can design a data the structure that works as a neuron, then create a network of these interconnected synthetic neurons, create a learning algorithm for the network, and apply it to the network.

Mathematical Modeling Basics of Artificial Neural Networks

(A) Artificial Neuron Model

A neuron is the smallest information processing unit that forms the basis of the performance of artificial neural networks. Synthetic neurons were designed to mimic the basic characteristics of biological neurons.

(B) Single-layer artificial neural networks

Although a single neuron only shows stimulus functions with a certain simple scheme, the main power of neural computation is due to the connections of neurons in the network. The simplest network is a group of regulated neurons, in one layer.

(C) Perceptron Multilayer Network

The artificial neural network consists of several computational units called neurons. The artificial neural network is a kind of parallel processor system and has the following characteristics: 1. Neural neurons are network processors. 2. Network connections have a particular weight that affects traffic signals. 2. Each neuron calculates its weighted sum of inputs and outputs it after passing the threshold function. The weights of network connections change during the training procedure according to the learning law, and after the realization of learning and the fixed weights act as network memory.

(D) Comparison of monolayer and multilayer neural networks

Multilayer neural networks have more capability than monolayer networks. Two-layer feedforward neural networks are able to approximate any function accurately, while single-layer artificial neural networks do not.

(E) Notes

Artificial neural networks are composed of several processors called neurons with cells and units operating in parallel. Neurons, in themselves, are test functions, but as a whole in the form of a network, they can solve complex issues ranging from estimating work progress rates to estimating the pre-stress of lift cables.

(F) Education

Learning capability means the ability to adjust network parameters with the passage of time as the network environment changes and the network experiences new situation. Most training algorithms are based on supervised training. The convergence of these algorithms is mathematically proven so that in the generalized Delta method, the first derivative of the total error is used to adjust the weights, which reduces the total error by applying this method.

(G) Generalization

The purpose of training is to make acceptable estimates within the optimal range of the problem. Factors affecting neural network generalization capability are a type of network and training algorithm, number and texture of middle neurons, number and distribution of training patterns. The power to extend artificial neural networks is based on interpolation. Despite this potential, extrapolation-based generalization is also important. It can be said that the network learns the function, learns the algorithm, or obtains the appropriate relation for some points in space.

(H) Several middle neurons

In general, there is no straightforward way to determine the most appropriate number for mid-layer neurons, and this is especially complicated when the number of mid-layers is increased. Unless there are very few intermediate neurons in the network, it will not be possible to obtain an accurate model of all forms of the response surface. In attempting to resolve this problem, a range of tissues of different midline neurons is usually considered and the tissue that works best is accepted.

(I) Network Evaluation

The value of the results needs to be determined before applying a neural network. The evaluation usually involves determining the effectiveness of the network on test issues that are not used in network training but are suitable for comparison. The test questions should be chosen so that they do not fit into specific educational patterns.

(J) The shape of data and how it is expressed

The input and output data of a network are usually continuous or discrete. However, sometimes they may be parametric with a combination of all of these.

(K) Network Training Method

The algorithms used can be varied depending on the type of problem discussed, but the form of training can be either by simply adjusting weights or by adjusting the network structure.

Nonlinear (probabilistic) simulation models

Stochastic simulation models combine deterministic and probabilistic models and allow the probabilistic properties of some system variables to be considered. Nonlinear simulation models usually use information such as the distribution of probable variables, the number of times they are repeated, and their mean and range of variation, and generally the simulation results are presented as probabilistic. One of the conventional methods of nonlinear simulation of water resources systems is the Monte-Carlo method. In this method, according to the statistical properties of the random variables of the system, value is randomly generated for each variable and the system is generated based on these values and the values of the definitive input variables are definitively simulated. By repeating this process, the statistical properties of the output variables can be estimated. Uncertainty is always a part of the planning process because many quantities of factors affecting the performance of water resources systems cannot be fully ascertained when designing and building the system. The uncertainty stems from the possible nature of atmospheric processes such as evaporation, precipitation and temperature (Hooshmandzadeh 2007).

1.4 The Importance and Necessity of Modeling and Simulation

As mentioned, modeling and then simulating it is one of the most important scientific methods because having the basic information about a process can use simulators such as temperature, pressure and flow rate. Access. In any existing process, computer-aided simulation enables it to be able to detect its behavior by changing operating conditions and also to find optimal conditions for reducing waste while increasing productivity (Sotudeh Karabagh and Zarif 2005). On the other hand, because we do not want to try and error repeatedly, it is possible to observe simulations at a lower cost and time if a series of events occurs or if a series of variables are introduced into the system. What will happen to the system, or what will happen if we do not make any changes and follow the same procedure? Or even if the system doesn't exist in real life and just guess what might be working on it and want to know if it does.

Optimization in some problems is one of the key steps in problem-solving, and since modeling and simulation are the basic stages of optimization, they are therefore very important in problem analysis.

Simulation is nowadays widely accepted. In 2006, the National Science Foundation (NSF) highlighted the potential of using technology and simulation techniques to revolutionize engineering in a report based on engineering science simulation. One of the reasons given for increasing interest in using simulation is that using simulation is generally cheaper, safer, and more ethical than real-world testing. For example, supercomputers are sometimes used to simulate the explosion of atomic devices and their effects in better support for preparedness in the event of an atomic explosion (Versky). Similar efforts have been made to simulate storms and other natural disasters.

One of the important applications of simulation is in industrial production. The simulation technique is a valuable tool for assessing the impact of massive investments on physical equipment or facilities such as factory machinery, warehouses and distribution and distribution centers used by engineers. Simulation can be used to predict the performance of existing or planned systems to compare alternative solutions to a particular design problem (Benedettini and Tjahjono 2008).

Another important aim of simulating production systems is to determine the quality of system performance. Types that are commonly measured in determining system performance are: (Banks et al. 2005).

1. System performance under medium and maximum load
2. System Time Cycle
3. Use of resources, workers and machines
4. Bottlenecks and blockages
5. Queuing in the workplace
6. Queuing and delays caused by systems and devices involved with the material
7. Staff Needs
8. System scheduling efficiency
9. System control efficiency

In modeling and simulation discussions, a model needs to be built before anything can be simulated. In this regard, modeling has the advantages that have become important. The benefits of modeling include:

1. More appropriate decisions.
2. Increase user insights by linking different problem variables.
3. A tool for better and more effective communication.
4. Save money.
5. Ability to investigate and predict the behavior of the system in the face of a variety of variables or parameters of a real situation.
6. Ability to analyze all possible combinations of potential factors using computer modeling.

The basic benefits of simulation discussed by Schmidt and Taylor (1970) and others are as follows: (Mahlouji 2010)

1. After building each model, it can be repeatedly applied to analyze the designs or policies proposed.
2. Simulation techniques can be used to help analyze any proposed system. However, incoming data is approximate and incomplete.
3. Generally, it is much less expensive to obtain simulation data than to provide real system data.
4. Simulation is usually easier to apply than analytical methods. Therefore, the number of potential users of simulation methods is much higher than the analytical methods.
5. While simulation models usually do not have such limitations. Using analytical models, the analyst can usually calculate only a finite number of system performance metrics, while the data generated from simulation models apply to the estimation of each performance metric.
6. In some cases, simulation is the only way to find a solution.

There have been many applications of simulation in a variety of contexts. Hayley and Lieberman (1980) provide the following examples to illustrate the broad capability of the simulation method:

1. Simulate operations at major airports by airlines to test changes in their policies and practices, such as maintenance and maintenance capabilities, passenger mounting and disembarkation facilities, auxiliary aircraft, and so on.
2. Simulate the transit of traffic junction with traffic lights with a regular schedule, to determine the best time sequences.
3. Simulation of maintenance operations to determine the optimal number of repair departments.
4. Simulate the uncharged flow of particles from the radiation path to determine the radiation intensity passing through the shield.
5. Simulation of steelmaking operations to evaluate changes in operation, capacity, and facility composition.
6. Simulate the economy of the country in terms of predicting the impact of economic policy decisions.
7. Simulating large-scale military battles to evaluate defensive and offensive weapon systems.
8. Simulate large-scale inventory distribution and control systems to refine the design of such systems.
9. Simulate all operations of each business enterprise to evaluate the wide range of changes in its policies and operations as well as provide the opportunity to simulate business operations to train managers.
10. Simulation of the telecommunications system to determine the capacity of the desired components to provide satisfactory service at the most economical level possible.
11. Simulate the performance of a developed river basin to determine the best combination of dams, power plants, and irrigation operations to provide the optimal level of floodwater flow and water resources development.

12. Simulation of production line operations to determine the amount of space needed to store materials under production.

1.5 Workflow and Flowchart

The basic steps of modeling and simulation include 10 steps as follows (Fig. 1.2):

1. **Problem Editing and Goal Setting**: To find the answer to the problem, you need to know what its purpose is.
2. So the first step in a simulation test (like any other test) is to determine the purpose of the test because it is the goal that determines how the test is done, the details needed, and the final results.
3. **System Definition**: At this point it is necessary to determine what methods and techniques can be used to study the system. The definition of a system is, in fact, the determination of the components of the system, the internal and external elements and factors of the system environment, and the parameters and variables of the system. Afterward, the relationships and rules governing the characteristics of the system and its variables are identified or formulated, then the system behavior is examined and the details of the variables change in the system are revealed.

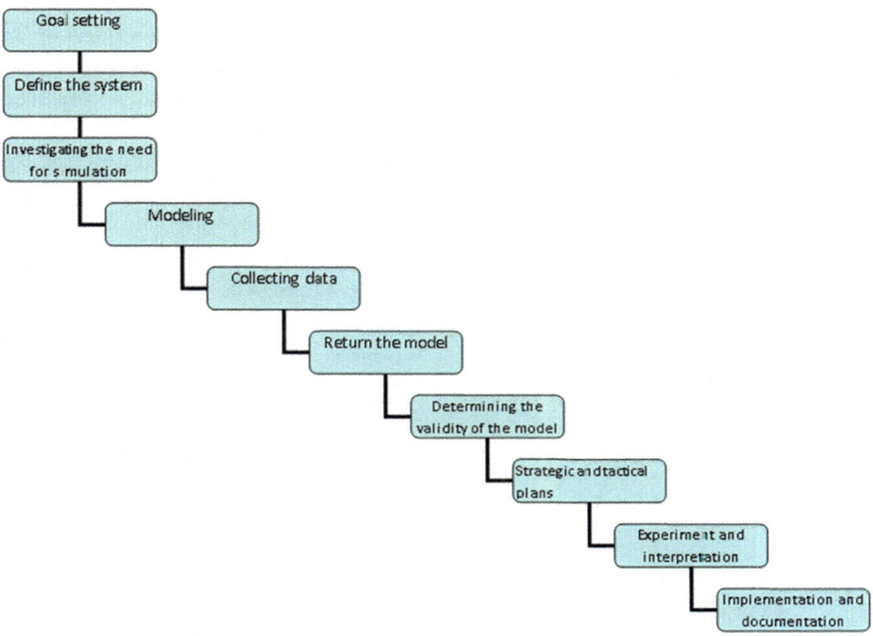

Fig. 1.2 Flowchart Modeling and Simulation Process

4. **The answer to the question**: should the simulation model be used in all decisions? If the actual conditions are not too complex and can be solved using analytical methods, naturally there is no need to use a simulation model, but if given the complex and high-risk conditions, only the simulation can be used, So the application of the simulation method will be.
5. **Modeling**: The art of modeling is the ability to analyze a problem, abstract its properties, select assumptions, and then complete and develop the model until a useful approximation of reality is obtained. The more complete the model, the more complex the situation reflects.
6. **Providing and collecting data**: Each study requires data collection. In a simulation model, the input data must be closely related to the information about the components of the system and the relationship between them. At this time, the analyst must decide what data is needed and how to collect it.
7. **Model Return**: Step 6 is removed by returning the model. At this point we need to describe a model of the computer system. Simulation models are very logically complex and have many interactions between system elements.
8. **Model validation**: This is the most important and most difficult step of the simulation process. Validation, namely, whether the constructed model accurately simulates and describes the behavior of a real system? So what matters is the reliability of the model, not the fact of its structure.
9. **Strategic and Tactical Planning**: Strategic planning means the test plan from which the desired information is obtained, and tactical planning means determining how each of the tests specified in the test plan is performed.
10. **Experimentation and Interpretation**: At this stage, planning errors and deficiencies will be identified, and the implemented steps will be reviewed.
11. **Implementation and Documentation**: The success of a simulation project can only be attributed to a researcher when the model is accepted, understood, and used. Accurate documentation of how the model is designed, developed, and operated can extend the useful life and chances of successful implementation.

1.6 Examples of Practical Simulating Methods

There are several ways to model and simulate the phenomena of water resources. Some of these important and practical methods are as follows.

1.6.1 Monte Carlo Method

The term "Monte Carlo Method" was coined in the 1940s by physicists working on a nuclear weapons project at the Los Angeles National Laboratory in the United States (beginning with the Monte Carlo method). Monte Carlo Method (Monte Carlo experience) is a computational algorithm that uses random sampling to calculate the results.

Monte Carlo methods are commonly used to simulate physical, mathematical, and economic systems. On the other hand, the Monte Carlo method is a class of computational algorithms that rely on random repeated sampling to calculate their results. Monte Carlo methods are often used when simulating a mathematical or physical system. Because they rely on duplicate calculations and random or random numbers, Monte Carlo methods are often configured to run on a computer. The tendency to use Monte Carlo methods becomes even more difficult when calculating the exact response with the help of deterministic algorithms is impossible or unjustified. Monte Carlo simulation methods are particularly useful in studying systems where there are many variables associated with the degree of pairwise freedom. These include fluids, highly coupled solids, and disordered materials and cellular structures. Monte Carlo methods are also useful for simulating phenomena with high uncertainties in their inputs, such as risk calculation in trade. These methods are also widely used in mathematics. An example of the traditional use of these methods is to estimate certain integrals, especially multidimensional integrals with complex boundary boundaries. There is not only one Monte Carlo method, but the term refers to a wide range of widely used methods. However, these approaches follow a certain pattern:

1. Define a range of possible inputs.
2. Generate random ranges from that range.
3. Using the inputs, perform a series of specified calculations.
4. Integrate the results of each computation into the final answer.

For example, we can calculate the value of p (l) by using the Monte Carlo method.

1. Draw a square on the screen, then insert a circle inside it. Next, spread several shapes of the same size uniformly (for example grains of sand or rice) across the square.
2. Then count the number of objects in the circle, multiply by four, and divide the number by the total number of objects in the square.
3. The ratio of intra-circle objects to in-square objects will be approximately equal to $\pi/4$ which is the ratio of the surface of the circle to the square. So you've got an estimate of π. Note how the estimate of λ follows a pattern specified in the Monte Carlo method.

We first defined a range of variables that was a square that surrounded our circle. We then randomly generated the inputs (distributing the grains uniformly in a square), then performed the calculations for each input (checking whether the grains were in a circle). Finally, we merged all the answers into the final answer. Also note that two other common features of Monte Carlo are:

Calculation relies on good random numbers

Gradual convergence towards better estimates when more data is simulated.

1.6.2 Las Vegas Method

The Las Vegas algorithm was introduced by Babaei in 1979 for the problem of isomorphism graph (isomorphism) as a dual Monte Carlo algorithm (Babai 1979; Grundy 2008). The Las Vegas algorithm is a random algorithm that always gives the correct answer, which means it always produces the correct answer or alerts us to the error. In other words, the Las Vegas algorithm does not gamble with the accuracy of the results, but gambles with the resources used to calculate it. A simple example is rapid random sorting in which the axis or the member influencing the sorting is randomly selected but the result is that we always have a ordered response. The Las Vegas algorithm has one limitation: it must always have finite runtime. In general, the Las Vegas algorithm can be used in situations where the number of solutions may be relatively low and where the validity of a candidate is a relatively easy solution. While the solution is complex.

Various mechanisms and conditions affecting groundwater resources cause changes in water reservoir volume and water table levels that need to be closely monitored and forecasted for proper planning, control and management. In recent years, the preparation of groundwater models and their use for simulating groundwater systems has been a major part of the projects related to the management, operation, protection and purification of groundwater. Groundwater models are often used in evaluating water resources to determine the long-term period of operation of regional or local aquifers. In particular, the flow model can provide useful information on hydraulic factors such as flow rate and flow direction. Also, subsurface conditions are not readily accessible and therefore models have become a useful tool for understanding, simulating, and predicting groundwater systems.

The following are some of the most important numerical solution methods used in groundwater:

- Border elements
- Dynamic planning
- Finite Elements
- Finite integral differentials
- Classic Limited Differences
- Random Walk Method

Among the methods mentioned above, the finite difference method is most commonly used to solve groundwater problems and is the most important numerical method for solving differential equations.

1.6.3 Finite Difference Method

The major difference between these methods and the finite elements is how the region is disrupted. In the finite difference method, the location of the points should be specified. However, various elements need to be determined in the finite element but the order in which the elements are formed does not matter. In the finite difference method, the study area is subdivided into square or rectangular elements, in which the nodes may be located within the elements or at the intersection of the grid markers. The segmentation and size of the network in the study area are determined based on the need and the degree of precision required and the shape of the area and other hydrological components that govern the area. If the study area is a homogeneous, rounded environment with a rectangular square shape and the stresses are not concentrated in specific environments, the element size is assumed to be constant, and small element size is assumed to be required in areas with high accuracy. The size of the network affects the accuracy of the results and the amount of computation.

When modeling groundwater, there is usually a simplified picture of the real world that is called the conceptual model. This model reflects the characteristics of the hydrogeological system. The most important goals of conceptual modeling are as follows:

Obtain accurate knowledge of the hydrogeological status of the area.

- Explain the problem of groundwater for a numerical model.
- Help in choosing a suitable numerical model.
- Logical simplification of the problem by appropriate assumptions.

Given that in simulation and complete reconstruction, there is usually never complete data to accurately describe a system, simplifying assumptions must be made in the conceptual model. The conceptual model should be designed to simulate the behavior of the system while being simple. One of the factors that makes the model predictions not accurate is the errors and deficiencies in the conceptual model.

1.7 Modeling Networking

To solve the partial differential equation, Environment divided into smaller components called cells. In the finite difference method, the study area is usually divided into several rectangular or square cells using two groups of perpendicular parallel lines, which, of course, the smaller the cell size, the larger the cell number, and the greater the accuracy of the calculations. In contrast, the run time of the model increases. Usually, in groundwater modeling in Iran, according to available information and statistics, cell dimensions range from 2 m to 5 km, which is either uniform or variable.

Due to the complexity and volume of computation of these methods and other methods, computers use these methods to be more precise in addition to speeding up simulation time. In this regard, various software for modeling and simulation of different phenomena have been created and accordingly, various software applications have been developed in water resources for different systems that are widely used today. Below we introduce and simulate several software applications of water resources systems.

Dams are one of the most important water resource systems. PLAXIS software is one of the software used in modeling underground dams. PLAXIS software is one of the most powerful and widely used software in the field of geotechnical engineering. Various versions of PLAXIS software have been released, including two-dimensional and three-dimensional versions, as well as tunnel versions of the software, and have different applications. One of the areas that use the two-dimensional version of PLAXIS software for analysis and analysis is the axial and symmetric strain analysis. Two-dimensional PLAXIS software is widely used in the fields of calculating the coefficient of reliability and analysis of stability states and using the software in simulation branches of the consolidation process, loading under load control modes and variable location control, analyzing and studying conditions Boundary water flow, pore water pressure analysis and analysis of boundary conditions of geometry in a particular problem have increased dramatically. The 3D version of PLAXIS software, known as PLAXIS-3D, is used to analyze issues in 3D. In the introduction of PLAXIS 3D software, it should be noted that the software has some limitations in terms of the features that it provides to the two-dimensional version of the software and its strengths are the three-dimensional problem analysis, Although PLAXIS-3D is capable of analyzing in three dimensions, loading analysis in this version of PLAXIS software can only be performed under load control conditions. The boundary conditions in the 3D PLAXIS software are defined as standard and the user is not able to change the boundary conditions while performing the PLAXIS project. It is not possible to check the pore pressure in this version of the software unlike its two-dimensional version. In PLAXIS-3D, the depth visibility is not defined, and the mesh defined in the PLAXIS 3D plan is the same at all depths. Simulation in the geogrid, tunnel and geotextile domains is not possible in the 3D version of PLAXIS software, but it is possible to analyze and model the plate items in which such items can be piles and Page pointed out. It should be noted that in the three-dimensional version of PLAXIS software, the c-phi reduction method cannot calculate the confidence factor.

The latest version of PLAXIS software to explain, is the 3D version of the tunnel, which is introduced as PLAXIS-3D TUNNEL and the software is capable of analyzing problems in two and a half dimensions. In explaining the concept of analysis in two and a half dimensions it should be noted that in the analysis with PLAXIS-3D TUNNEL, the software will draw one section of the tunnel and repeat the other section. The PLAXIS software makers have identified and reviewed these vulnerabilities, and the software panel has made significant improvements in all its versions compared to the past. The possibility of 3D analysis in many areas of science has been provided in the 3D version of PLAXIS software, and in parallel with the aforementioned possibility, the processing speed in this software has also increased significantly, which has led to the 3D version of the software being analyzed. And to study many scientific areas of soil and subsurface applications in civil engineering (Sharif Consultants 2016).

Another software used to simulate the failure of dams due to floods is the Mike11 software. This software was developed by the Danish Hydraulic Institute (DHI) and is capable of simulating one-dimensional flow, sediment transport and water quality in unstable rivers, estuaries and irrigation networks. The program uses the finite difference method to solve one-dimensional governing equations of flow, sediment transport, and water quality. The hydrodynamic model is actually the underlying element of all the systems mentioned (Komasi et al. 2015).

In water distribution and distribution systems, WaterGEMS software is one of the most widely used and simplified software for modeling and simulation of these systems and has the ability to run in Arc GIS, AutoCAD, microwave or separate environments. This program has many capabilities that can be mentioned: calculation of speed, pressure and other hydraulic parameters, simulation of fire state, simulation of water quality (pollution modeling in distribution networks [WaterGEMS], concentration calculation Pollutants after entering the grid and at specific times), energy cost calculations and more advanced topics such as designing and optimizing the water distribution network using genetic algorithm, finding the location of water leakage in urban water distribution networks and so on. WaterGEMS software is a full-fledged version of WaterCAD software and has the added features of WaterCAD software. Features of WaterGEMS more than WaterCAD include the Skelebrator module, Darwin tools module and SCADA Connect module. WaterGEMS software also has the ability to integrate and sync with AutoCAD and ArcGIS software, which WaterCAD software is unable to do. It should be noted that all files created by this two software are easily executable in one another, without any interference. Just have two versions of one software (Water GEMS Iran).

Also, EPANET is one of the most powerful water distribution network analysis software developed by the US Environmental Protection Agency (EPA) for free. EPANET is a computer program that simulates the hydraulic and qualitative behavior of water inside a pressure pipe network. A network consists of pipes, nodes, pumps, valves, and storage tanks or tanks. EPANET simulates the water flow in each pipe, the pressure in each node, the height of the water in each tank, and the concentration of one type of chemical throughout the network over a period. Typically, this software cannot design (determine pipe diameters), only with engineering knowledge and trial and error can the diameters of the distribution network be determined to estimate pressure and speed constraints. In branch networks this is possible, but in networks where the number of loops is high, it will not be possible to determine the diameter of the pipes by trial and error. One of the popular and widely used versions of this program (and of course unfamiliar in Iran) is the version of WaterNetGen which is one of the important features added by specifying the network constraints (speed and pressure) and specifying the diameter The available pipes and the Heizen-Williams coefficient of each design the grid program to satisfy the constraints (Parsian Modern Training Center).

Nowadays there are several software for simulating the model of groundwater, the most important of which is GMS and MODFLOW. GMS supports a variety of models and provides a great deal of information sharing between different models and data types. This software is a comprehensive and graphical environment for groundwater modeling. The GMS contains an interface, cartographic and several codes used in modeling such as UTCHEM, FEMWATER, SEEP2D, MODPATH, SEAM3D, RT3D, ART3D, MODEM, MT3DMS, MODFLOW, PEST, and UCODE.

The GMS software was developed by the Environmental Modeling Research Laboratory of Brigham Young University. MODFLOW is a three-dimensional model of saturation, finite difference, and block axis developed by the US Geological Survey and used in steady-state and unsteady-state analyzes. GMS is regarded as a very powerful pre-processor and post-processor for the MODFLOW 2 code. Input data for MODFLOW are provided by GMS and stored in files that are called by MODFLOW when GMS is launched. MODFLOW can be networked in both software and cellular center (One Search).

MODFLOW software was first introduced in 1984 as a three-dimensional finite-difference model. The MODFLOW code provided to simulate three-dimensional groundwater flow includes the main program and several sub-programs that are subdivided into several stand-alone software. Each software package is used to simulate one of the hydrological systems such as river feeding, drainage, water abstraction by wells, or to solve linear flow equations by a specific method. Splitting the code into several sub-programs makes it easy for the user to simulate hydrological aspects. On the other hand, it is possible to add new features and parts without having to correct the existing ones. The code is in FORTRAN 77 language and applies to many computers with FORTRAN 77 compilers. The benefits of this model are:

- Solves the equation using a finite difference method that is easy to understand.
- Works on many different computers.
- Used in one-dimensional, two-dimensional, half-dimensional, and three-dimensional modes.
- Its simulation aspects have been thoroughly tested.
- There is a lot of material and publications about it.
- Can simulate various effects of an aquifer, which includes: pressure and free aquifer, reservoir changes, bedrock and areas that are outside the aquifer but affect water flow, rivers that Aquifers are in exchange for water, drainage and springs that discharge water from the aquifer, seasonal springs, reservoirs that exchange water with the aquifer, rainfall and irrigation, evaporation and perspiration, and feed or drain wells.

The model inputs must include the aquifer properties for each cell. Also, if we have other information about other tolls, including wells, rivers, drainage, flow barriers, etc., we must include them. At the model outputs, after solving the equations by the model, parameters such as groundwater head at different time steps, groundwater level alignment curves, water balance and flow rate for each cell are extracted (concepts and models). Groundwater).

The SWAT model is currently widely used in Europe and the US to estimate and assess the impacts of global climate change on water resources and their quality. In the late 1980s, the American Agricultural Research Institute recognized the need for a model to simulate river flow larger than 1000 km^2. The SWRRB model was only responsive to basins up to 100 km^2. This model was only able to calculate the sub-basin parameters with 10 sub-basins, thus requiring the ROTO model.

The model captures and connects multiple SWRRB simulation outputs. Due to a large amount of input and output data, the two SWRRB and ROTO models were combined and replaced with a single SWAT model. Various versions of this model including SWAT 94/2, SWAT 96/2, SWAT 98/1, SWAT99/2 and SWAT2000 have been developed over time. SWAT 94/2 was developed based on several hydrological response units (HRUs). In SWAT model 2.96, agricultural irrigation and fertilizer parameters were added to the model as management components. This was the first edition in which the CO2 increase parameter was added to the model to model plant growth due to climate change. In this edition, the evapotranspiration equation was added to the model using the Penman-Montez method, lateral flow calculations, nitrogen content equations and pesticide estimation. In the SWAT version 98/1 the model was modified for use in the Southern Hemisphere. Other refinements of this edition include the refinement of the snow melting equations and the refinement of the nitrogen cycle calculations. SWAT 99/2 modified the nitrogen cycle process, modified the wetland calculation process, added nutrient drainage parameters to the wetland, reservoir, and lake. Also corrective and adaptive formulas with urban areas were added to the model. In SWAT 2000 editing, adding bacterial transfer formulas, adding Green and Ampt equations to the model to calculate infiltration, modifying production data and reconstructing climatic data, the possibility to enter radiation parameters, relative humidity, wind speed, evaporation, and transpiration was done by the user or simulated by the model. In addition to all the changes made to this model editing, in order to make it easier for users to interact with the model, an interface with Windows was developed through ArcView software. This model is able to simulate various parameters in great detail for large scale watersheds with low cost and short time. SWAT can simulate the long-term effects of the parameters in the basin and under different scenarios. This model is time-dependent and long-term modeling and is not designed to simulate single flood events. The SWAT model has a physical basis and can be used in watersheds that do not have regular inventory. This model uses easily accessible input data. The SWAT model is an efficient computation that performs simulations of very large basins with different management solutions with very low investment in the shortest time and the user will be able to study the long-term effects. In the tenth (2000) edition of AVSWAT, using ArcView software is a graphical environment for the SWAT model. SWAT is a model for simulating river basin parameters and predicting management plans, sedimentation, large-scale agricultural chemical parameters, and basins with different diversity in soil type, vegetation type and different land use as well as for different management conditions in the length of the courses are long. This model is a physical model that enables the user to study and compare the desired effects over a long period of time with different inputs. The model can also be studied in different contexts and in different basins for modeling water quality parameters. In the SWAT 2009 edition, bacterial transfer formulas have been further developed and various weather forecasting scenarios have been added. A generator has also been added to produce rain data in less than a daily time. Also the protection parameters used in daily CN calculations depend on the soil water content or evapotranspiration of the plant and in this edition the calculation of dry and wet nitrate and ammonium sediments has been developed.

Flow 3D software is one of these 3D computer numerical models and a product of Flow Science, which was launched in 1980. Flow 3D software is powerful and highly accurate for Computational Fluid Dynamics (CFD) that gives engineers valuable insight into many of the physical processes that have occurred in the fluid stream. The ability to accurately estimate fluid free surface has made Flow 3D software an ideal choice for design purposes. In CFD methods, the solution field (flow and problem range) is subdivided into small grids and the fluid-governing equations (Neuer-Stokes equation and continuity equation) are discretized and solved for each cell of the network. Flow 3D software can model a variety of streams. The governing equations are solved in three main directions X, Y, Z and allow the user to study the flow behavior carefully. Due to the capabilities of Flow 3D software, it can be used by many engineers in various fields such as civil, water, mechanical, aerospace, etc. for design and research purposes. The relationship between Flow 3D software and civil engineering are discussed below. Civil engineers face many issues related to fluid (often water) environment. Trends in water engineering, hydraulic structures, offshore, river engineering, environment, structures and soil mechanics are inevitable in many ways modeling fluid behavior or its effect on structures. Some of the issues that civil engineers can solve using this model are as follows:

Combine shallow water model and 3D flow around 3D structures such as bridges and dams.

- Modeling of moving solid objects in a fluid environment.
- Modeling of scouring and sedimentation, which is very important in the design of bridges, dams and reservoirs.
- Stress modeling and deformation of solid bodies under load-induced currents.
- Modeling events such as dam failure, tsunami, flood.
- Modeling of dams and overflows, determining the maximum possible flood rate and determining the location of cavitation and pressure on the valves.
- Design of fish stairs.
- Design of water and wastewater treatment equipment, analysis of tanks, control structures and pumping stations.
- River hydraulics can be used to study the complex dynamics of a river and its behavior, especially by using free water levels that are of interest in river engineering issues.

Given these capabilities, Flow 3D software is a very useful and essential tool for civil engineers. Design of hydraulic and offshore structures, hydraulic design and study of flow behavior in canals and rivers, analysis of stresses and forces applied to structures in the vicinity of streams, etc. Consultants or anyone looking to work in the future are bound to become familiar with Flow 3D software. Due to the increasing development of numerical models and the application and replacement of physical models with numerical models in research, postgraduate and doctoral studies researchers are also involved in their research projects related to flow and turbulence studies, sedimentation and scour, hydraulic structures, and so on. At these points, they need to master software such as the powerful Flow 3D software (Iranian Computational Science Database 1986). The hydrological modeling system (HEC-HMS) is designed to simulate the full aqueous processes of dendritic basin systems. The software includes many water analysis methods such as infiltration event, unit hydrograph and routing. The HEC-HMS also includes steps required for continuous simulation, including evapotranspiration, snowmelt and soil moisture calculation. Advanced capabilities are also provided to simulate grid runoff using the ModClark quasi-linear runoff distribution. Complementary analysis tools are provided for parameter estimation, in-depth area analysis, flow forecasting, sediment erosion, and transport and water-soluble material quality. The features of this software interface are fully integrated, including a database, data entry section, computation engine, and reporting results tool. A graphical interface or interface allows the user to seamlessly switch between different parts of the software. The simulation results are stored in the HEC-DSS (Data Storage System) and can be used in conjunction with other software for studying available water, urban drainage, flow forecasting, future urbanization impact, tank overflow design, damage reduction Flood, storm and floodplain regulations and basin system operations, to be used. Currently, the development of maximum possible rainfall is a key requirement for dam safety studies and other studies where it may pose a significant risk to human life. The standard for developing the storm in the United States is the HMR52 weather report. CEIWR-HEC is currently developing a new sedimentation method implemented in the HMR52 method. This new sedimentation method calculates the outline of each sub-basin, the pattern of the hurricane, and other storm information quickly and calculates precipitation for each sub-basin. The HEC-HMS basin map section is now modernized. This allows users to display new data formats as map backgrounds. The new section of the watershed map improves sub-watershed display and enables mapping, repository elements as well as easy entry of mapping data for elements.

HEC-RAS software is one of the models of the Hydrologic Engineering Center that can be used for streamflow in the river. This model is very simple yet practical. The above model performs river routing in both steady and unsteady state. The canalization of canals can also be defined in this model. Also, in this model, if any aquatic structures including bridge, dam, dam, culvert, etc. can be added to the model and its impact on routing can be observed. Compared to the MIKE11 model, one of the components is hydraulic modeling, it is much simpler and more practical. Although the MIKE11 model is somewhat more accurate than HEC-RAS, it can still be trusted. The outputs of the above model can be related to changes in water surface profiles in discharges with different return periods in the desired intervals in the river, flow rate values, normal depth, critical depth and hydraulic properties and parameters in the river. Model inputs include waterway cross-sections, roughness coefficients (in this section different roughness coefficients can be defined in a cross-section depending on the depth and shape changes of the sections), and design discharges at different return periods and the gap between sections. (Faramarz 2009).

Although the MIKE11 model is somewhat more accurate than HEC-RAS, it can still be trusted. The outputs of the above model can be related to changes in water surface profiles in discharges with different return periods in the desired intervals in the river, flow rate values, normal depth, critical depth and hydraulic properties and parameters in the river. Model inputs include waterway cross-sections, roughness coefficients (in this section different roughness coefficients can be defined in a cross-section depending on the depth and shape changes of the sections), and design discharges at different return periods and the gap between sections (Faramarz 2009).

One of the applications in dynamic systems is Vensim software. This software is a modeling tool for water resource issues. Vensim is capable of visualizing, processing, simulating, analyzing and optimizing models of dynamic systems in water resources. Vensim provides a simple and flexible way to simulate loop and flow, diagram models. This software introduces and defines the relationships between system variables by connecting words to arrows. After defining the above relationships and building the model, all aspects of the behavior of the system will be able to be simulated. This software is a product of the Ventana Systems Institute. Other software in the field of water resources includes WEAP software, RIBASIM, MODSIM, etc.

The steps of simulating the water resource system using Vensim software are: (Vensim)

1. Model Structure: Defining system variables and the relationships and relationships between these variables (including input, output, and auxiliary variables)
2. Preparation of the file containing the input variables: This is a.xls file in Excel software that can be defined by Vensim.
3. Model implementation and simulation.
4. Transfer model output from pdf mode to.xls format and analyze output results.

SPSS software, or "social science statistics package", is a summary of a Windows program or software that receives, analyzes, and provides various information (e.g., information on a questionnaire). The first version of the software was released six years after the establishment of SoftNe. The software company was acquired by IBM on July 7, and IBM gave it the new name PASW. But strangely enough, again in version 2 IBM decided to name it SPSS Statistics.

SPSS is divided into three general stages:

- Import information into the software
- Select the type of analysis
- Getting output
- Applications and types of SPSS

The first application of this software is quite clear. Different students or researchers use this software to analyze the information of the questionnaires they produce and to present it in graphs and tables. SPSS is one of the software widely used for statistical analysis in the social sciences. This software is used by market and market researchers, health researchers, mapping companies, government agencies, educational researchers, marketing organizations and more. For example, you will be researching how much women are satisfied with the city's lighting situation. Questionnaires are prepared by women, with questions that meet all of the standard criteria of a questionnaire (reliability and validity), then enter the information into SPSS.

At the analysis stage it tells you what the information in your questionnaire represents (relative satisfaction, dissatisfaction, etc.) and then explain it in simple language for the article or other students. You make it in the form of a chart or a table. In general, this software has the following applications: (SPSS Software Comprehensive Training).

- Provide statistical summaries such as graphs, tables, statistics, etc.
- Types of mathematical functions such as absolute magnitude, sign function, logarithm, trigonometric functions, etc.
- Preparation of custom tables such as frequency tables, cumulative frequency, frequency percentage, etc.
- Types of statistical distributions, including discrete and continuous distributions
- Providing a variety of statistical designs
- Perform one-way, two-way, multivariate analysis of variance and covariance analysis
- Time series analysis techniques
- Creating random and continuous data
- Calculate types of descriptive statistics
- A variety of tests related to comparing averages between two or more independent and dependent communities
- Ability to exchange information with other software
- Processing different types of regression

Human activities in agricultural, urban, and industrial lands produce significant amounts of nutrients and organic matter that they contaminate when they reach rivers. Computer models such as the Qual2kw model are widely used for river water quality management. The model, which simulates the river in a one-dimensional, non-uniform continuous stream, was developed by the US Environmental Protection Agency and developed by Qual2e.

Water quality models provide a practical framework by simulating important physical, chemical, and biological processes. The Qual2kw model can simulate several parameters including temperature, pH, biochemical carbon demand, the oxygen content of sediments, dissolved oxygen, organic nitrogen, nitrogen, ammonia, nitrite and nitrate, organic phosphorus, inorganic phosphorus, total nitrogen, phosphorus. Whole, phytoplankton, and algae are the floor. The main equation that this model solves is the one-dimensional displacement/diffusion equation that is used to simulate all parameters except foam algae. The effectiveness of this model has been proven in various researches in Iran, including in Karun, Zayandehrood and Kainesaras rivers of Zanjan province. The model can simulate different parameters along the river and at different times of the day and can be used as a reliable management tool (Taheri Soodjani et al. 2015).

1.7.1 Describes Examples of Modeling and Simulation

Suppose we intend to measure the success rate of air versus a car in the design phase of a car. One way to do this experiment is to build a real car, drive it and then measure the air resistance, which, although it does cost us a lot of money because it has a high cost because it has to be built first, then tested. If the test fails, then we need to redesign the design and repeat the test after building another real prototype, and the process will continue until the design is perfect for the car with the features. We find that such a method is very costly and involves both economic costs and time costs, as it will take time to build a high-cost car at each test stage. In this case, the experts turn to the model for such an experiment. That is, a small physical object with the aerodynamic characteristics of the car's design is constructed and placed in a wind tunnel to simulate the movement of the car in real space and thus measure the amount of air resistance.

The points of interest in this modeling are one side of the model and the other features. The model is very simple and small and only car aerodynamics are included in the model, since our purpose is to study only the aerodynamic characteristics of the car and the model must not necessarily resemble the real car in other respects. For example, in the construction of such a model, no consideration is given to the robustness or beauty of the model, as examining such properties is beyond the scope of this particular modeling.

Another example of water resource simulation is a study in which the geological, hydraulic and hydrological information is used to numerically simulate the groundwater flow path of the Hamadan-Bahar aquifer using GMS software. In this research, first, the three-dimensional hydrological model of Hamadan-Bahar plain was prepared and then the flow in the plain was simulated with MODFLOW numerical code. After simulating the flow, the model was calibrated by trial and error. Groundwater flow path estimation was also performed using MODPATH numerical code. Finally, the areas of congestion of the wells studied were plotted for the movement of groundwater with different movement times. After the simulation, it was observed that the observed and simulated water level difference was within the permissible range of 35 $\pm$ m as the optimal range. Groundwater flow path estimation with MODPATH numerical code showed that in forward movement, the longest flow direction of 43400 m and in the backward movement of 674/8270 m were identified as the longest flow direction. The simulation results of the groundwater flow transmission simulation revealed the fact that the flow from the aquifer boundary to the center and the aquifer outlet was aligned with the hydraulic gradient. The overall result of this study was that, overall, it can be stated that the ongoing process of groundwater flow will increase the level of aquifer contamination that will cause irreparable damages to the groundwater body (Bayat Varkashi et al. 2018).

1.8 Summary

The main purpose of this chapter is to provide an overview of the modeling and simulation topics focusing on water resources management. In this regard, firstly, a review of studies on modeling and simulation of water resources was made and a summary of them was provided. Present paper describes various water systems were modeled and simulated. Then a definition of the keywords and terms in the text is provided to clarify their general meaning. Subsequently, the basics and logic of the subject were studied to determine its basis. The varieties of models, concepts and logics were mentioned, and their nature and structure were identified. Finally, it is mentioned that the ultimate goal is a model is to model and identify the phenomenon. Furthermore, the importance and necessity and the reason for modeling and simulation was stated and their benefits were also identified. The study discussed the method of doing so and the process of its implementation. Moreover, the Flowchart presents the process of modeling and simulation and then different and applied simulation methods are mentioned, among which the Monte Carlo Method is the most important and practical one. Computer and software are generally used for modeling and simulation because computers allow this to be done in the least time and with the most precision. Also, each software is related to the simulation of a particular system so they use simulation methods specific to the same system so that the user only inputs the information to the software and it delivers output, so the user does not need to learn how to simulate different systems. Important and useful software for each of the aquatic systems with their logic, application and generality were mentioned. Finally, after providing all the relevant material needed to clarify the topic for the reader, two examples were outlined: the first is about general modeling and simulation and the second is about modeling.

Acknowledgements The authors thank Iran's National Science Foundation (INSF) for the support of this research.

References

Abde KolahchiA, Majidi A, Firoozbakht M (2010) Simulation of the underground dam using PLAXIS software. Civilization and retrofitting

Azari H, Matkan AA, Shakiba A, Pourali H (2009) Simulation and warning of floods by combining hydrological models in GIS and estimating precipitation by remote sensing. Iranian Geol Q 3(9):51–59

Babai L (1979) Monte-Carlo algorithms in graph isomorphism testing. Université tde Montréal Technical Report, DMS, pp 79–10

Banks J, Carson J, Nelson BL, Nicol D (2005) Discrete-event system simulation, 4th edn. Pearson Prentice Hall, Upper Saddle River, NJ. ISBN 978-0-13-088702-3

Bayat Varkashi M, Fassihi R, Zare Abyaneh H (2018) Numerical simulation of the groundwater flow of Hamedan plain spring. Health Environ 11(1):49–62

Belaineh G, Peralta RC, Hughes TC (1999) Simulation/optimization modeling for water resources management. J Water Resour Plann Manag 125(3):154–161
Benedettini O, Tjahjono B (2008) Towards an improved tool to facilitate simulation modeling of complex manufacturing systems. Int J Adv Manuf Technol 43(1/2):191–199
Bozorg-Haddad A, Saifullah Aghmoni S (2013) Introduction to analysis of uncertainty in water resources systems. University of Tehran Press, First Time, Iran, Tehran
Cartwright N (1983) How the laws of physics lie. Oxford University Press
Faramarz M (2009) Water science site. http://waterscience.blogfa.com
Grundy D (2008) Concepts and calculation in cryptography. University of Kent, PhD thesis
Hacking I (1983) Representing and intervening. introductory topics in the philosophy of natural science. Cambridge University Press
Hajkowicz S, Collins K (2007) A review of multiple criteria analysis for water resource planning and management. Water Resour Manage 21(9):1553–1566
Hooshmandzadeh M (2007) Simulation of water resources systems - a case study of the century dam. In: 3rd Conference on applied geology and environment, Islamshahr, Islamic Azad University, Islamshahr Branch
Khayami M, Danesh Sh, Khodashenas SR, Arbitration K (2007) Simulation of qualitative conditions of dam reservoirs (Case study: Torgh Dam reservoir). J Soil Water (Agric Sci Ind) 1(23):29–37
Khosh Fetrat A, Asghari K, Karimi Jashni A (2006) Simulation of the operation of dam reservoirs as a multi-barrier system. Case study: Zayandehroud, Koohranak and Cheshmeh Langan multi-barrier systems. In: The first regional conference on the exploitation of water resources in the Karun and Zayandehrud basins
Komasi M, Kohzadi A, Hosseini A (2015) Hydraulic stimulation of the Dez Dam failure phenomenon by MIKE computer model
Loux DP (2008) Water resources management models. https://www.nae.edu/19582/Bridge/V38N2/WaterResourceManagementModels.aspx
Mahlouji H (2010) Simulation of discrete systems, 9th edn. Sharif University of Technology Publishing Institute, Tehran, Iran
McKinney DC, Cai X, Rosegrant MW, Ringler C, Scott CA (1999) Modeling water resources management at the basin level: review and future directions. Int Water Manag Inst, SWIM, Colombo, Sri Lanka
Mir A, Mons-Hernandez A (2009) Integrated water resources optimization models: an assessment of a multidisciplinary tool for sustainable water resources management strategies. Geogr Compass 3(3):1176–1195
Moradi J, Shahnazari A, Darzi A, Jafari B (2015) Simulation of existing conditions of resources and uses of Tajan Irrigation and Drainage Network using the Vensim model and providing executive solutions. In: 3rd International conference on applied research in agricultural sciences, Comprehensive University of Applied Sciences, Tehran, Iran
Oberkampf WL, DeLand SM, Rutherford BM, Diegert KV, Alvin KF (2002) Error and uncertainty in modeling and simulation. Reliab Eng Syst Safety 75(3):333–357
Radmanesh F, Golabi MR, Khodabakhshi F, Farzi S, Zeinali M (2020) Modeling aquifer hydrograph: performance review of conceptual MODFLOW and simulator models. Arab J Geosci 13(5):1–9
Salavi Tabar A, Bandari R, Hassani Kh, Caution M (2011) Integrated simulation of surface and groundwater resources case study: Haraz River Basin. In: Second national conference on applied research in Iranian water resources, Zanjan Regional Water Company, Zanjan, Iran
Samadyar H, Samadyar H (2008) Water resources modeling process. In: 2nd Specialized conference on environmental engineering, University of Tehran, Tehran, Iran
Sotoudeh Qarabagh R, Zarifi MA (2005) Computer-assisted process simulation for chemical engineers and the need for training. Iranian J Eng Educ 7(25):44–47
Taheri Soodjani H, Barati K, Shayannejad M (2015) Introducing the basic equations of the Qual2kw model and the practical guide of the model. J Water Sustain Dev (No. 2):45–35
Tolk A (2015) Learning something right from models that are wrong – Epistemology of Simulation. In: Yilmaz L (ed) Concepts and methodologies in modeling and simulation. Springer, pp 87–106

Torabian A, Ashemi SH (2002) Qualitative modeling of surface-kinetic waters, constants, and rates. Tehran University Press, Iran

Zeinali M, Azari A, Heidari MM (2020a) Multiobjective optimization for water resource management in low-flow areas based on a coupled surface water-groundwater model. J Water Resour Plann Manag 146(5):04020020

Zeinali M, Azari A, Heidari MM (2020b) Simulating unsaturated zone of soil for estimating the recharge rate and flow exchange between a river and an aquifer. Water Resour Manag 34(1):425–443

Chapter 2
Optimization in Water Resources Management

Masoumeh Zeinalie, Omid Bozorg-Haddad, and Hazi Mohammad Azamathulla

Abstract The environmental crisis and the destruction of natural resources is one of the problems of twenty-first century, due to the available resources, in terms of limitations, have forced humanity to think of ways to combat this trend. Therefore, the water resources planning and management, as well as sustainable development and optimal use of natural resources, has led to the human use of optimization methods and techniques. Water crisis, increasing water demand, and the occurrence of intermittent droughts, saving water consumption, and efficient use, it is necessary to use appropriate optimization techniques can be helpful in this regard. Numerous studies have been conducted to achieve effective use of available resources without harming the assets and resources of future generations using optimization methods. Limited freshwater resources, on the one hand, and growing demand, on the other, are exacerbating people's concerns about water resources. Therefore, the importance of planning the optimal use of surface and groundwater is increasing. The unnecessary and inappropriate groundwater harvesting has occurred over the last few years it is a serious threat to water resources, which drastically affects the environment, especially in dry and semi-arid areas. Using the concept of optimization, which is the most appropriate output value of a system due to its constraints, it can be

The original version of this chapter was revised: For detailed information, please see Correction. The correction to this chapter is available at https://doi.org/10.1007/978-981-33-4295-8_14

M. Zeinalie
Department of Irrigation & Reclamation Engineering, Faculty of Agricultural Engineering & Technology, College of Agriculture & Natural Resources, University of Tehran, Karaj, Iran
e-mail: masoumeh.zeinalie@ut.ac.ir

O. Bozorg-Haddad (✉)
Distinguished Professor, Department of Irrigation & Reclamation Engineering, Faculty of Agricultural Engineering & Technology, College of Agriculture & Natural Resources, University of Tehran, Karaj, Tehran
e-mail: OBHaddad@ut.ac.ir

H. M. Azamathulla
Department of Civil and Environmental Engineering, University of the West Indies St. Augustine, St. Augustine, Trinidad and Tobago
e-mail: Hazi.Azamathulla@sta.uwi.edu

O. Bozorg-Haddad (ed.), *Essential Tools for Water Resources Analysis, Planning, and Management*, Springer Water, https://doi.org/10.1007/978-981-33-4295-8_2

concluded that optimizing water resources is one of the best ways to conserve water resources.

Keywords Water resources management · Optimization · Water crisis

2.1 Introduction

Optimization is widely used in water resources management today. Rao (1984) divided optimization methods into two main categories. The first category is the classical mathematics-based method, and the second is the search or numerical methods, which include direct and indirect search. In classical and indirect search methods, the objective function must be continuous and derivative, and indirect search methods such as random jump, various algorithms are used to search point by point in the space of variables (Rao 1984). In many engineering problems, the target function has several local and global optimal point of reference that the classical methods are not able to distinguish between them and find the optimal national point easily. Also, in the form of practical problems, the target function may be separate or accompanied by sudden changes that classical and indirect methods are generally unable to solve. Direct search methods may also be useful in small problems with a limited number of variables. However, in practice, they will not work well when there are many decision variables, and their scope is broad (Rao 1984). To this end, in recent years, many researchers have turned to artificial intelligence to solve this problem. These methods include evolutionary calculations such as Anil simulation algorithm (SA), GA, and PSO. Inspired by nature, these methods consider a set of points or populations in the space of answers and lead the model in different directions to find the optimal answer. To this end, in recent years, many researchers have turned to artificial intelligence to solve this problem. These methods include evolutionary calculations such as Anil simulation algorithm (SA), GA, and PSO. Inspired by nature, these methods consider a set of points or populations in the space of answers and lead the model in different directions to find the optimal answer. Although water resources are renewable, their volume is constant, and in contrast to human demand for it is increasing. Over the past 100 years, global water demand has increased six-fold. Unfortunately, pollutants, including industrial effluents, agricultural wastewater, and urban and rural wastewater, pollute water resources and exceed consumption standards (Ghanbari et al. 2014). In water resources management and planning, scientific decision-making is the key to future scientific action. For this purpose, optimization is one of the appropriate tools in the field of water resources. In mathematical sciences, optimization provides the best answer from the set of available answers with the presence or absence of different constraints or constraints (Bozorg-Haddad et al. 2016). Optimization is one of the applied sciences in mathematics, economics, and management, and its tools have been widely developed in various fields. Also, due to the increasing development of optimization methods in the field of water resources management, there is a need for modern methods regarding the optimization of water resource systems (Rao 1984). Belaineh et al. (1999) examined the simulation and optimization of a hypothetical area model with a greater focus on joint water management and reservoir management under different scenarios. The results showed that more details are used in the model, the better the joint water management would be, so that in a model with more detail, 13% more water was provided. Shangwan (2001) study revealed the use of appropriate techniques in optimization could save up to 50% of water and minimize damage. The study aims to compare the various techniques of linear programming optimization (NLP), collective intelligence (PSO) and genetic algorithm (GA) in managing agricultural water allocation in drought conditions by maximizing income. Kumar and Reddy (2007) used GA, EMPSO, and PSO methods to derive operating policies in multi-purpose tanks. This technique was used for the Bhadra multi-purpose tank in India. According to the objectives of this reservoir, the goal is to maximize electricity generation and minimize water and PSO shortage compared to EMPSO irrigation. The results showed that it has better results in extracting reservoir exploitation policies. Mayer and Muñoz-Hernandez (2009) reviewed the optimization of aggregated models of water resources. Matsukawa et al. (1992) presented a simulation-optimization model for a strategic plan to develop the exploitation of the Mad River in California. The river's watershed consisted of a multi-purpose pond and a free aqueduct that are hydraulically connected. Nikkami et al. (2012), applied multi-objective linear programming model, found that the total profitability of the Abol Abbas Basin before optimization was 127601 million Rials and after optimization of land use optimization with 3687 percent increase has reached 299.89 million Rials. Azamathulla (2013) used soft computing techniques Water resources study, namely radial radiant performance (RBF), adaptive neural-fuzzy inference system (ANFIS), expression gene programming (GEP), and linear genetic programming (LGP) in engineering. The scope of application is more in optimizing the prediction of cascading phenomena that occur in pipelines placed under the sea or river, bridge or waterfall piers, ponds, and overflows. Zeinali et al. (2020) focused on unknown sorting genetic algorithm II with a coupled surface water–groundwater model in the southwestern region of Iran. The advantage of this structure is that considering various limitations, it achieves and maintains a balance between surface water output and groundwater. The optimal outflow of surface and groundwater in humid and dry areas is achieved by linking the optimization algorithm with the Cooper model. Fatehi and Beauty (2010), in their research results in a case study in the Firoozabad region, Pakistan, showed that using multi-objective linear programming, farmers' profits, and water abstraction from underground aquifers can be optimized simultaneously.

2.2 Definitions and Terms

In the Water Resources Management System, some of the following term is used to refer (Gleick 1993).

Management: The process of effective use of human and material resources is used in planning, organizing, resources and tools, guiding and controlling the system to achieve specific organizational goals or objectives.

Planning: Planning is organizing various activities, such as "deciding on the time and place of activities, visualizing the desired status of the system in the future and creating appropriate tools and methods to achieve this desired situation."

Analysis: To be analyzed, the system is first divided into smaller components, and after reviewing and, analyzing the problem, the conclusion is made on well-defined system. This comprehensive approach to all the components of the system is done with the help of analysis so that all the components of the system can be reviewed.

Decision: Decision-making in water resources systems is scientifically possible to select the best option for management and planning purposes.

The objective function: The objective function is a goal that is considered through the optimization process, or in other words, finding the optimal value for the objective function is defined. Such as maximizing profits, minimizing costs, maximizing the reliability efficiency of the system, he pointed out. In a numerical solution, a problem may have several different answers, and to compare them and choose the optimal answer, a function is defined as the objective function. Therefore, it can be said that choosing the optimal answer through the objective function is one of the most important and basic optimization steps.

Decision variables: These types of variables are variables whose value determines the objective function. These variables can be controlled and managed in the system to achieve optimal optimization in the system. It is important to find the optimal value for decision variables so that the objective function is maximized or minimized.

Mode variables: These types of variables are dependent variables, the value of which determines the overall status of the system, and the value of which is determined by independent decision variables and by the system simulation method.

Decision space: All the answers to an optimization problem are called decision space. The space of decision itself is divided into two parts, the space of final and infinite decision. The finite decision space is created by the intersection of the allowable value of the decision variables of that problem, whereas in the infinite decision space there is no allowable value for one or more variables in the domain.

Justified and unjustified answers: Justified answers are answers for which the limitations of the problem are met. But unjustified answers are called violations of the limitations of the problem.

Optimal Solutions: Optimal Solutions are values that are in the possible decision-making space. The optimal Solutions is divided into three parts: absolute optimal, local optimal, and near-optimal.

Absolute optimization: If the response to an optimization problem reaches its best value, it is called the absolute optimal Solutions. The point in the decision space that for each value of that point, the target function occupies the peak (maximum value) and the door (minimum value). So that there is no other peak above or no other valley below it. In fact, in maximization problems, the value of the target function for the absolute optimal answer is greater than or equal to all the values of the target function for different decision variables, and in minimization problems, the value of the optimal objective function for the absolute optimal answer is the highest. The values of the target function are less than or equal to the variables of different decisions.

Local optimization: Local optimization is solutions the best answer for the objective function around its neighborhood. Optimal localized responses are superior to those around them but will lose their superiority if performed on a larger scale.

Close to optimal: In most optimization problems, there are situations where there is no absolute optimal Solution to the problem and the probability of achieving the absolute Solution is very low, in this case the near-optimal answer is considered the best value for the objective function. It is worth noting that there is a very little gap with the absolute answer to the problem. For this reason, in many optimization methods, the search process stops after reaching the Solution close to the problem.

Optimization: Selects the desired options from a set of available options. Or rather, finding the most appropriate possible values for the desired variables in a problem, so that for the amounts obtained, the desired value is achieved for the problem.

Modeling: A natural phenomenon in the form of physical components or mathematical relations, which includes two types of physics modeling and mathematical modeling. Physical modeling can be implemented in cases where there is the possibility of laboratory construction of components and systems related to that problem. But in physical water management, physical modeling is not very practical because building such a model in a laboratory requires a lot of money and is still time-consuming. In modeling, the greater the number of components, the more accurate the modeling.

Mathematical modeling: It is the expression of a system and or a system in mathematical language and its theorems and symbols. Mathematical modeling is an attempt to develop a mathematical model for a specific system.

Simulation: Is one of the methods to solve programming models when it is not possible to use algebraic analysis methods and not possible to experiment in the real world. Simulation relies on test and error to examine the effect of different conditions on the system and evaluate its results.

System: The system relates each building, device, method, or method of actual operation to a given time, input, cause, or stimulus of material, energy, or information to output, effect, or response to the form of information, energy, or material. In other words, the system-defined a set of related components that convert multiple inputs to multiple outputs. But all the said compliments have been given to different definitions of scholars, which have common points in connection with the compliments of the system, examples are given below (Modarresizdi and Asif Vaziri 2003).

1. System components are independent of each other but may be interconnected.
2. A system is a combination of the logic of its constituent components.
3. The system establishes a connection between output and input, or cause and effect.
4. The input and output of a system do not necessarily have the same nature.

2.3 Fundamentals and Logic

We always been looking for the best option and path. The concept of optimization derives from this concise phrase. we are looking for the best route or the best option for different life issues. The goal of optimization is to find the best acceptable solution, given the limitations and needs of the problem. For example, a farmer's target is to increase the economic profit from the sale of products (target function) according to restrictions such as cultivation area, amount of water resources, number of workforces, planting costs, and so on. That we can maximize product sales, given existing constraints, is to optimize our target function. Optimizing water resources systems is one of the practical solutions to improve the status of existing water resources. Being able to minimize significant profits in the agricultural sector by minimizing water resources (target function) means optimizing (the simple concept of optimization). In general, optimization occurs when one or more available resources (which can be water, land, financial resources, labor, etc.) limit development. In the presence of abundance in terms of all the resources needed for development, the importance of optimal use of resources is not so much discussed. Therefore, optimization often occurs when it is not possible to use one or more resources freely and without worries, or when harvesting from a certain amount can cause severe damage to the water system or reduce the economic efficiency of the project. In Iran, water is a limiting source of development in some areas. In most parts of Iran, water shortage is a severe problem, and the optimal use of water resources is a national issue, so using optimization methods is a way to manage water resources better.

The choice of the appropriate optimization method depends on the circumstances and characteristics of the problem under consideration. Each of the optimization methods, depending on the characteristics of each problem, is used for a specific type of problem. Therefore, it is necessary to acquire enough knowledge of the problem as well as the existing solution methods to choose the right optimization method.

Types of optimization methods:

(A) Optimization by linear programming method
(B) Optimization by nonlinear programming method
(C) Optimization with a definite discrete classical method
(D) Continuous optimization (linear stochastic programming)
(E) Optimal discontinuous optimal optimization (dynamic stochastic programming)
(F) Optimization by modern methods (evolutionary and meta-exploration algorithm).

Optimization by linear programming is one of the most common mechanisms for formulating a wide range of problems. A linear programming problem consists of several linear relationships describing the structure of the problem and its physical characteristics. Decision) and indexes and constraints are a set of relations of equality or nonlinear inequality of unknowns and indexes (Bozorg-Haddad et al. 2014).

The word line means that all mathematical relationships (objective function and constraints) of this model are necessarily linear functions, so LP means planning activities to obtain an optimal result.

It is worth noting that the linear method is one of the most practical optimization methods in the field of water resource planning and management, the popularity of which is mainly due to the available software packages that can be used to solve the desired problems easily.

Here is a standard and classic form of a linear optimization problem according to Eq. (1.1).

$$
\begin{aligned}
\text{Maximize F } &= c_1x_1 + c_2x_2 + \cdots + c_nx_n \\
\text{Subject to } &: a_{11}x_1 + a_{12}x_2 + \cdots + a_{1n}x_n \le b_1 \\
&a_{21}x_1 + a_{22}x_2 + \cdots a_{2n}x_n \le b_2 \\
&a_{m1}x_1 + a_{m2}x_2 + \cdots a_{mn}x_n \le b_m \\
&x_1, x_2, \ldots x_n \ge 0
\end{aligned}
$$

in which, F = the objective function of x_n, = the decision variables of c_n, = the corresponding coefficients x_n in the objective function of a_{mn} the coefficients in the constraints and b_m the negative values to the right of the constraints. The number of m is the first constraint (ie, those with the $a_{m1}x_1 + a_{m2}x_2 + \cdots a_{mn}x_n$ function) are called functional constraints. The constraints of the $x_n \le 0$ constraints, the variables x_n, the decision variables and the fixed data c_n, b_n, a_{mn} are the model parameters (Bozorg-Haddad et al. 2014).

Assumptions of linear models: Linear models, like any other modeling, have predetermined assumptions. These assumptions include:

Assumptions are proportionality, aggregation, separability, and definiteness.

Proportionality assumption: The proportionality assumption means that the amount of consumption from each source is independent of the amount of consumption from other sources.

Conclusion Assumption: The assumption of appropriateness alone is not enough for the linearity of the objective function and the definition of constraints. The assumption of aggregation means that the total consumption is equal to the sum of the consumption values from each source.

Assumption: Decision variables Some physical problems are just integers. However, the answer to n from LP is not necessarily integers. Therefore, the assumption of forgiveness means that each unit of activity can be divided into any desired number. However, even when the correct answer is necessary, LP can still be used sequentially (Hillier and Lieberman 1980).

In this case, if the answer is not an integer, they can be rounded to the nearest integer. This usually does not cause problems when the decision variables are significant numbers. However, in some cases, there are problems with rounding up the answers. In such cases, linear programming with the correct number must be used.

2.3.1 *Categorization of Optimization Methods*

The choice of each of the following methods depends on the conditions and characteristics of the optimization model in each problem, and each of the options is briefly described (Hillier and Lieberman 1980).

1. Linear methods—nonlinear methods
2. Definitive methods—Random (probable)—Stochastic
3. Continuous methods
4. Explicit fuzzy methods
5. Modern Classical Methods
6. Simulation and search methods
7. Random sampling method
8. Targeted test and error method
9. How to solve the device of equations
10. The method of using partial derivatives of the objective function.

In Linear models decision space are polygons surrounded by constraints (direct lines) and the objective function moves as a straight line within it, but in nonlinear models of decision space, the form of enclosed space is with constraints or lines. The curve or space of the linear enclosure is a function of the nonlinear target or a combination of both.

There are several ways to solve such (linear) models, such as drawing methods and the Simplex method, but in the nonlinear method it is possible to have an optimal answer at any boundary point, and the derivative of nonlinear curves can be used. It should be noted that due to the complexity of some models, it is sometimes not possible to calculate the models by derivative method, so we use the numerical calculation method.

2.3.1.1 Definite-Random Method

Using a definite method to solve an optimization problem will certainly ignore the possible nature of the problem phenomena and may not lead to logical and correct answers to the problem. But stochastic linear programming (SLP) and stochastic dynamic programming (SDP) methods that help us find optimal solutions to the problem.

2.3.1.2 Classical-Modern Methods

The classical method of optimization is often based on mathematical concepts, and solving them numerical solution methods used to achieve the absolute optimal method. In some optimization problems, when the details of the problem increases, it leads to a more complex mathematical model, which sometimes makes it impossible to solve engineering problems with these methods.

Modernization and optimization, such as evolutionary and meta-exploration algorithms in the engineering sciences, are widely used to cover the weaknesses of the classical method. The basis of this method and the mentioned algorithms are the use of simple processes and concepts in nature, and by performing many attempts and errors, they achieve the optimal solutions. It is worth mentioning that in the classical method, it is possible not to have access to the absolute optimal answer, therefore, by solving the modern method, we will be able to access the answer close to the optimal, and therefore it is superior to the classical method.

2.3.1.3 Simulation and Search Methods

The basis for this method is the consecutive selection of points from the whole decision space, simulation of each point, control of the experimental answer and control of the presence of the desired point (answer) in the justified decision space.

2.3.1.4 Random Sampling Method

This method is presented to address the problem of reliability between achieving the optimal solution and the time required to solve optimization problems. Sampling can be done both from network points and all available points. However, sampling from network points increases the calculation speed and reduces the volume and accuracy of the calculation, and in sampling from all existing points, the calculation speed decreases, and the volume and accuracy of the calculations increase. In this method, one point is randomly selected each time, so it is possible that in subsequent sampling, one of the measured points will be re-selected, so two solutions are created:

In the first approach, the new answer is measured regardless of the previous choice, and in the second approach, each new answer can be compared with all the previously measured responses. We will pay for another new one. Therefore, the second approach has an advantage over the first approach due to the reduction of the volume and amount of repetitive simulation and the time of calculations is less (Bozorg-Haddad et al. 2014).

2.3.1.5 Targeted Trial and Error Method

In the trial and error method, the next point is selected in the direction and a path that the user is likely to get a better answer in that direction. In other words, explaining the problem, the designer chooses the next examples in a direction that is likely to occur if the problem is solved to the most positive change in the value of the target function. will give. Reaching the optimal answer in this way is entirely based on the decision of the problem designer and experience in this field.

Of course, such methods are often referred to as user-friendly, and the final answer obtained from problem-solving is completely in the direction intended by the designer, but there is no always a guarantee that the final answer is optimal or even optimal for moving the chosen path.

2.3.2 How to Use Minor Derivatives of the Objective Function

This method complements the targeted test and error method. In this method, trial, and error used, except that identifying the direction of movement in the decision space has a mathematical logic. There is no guarantee that the selected direction will be optimal in the target test and error method, so we use partial derivatives to eliminate this defect and problem.

In this method, according to the gradient and slope of the target function and the path to be moved and tested, the next point is selected and the direction of the highest slope facing the objective function is at the current point. In this way, it can be claimed that the selected path to the cut will have the most positive effect on the objective function.

The problem with this method is that it calculates the partial derivatives of the objective function, which may be easy to calculate in this partial derivative and problematic in some cases according to the parameters of the problem. To solve this problem, it is possible to change the shift margins of each of the decision variables.

2.3.3 Challenges and Opportunities in Using Optimization Models

One of the basic pillars of water resources management in present situation is the optimal use of available water resources. To this end, considering the diverse dimensions and complexities of water resource systems, managers and planners today have resorted to using optimization models as an efficient tool to achieve optimal decisions (Navid Nejad 2014).

The development of human knowledge and the creation of modern tools and their combination with existing optimization models, new opportunities for "better decision making" have been provided in water resources planning and development." (Soltani 1994).

Create tools and access to modern technology, such as the World Wide Web and geographic information systems. Value of new data and information due to behavioral changes in watersheds and the importance of using this information quickly for analysis, design and information in emergencies, as well as public education for consumption. And water-saving has created new coordinates in water resources management (Zahraei 2016).

Nowadays, with the rapid expansion of information technology, a new space has been created for the development of new tools for analyzing, planning and managing water resources systems. By accessing the Internet, infinite space of information and science is provided to the user, which can be used to make decisions as effectively as possible. Useful tools that have been widely used include remote sensing and geographic information systems. Using remote sensing, quantitative and qualitative variables can be identified, and effective parameters about them can be measured and interpreted. For example, using sensors on aircraft and satellites, valuable information such as soil moisture, snow cover and flood expansion can be obtained, which is of great importance in water resources management (Zahraei 2016).

2.3.4 Applications of Optimization in Water Resource Management

To understand the problem of water resources management and the purpose of which is predetermined, after explaining the conceptual and mathematical model of the problem, optimization methods can be used to solve and find the optimal solution For this purpose, a specific definition of the system must first be provided. A system is a set of related sections and components that pursue a specific goal, and specific inputs lead to access to the corresponding outputs (Bozorg-Haddad et al. 2016).

A system feature can be divided into inputs and outputs, which can be controlled or uncontrollable. Also, system inputs include primary data, relationships are the constraints of constraints, and decision variables are needed to determine the status or performance of a system.

Figure 2.1 shows an overview of the components of the blue system, in which each system (S) consists of a set of relationships and equations with some data and indexes that are given as inputs (U, P, L, D, O), the output (R) is obtained.

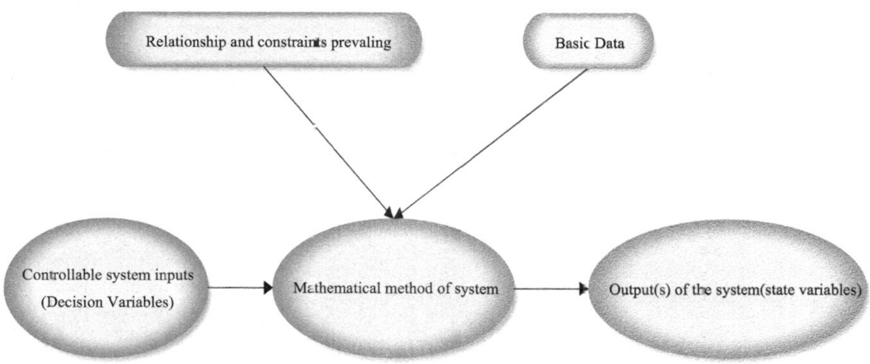

Fig. 2.1 General outline of the components of water resources systems (withdrawal from Bozorg-Haddad et al. 2014)

Uncontrollable inputs to a set of data related to natural phenomena and processes in a system whose selection and change of values is beyond the control of the system analyst. In this system, there are uncontrolled inputs (UP). In mathematics, computer science and economics, optimization selects options from a set of justifiable or unjustifiable selections. In general, in the optimization process, assuming that R is unknown as the output of the water resources system, the overall goal is to find the best value for R by determining the values for the inputs and controllable variables. According to the issues raised, in optimizing water systems, the system should be introduced by developing a suitable conceptual and mathematical model (Bozorg-Haddad et al. 2014). The simplest and most common problem of water resources management, which can be solved using a variety of optimization methods, is the issue of dam operation.

Example: Consider a barrier that is exploited during a monthly T-period. The operation of this dam in different months depends on the volume of needs in downstream agricultural fields, the amount of volume required for each period is determined every month during a period of T months. If the shortage index is defined as the difference between the release of the dam and the volume of downstream needs, the purpose of solving this problem is to manage water resources management to achieve the maximum total supply of downstream needs or to achieve the minimum total deficit volume in the T-period in planning. Therefore, the decision-making variables in solving this problem are the monthly release volumes during the operation period.

Figure 2.2 shows the different parts of the dam system for optimal operation.

It should be noted that a dam can be exploited for different purposes such as industry or agriculture, maximizing hydropower production, minimizing the shortage of drinking, shipping, and environmental needs on a daily, monthly and annual basis. As mentioned, each optimization problem has two main components, the objective function, and the constraints, which are identified in the modeling process, and their

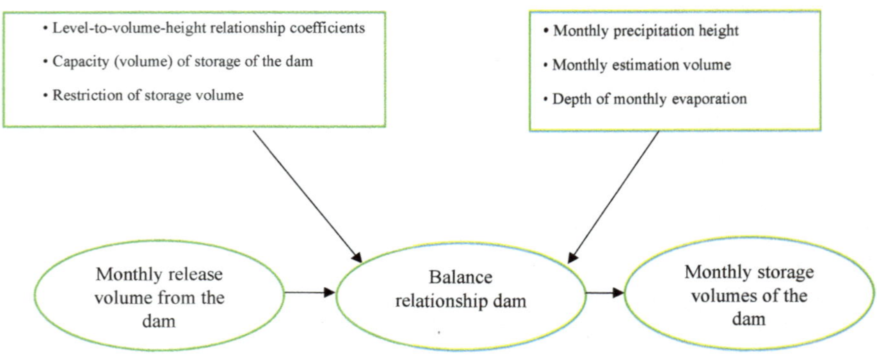

Fig. 2.2 General diagram of the components of the operation model of the dam system harvesting from the great (Bozorg-Haddad and Saifollahi-Aghmioni 2013)

mathematical structure is formed. The goal of any optimization problem is to achieve the desired value for the objective function by observing all the conditions of the problem and identifying the absolute optimal solution among the justified answers to the problem.

2.4 Importance and Necessity

Due to the breadth and complexity of water resources systems, policy determination, selection of options, and appropriate strategies for the proper functioning of water resources systems in different situations are necessary and require a comprehensive optimization process.

The World Water Commission predicts that water use will increase by about 50 percent in the next 30 years. Therefore, the use of simulation and optimization techniques in decision-making and management strategies for the principled use of water resources will be inevitable.

2.4.1 Significance of Consumption Optimization and Technological Strategies for Water Consumption Optimization

Rapid population growth and urban development in the third millennium, especially in developing countries, have significantly increased water demand to meet population needs. This is especially true in areas of the world that are naturally plagued by water scarcity. In our country, too, the situation of water resources is not very favorable, Per capita water divides the world's countries into five categories, which are shown in Fig. 2.3.

- The first group of countries with a per capita capacity of more than 5,000 cubic meters per year, most of the countries of South and North America, parts of Europe, Africa and Russia
- The second group of countries with per capita water between 1700 and 5000 cubic meters per year, China, Afghanistan, Kyrgyzstan, Tajikistan, Turkmenistan, Iran, Iraq, Turkey, and Syria.
- The third category of countries that have a per capita renewable water between 1000 and 1700 cubic meters per year, India, Pakistan, Ethiopia, Somalia.
- The fourth category of countries that have a per capita renewable water of 500 to 1000 cubic meters of water per year, countries such as Sudan, Egypt, South Africa, and Morocco.
- The fifth category of countries that have a per capita renewable water of less than 500 cubic meters per year, Saudi Arabia, Oman, Yemen, Libya, Algeria, Tunisia.

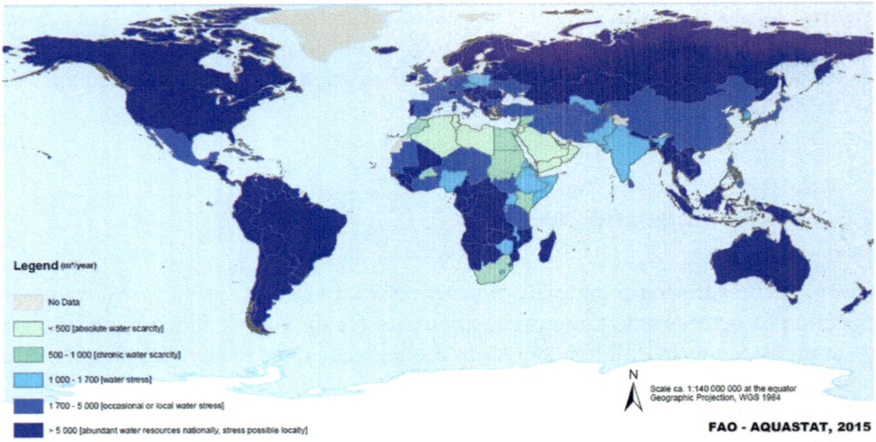

Fig. 2.3 Per capita renewable water in the world (cubic meters per year)

2.4.2 Modernization and Optimization of Irrigation Systems

Modernization and optimization of irrigation systems in the management of irrigation systems are often used as a tool to improve irrigation efficiency, produce more agricultural products with less water consumption. The options available to solve the problem of water scarcity in the community include the following:

1. Using more virtual water
2. Improving the economic efficiency of water
3. Improving the technical efficiency of water. Allan (1997, 1999).

Discussions on modernization and optimization activities can be considered in both the farm and irrigation areas.

In farm water management, the selection and use of irrigation equipment and irrigation planning are influenced by a person's decision.

Much research has been done to evaluate the quantitative impact of irrigation modernization and optimization programs on the use of surface water in the basin. However, it is necessary to assess the effects on the socio-economic stability of agricultural communities and water quality in the river basin, so that analysis Suitable for the benefits and costs of improving water efficiency (Playán and Mateos 2006).

2.4.3 Lack of Water Resources and Compensation Strategies

Providing sustainable freshwater resources has become a problem in many countries around the world. World population growth and industrialization, on the one hand, and changes in global warming, which has led to increased droughts, storms and floods around the world, are fueling water shortages. Despite these shortcomings, it seems that we need to look for different ways to compensate for these shortcomings. Water shortage compensation strategies can be divided into three main categories:

2.4.4 Wastewater Treatment and Gray Water

Wastewater treatment is a process in which wastewater is converted into water that can be reused or left in the environment without harm or danger.

Clean and healthy water is one of the most basic human resources for life on Earth. One way to access safe water is to treat wastewater. Sewage is one of the most valuable and essential sources of water today. In the past, sewage was considered a useless waste for various reasons.

One way to solve the water shortage problem is to recycle gray water. Gray water refers to recyclable water from sewage from homes or office buildings, which includes effluents other than sanitary water.

Due to daily activities such as washing clothes, washing dishes and bathing, it can be recycled and reused.

2.4.5 Water Desalination

Given that more than 95% of the world's water is in the form of saltwater, water supply from this source has always been of human interest. Accordingly, various methods have been developed for water desalination. The main methods in the process of desalination are saltwater, reverse osmosis and thermal methods.

2.4.6 Optimizing Water Consumption

The daily increase in water consumption has had consequences such as increased demand, reduced supply, drought, floods and pollution of human resources.

To increase the demand for water, we can pay attention to reducing consumption and increasing resources. Meanwhile, reducing water consumption not only helps to maintain limited resources but also reduces the need for water treatment. Water consumption can be classified into three categories: household, industrial and agricultural, depending on the source of consumption. In the following, we will review some new technologies in optimizing water consumption.

2.4.7 Optimizing Domestic Water Consumption

Research by Angelin has shown that the more customers are aware of their consumption in different parts of the home, the more controlled their consumption pattern will be. This helps consumers consciously reduce water consumption.

In this regard, the company tried to provide a new solution to inform and increase customer awareness of its consumption pattern. Recently, Angelin, in partnership with Advisor, a behavioral software company, has tried to study the water consumption strategy of its customers.

2.4.8 Optimizing Water Consumption in the Industry

One of the places where we need to pay attention to the control and optimization of water consumption is an industry. The use of water is common in all industries, and without the presence of this vital substance, it will be impossible to carry out industrial activities.

The increase in the price of water used by industrial processes on the one hand, and the application of environmental standards on the permissible concentration of pollutants during effluent, on the other hand, have doubled the importance of controlling water consumption in industries. Accordingly, various strategies for optimizing water consumption in industries are recommended, and we will examine some of them in the following.

2.4.8.1 Dry Machining

Ford Machinery has found a new way to reduce water consumption by focusing on the amount of water consumption in the machining sector. The basis of this method was later considered as the basis of dry machining. In this method, a composite lubricant containing some oil and a small amount of water was used, which was sprayed directly on the tip of the tool. Studies have shown that this method, in addition to having the necessary efficiency and uniformity with conventional solutions, has drastically reduced the amount of water consumed.

2.4.8.2 Integrate Steps in the Color Line

In the painting process, it is necessary to apply a phosphate coating on the metal to increase the adhesion of the original color to the metal. At this stage, a large amount of water is used to regulate humidity as well as washing. The painting process at Ford consisted of three main steps. In the first stage, the phosphate coating was applied, in the second stage, the underlayer paint was sprayed, and finally, the main paint was applied to the part in the third stage.

By changing the design of the painting process, Ford was able to integrate the first and second stages, reducing water consumption to one-third.

2.4.8.3 Sewer Recycling and Treatment

It is important to note that reducing water consumption can also help reduce energy consumption. There are several ways to reduce water consumption in the industry. One of the issues that need to be considered more in the industry is that in many heavy industries, the quality of water consumed does not have to be the same as that of drinking water. One way to reduce water consumption is to use purified and recycled water.

Cascade Tishou Company has been able to purify the consumed water as much as possible by using the filtration process. Recycled water is used instead of freshwater for various processes. This has led to a significant reduction in water consumption. The result of these processes is a reduction of 9.84 billion liters of water consumption and $ 24 million in annual storage for the company.

2.4.8.4 Water Recovery from Raw Materials

Using the recycling process, Nestlé has been able to supply its water from milk. In this factory, fresh cow's milk, which contains 88% water, is heated at low pressure to remove some of its water. As a result of this process, the water vapor produced in the next stage is subjected to the condensation process and after a series of preparation processes, it can be used for steam washing and cleaning machines. The effluent from this stage returns to the cycle after reassembly and is used a second time. This water is used for agricultural irrigation and washing processes. Nestlé Company receives 1.6 million liters of cow's milk per day, which has been used to reduce the amount of water consumed annually by 15% by designing these processes.

2.4.8.5 Use Recycled Materials

One of the processes that reduce water consumption in the paper industry is the use of recycled paper. Recycled paper, due to its fibrous structure, consumes less water than fresh and ready-to-use paper.

2.4.9 *Optimizing Water Consumption in Agriculture*

Global statistics show that 70% of the world's water is used in the agricultural industry, which is a high volume. Rising demand for food and water, on the one hand, and population growth, on the other, are forcing us to control the amount of water consumption in the agricultural industry. To do this, we can use different systems, some of which are discussed below.

2.4.9.1 Variable-Rate Irrigation System (VRI)

Irrigation is the most important issue of water consumption in agriculture. One of the most common methods of irrigation is the use of a central control system in which water is distributed 360 degrees everywhere. The problem with this method is the blind irrigation of agricultural land, which means that all parts of the land are irrigated equally, regardless of its conditions.

2.4.9.2 Water Pressure Control

By reducing the water pressure in a irrigation line, more control can be exercised over the amount of water consumed in that area. Farmers can use low water pressure to keep moisture close to the surface and reduce soil erosion. Also, with this operation, the irrigation line is controlled and does not leave the agricultural land.

2.4.9.3 Irrigation Planning

The use of soil moisture control sensors makes it possible to monitor wireless systems. By monitoring the amount of soil moisture depending on the day and time, irrigation can be planned completely automatically. By adding these features, the group is trying to reduce and optimize water consumption as much as possible.

2.4.9.4 Use of Intelligent Irrigation Systems

Smart Druplt is a small robot used to irrigate small gardens and farms. The robot can control the irrigation schedule based on the type of plant, the weather and the conditions and type of soil. The operating range of each robot is about 10 meters and it can supply any pot, plant or tree that is in this range.

2.4.9.5 Web Automation Systems

The system provided by Rachio is a smart web and automation software. This software allows you to control the irrigation process through smartphones and smart personal assistants such as Alexa (Amazon's smart assistant). This software can regulate the amount of irrigation based on rainfall or drought by collecting climate information in the area. One of the accompanying systems of this software package is the wireless water flowmeter. By measuring the amount of water passing through the pipes, this flowmeter controls and manages the leakage or bursting of the pipes online.

Water resource management activities can be divided into three categories: developing water resource management policies, managing measures to achieve policy goals, and evaluating their effects. The macro policies of water resources management determine the relationship between development and how these resources are exploited for national development purposes. The first step in formulating macro-policies for water resources management is to propose different options based on the pervasive and comprehensive limitations and goals of water resources development and management. Nowadays, due to the growing need for water resources, which is the result of a growing population and agriculture, it is not possible to plan only with enough and unreliable water resources. For this reason, we will briefly examine some of the necessities of this system.

1. Restriction of available water resources and heterogeneous distribution of resources
2. Increasing population growth
3. Increasing the vulnerability of various water supply systems due to their more complex nature
4. Increasing the per capita water consumption due to the high level of welfare and health of the people
5. Increasing the need in various industrial and agricultural sectors according to the development process of these sectors
6. Climate change and the need for long-term forecasts
7. Pollution of water resources
8. Destruction of resources, especially groundwater, due to over-disposal and lack of proper management.

2.5 Practical Examples

Due to the multiplicity of optimization methods used in water resources management, and the breadth and complexity of water resource systems and the need for optimization in its various sections, successful applications of genetic algorithms are examined in this section.

Genetic Algorithm: The genetic algorithm (GA) is one of the first and most successful evolutionary algorithms, and its successful applications have been reported in various sciences. GA is inherently suitable for solving specific predisposition problems and can be applied by making changes to solve continuous optimization problems and nonlinear combinations under nonlinear constraints of equality and inequality (Halland 1975; Goldberg 1987). To improve the performance of this algorithm, several improved versions have been invented so far, the most important of which are series genetic algorithms, turbulent genetic algorithms, hybrid genetic algorithms, self-organizing algorithms, genetically modified zygomatic (1999) and algorithms.

Structure of genetic algorithms: GA includes coding, evaluation, selection, composition, mutation, and decryption operators. First, the problem variables are coded and the algorithm deals with its encrypted form. In the evaluation stage, the evaluation function from the objective function evaluates each string with a numerical value and determines its quality. The higher the quality of the answer string, the higher the value of the answer, and the higher the possibility of participation for the next generation. In the selection process, a pair of chromosomes are selected to be combined.

The interface is the choice between two generations. The selection process is random and is the criterion for selection. In practice, the combination of the old generation of chromosomes is mixed, and the new generation of chromosomes is formed. The combination operator in the genetic algorithm eliminates the scattering or genetic diversity of the population and allows good genes to be combined. The mutation generator produces other possible responses. In the operator, the mutation is applied with a certain probability to the strings of the previous generation mating stage, and the strings obtained from the next generation strings are added. The result of the mutation operation will increase the likelihood of escaping from local optimal points. Mutations lead to the search for intact spaces in the problem, and its most important task is to prevent convergence in the local optimal. Mutations occur based on the probability of a mutation, and if the mutation steps are significant, the search for the genetic algorithm will be completely random.

The image decoding operator is the coding operator. After the algorithm provides the best answer to the problem, the decryption operation is applied to the answers so that the answer appears as a real function. Usually, the share of each of these operators in the production of a new series of strings is determined before the implementation of the program. After generating each new set of binary strings, the appropriateness of each string is calculated, and the condition required for the termination of the algorithm is examined. In case of non-fulfillment of the condition considered for the implementation of the algorithm, by repeating the above steps, new sets of fields are produced, and the corresponding proportional values are calculated. After obtaining convergence or providing the termination condition, the algorithm is stopped, and the best string obtained in the last generation is introduced as the optimal answer. The flowchart or trend of this optimization method is as shown in Fig. 2.4.

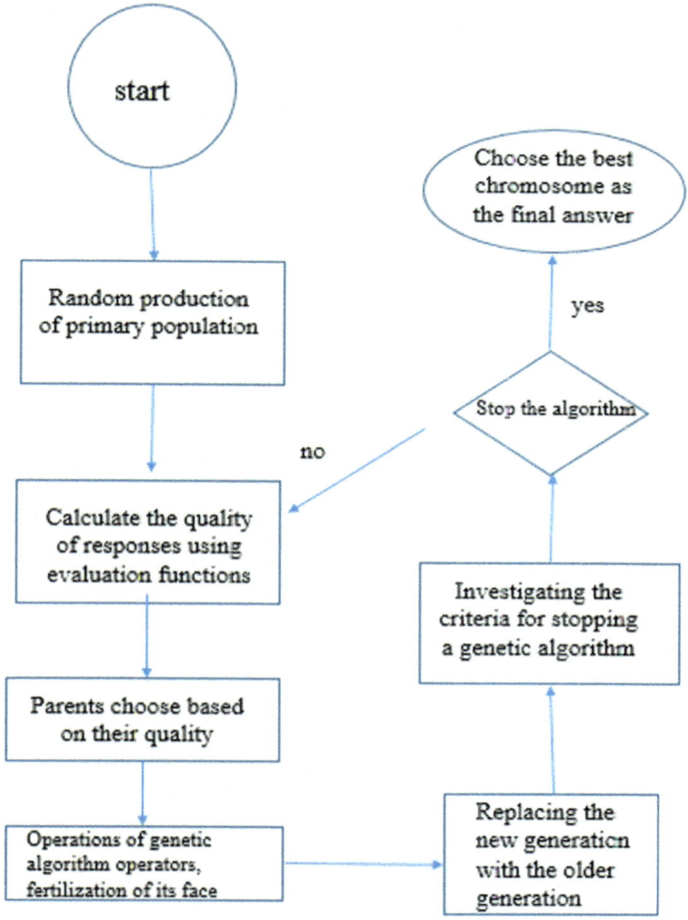

Fig. 2.4 Flowchart of the genetic algorithm method

Some examples of optimization are as following:

2.5.1 First Example

2.5.1.1 Pumping Program Problem

The purpose of this problem is to find the most economical pumping program from different pumps to meet the needs of different uses. Suppose there is a pumping of water by n different pumps and the cost of electricity consumption of pump I am at the unit price. Also, the need to form different consumption daily equivalent is bj to j for each consumption. Finally, it is assumed that the amount of supply from each pump I to j consumption is equal to said.

If xi is the pumping equivalent of each of the I pump, The problem would be to find the best values of xi to achieve the lowest cost and satisfy the various uses as follows:

$$\begin{aligned} \text{Maximize F } &= c_1x_1 + c_2x_2 + \cdots + c_nx_n \\ \text{Subject to } : a_{11}x_1 + a_{12}x_2 + \cdots + a_{1n}x_n &\leq b_1 \\ a_{21}x_1 + a_{22}x_2 + \cdots a_{2n}x_n &\leq b_2 \\ a_{m1}x_1 + a_{m2}x_2 + \cdots a_{mn}x_n &\leq b_m \\ x_1, x_2, \ldots x_n &\geq 0 \end{aligned}$$

As can be seen, all relationships are linear. Therefore, this is an example of linear programming. Of course, it is assumed that the mathematical model of the problem is not in the form of standard linear programming, because the problem is minimization. Also, the constraints are even more significant, and this is the case with many linear programming problems.

2.5.1.2 Drawing Method for Solving Linear Optimization Problems

This method can be used to solve linear optimization problems with two or at most three decision variables as well as with a linear target function. The advantages of this method include the ability to visualize the visual and tangible presentation of concepts and definitions of linear issues in it. Also, helping to quickly understand the process of solving the model and understanding many points that do not exist in linear programming are other advantages of the drawing method. In this method, by drawing the model constraints on the coordinate system, the justified (available) decision space is determined, and the optimal answer can be selected from within this space according to the value of the target function. The steps of the drawing method are as follows:

Step 1: According to the decision variable sign, the desired quarter is selected in the coordinate system.

Step 2: Each of the constraints is considered equal and is drawn as an equation and as a line.

Step 3: For unequal constraints, a point is selected at the desired end of one of the two sides of the line drawn for that constraint, and its coordinates are placed in the inequality. In the event of an unbalanced setting, the justified level is the same direction as the selected point on the other side.

Step 4: For all the constraints, step (3) is performed, and the standard level of the selected (justified) levels for each constraint is considered as the justified level of the problem.

Step 5: The objective function is plotted linearly for each of the hypothetical values of the decision variables.

Step 6: The target function line is moved in the right direction according to the type of optimization problem (maximization or minimization). The point of intersection of the justified surface, which is the last point of contact of the line of the target function moving in the right direction with the justified surface, is the optimal point.

Step 7: The equations of the constraints whose optimal intersection point is formed are considered in the equation device, and the optimal point coordinates are calculated. The following example clarifies the above description and drawing method (Bozorg-Haddad et al. 2014).

2.5.2 Second Example

It is essential to properly manage water and reservoir systems, plan to design and operate the reservoir, and optimize to maximize productivity and maximize profits. To this end, in 2008 to use the tank in real-time in a chiller tank system in Madhya Pradesh, India, two models of genetic algorithm (GA) and linear programming (LP) to obtain the appropriate time and amount of water released from the tank for Different goals, such as downstream irrigation, etc., have been developed and compared and analyzed. Finally, the results showed that the GA model is superior to the LP model (Azamathulla et al. 2008).

2.5.3 Third Example

Consider the following optimization: advise must take from a consulting engineering company to build or not to build a dam. The consulting engineer has taken over the project on 100 rivers. They know on which rivers the dam will be built or not, if it has a maximum economic benefit, maximum water storage, (Multi-Objective). When we consider all of this, we define an optimization problem.

However, how many choices do we have? In other words, the number of selected cases in this optimization problem is to decide which river the dam will be built on or not. If we have 100 rivers, we are faced with 100 decision-making options, which are called decision variables.

However, what does the answer to this optimization mean? That is, the results are obtained after those decisions, such as the maximum amount of economic profit or the maximum amount of water storage, and so on, which are called the answers to the objective function. If we have a function of a goal, our problem will be a single goal, and if we have several goals, there will be several goals.

2.5.4 Forth Example

Suppose that the design diameter of the pipes of an urban water distribution network is considered. In this example, finding the most appropriate values for the diameter of the network pipes is such that in addition to complying with the hydraulic constraints, the cost of running the network is also minimized. If we consider a network that has only 20 pipes and ten commercial diameters for it in the market, 1020, different modes or options for designing this network are possible.

The question is that if this optimization problem is solved by trial and error, there is no choice but to search among the 1020 different design modes. In each step of this search, in addition to calculating the cost of the pipes used in the construction of the network, it is necessary to consider the compliance with the hydraulic limitations of the network. So, if it takes only one second for each search to perform the calculations, it will take several centuries for the overall assessment of the possible states and the assurance of a possible optimal answer to this optimization problem. So, what can be done to solve this optimization problem?

Optimal answers to this example are easily obtained using a variety of evolutionary and meta-exploration optimization methods.

2.6 Summary

The primary purpose of this chapter is to review Berkeley and the details of the optimization theme with a focus on water resources management. In this regard, first the introduction and review of the studies conducted in the field of optimization in water resources, then a definition of keywords and terms related to optimization was presented to determine their general meaning. In the following, the basics and logic of the subject were mentioned in order to determine its basis. In this section, the types of models and the concept and logic of each of them were mentioned, and their nature and structure were determined. the importance and necessity of the issue were pointed out, and the reason for optimization in water resources management was stated, and its advantages were identified. the method of optimizing the genetic algorithm, which is one of the practical methods of optimization in water resources management, was examined a flowchart of the optimization process was presented. Finally, after presenting all the essential and required materials to clarify the issue for the readers, four examples in the field of optimization were mentioned, and it was shown that optimization in sustainable management. Water resources are a fundamental and essential factor in decision-making and policy-making that cannot be ignored.

Acknowledgements The authors thank Iran's National Science Foundation (INSF) for the support of this research.

References

Allan T (1997) Virtual water: a long term solution for water short Middle East economies? British Association Festival of Science, water and development session. University of Leeds, United Kingdom

Allan T (1999) Productive efficiency and allocative efficiency: why better management may not solve the problem. Agric Water Manag 40:71–75

Azamathulla HM (2013) A review on application of soft computing methods in water resources engineering. metaheuristics in water, Geotech Transp Eng:27–41

Azamathulla HM, Wu FC, Ab Ghani A, Narulkar SM, Zakaria NA, Chang CK (2008) Comparison between genetic algorithm and linear programming approach for real time operation. J Hydro-Environ Res 2(3):172–181

Belaineh G, Peralta RC, Hughes TC (1999) Simulation/optimization modeling for water resources management. J Water Resour Plann Manag 125(3):154–161

Bozorg-Haddad O, Saifollahi Aghmioni S (2013) Introduction to uncertainty in water resources systems. Tehran University Press

Bozorg-Haddad O, Farhangi M, Fallah-Mehdipour E, Mariño MA (2014) Effects of inflow uncertainty on the performance of multireservoir systems, Journal of Irrigation and Drainage Engineering, 140(11):04014035. https://doi.org/10.1061/(ASCE)IR.1943-4774.0000756

Bozorg-Haddad O, Janbaz M, Loáiciga HA (2016) Application of the gravity search algorithm to multi-reservoir operation optimization, Advances in Water Resources, 98, 173–185. https://doi.org/10.1016/j.advwatres.2016.11.001

Fatehi F, Beauty M (2010) Factors influencing the management of groundwater utilization using a multi-objective planning model: a case study of Firoozabad plain. J Agric Sci Technol Nat Resour Water Soil Sci 53

Ghanbari R, Nemati Z, Borzoi A (2014) Methods of using water resources management in agriculture (Challenges and Strategies). In: Second national conference on planning, environmental protection and sustainable development, Tehran, Permanent Secretariat of the Conference, Shahid Beheshti University, Iran

Gleick PH (1993) Water in crisis: a guide to the world's freshwater resources

Goldberg DE (1987) Genetic algorithms in search, optimization, and machine learning, Addison Wesley Publishing Company Inc

Halland J (1975) Adaptation in natural and artificial systems. MIT Press Cambridge, p 228

Hillier FS, Lieberman GI (1980) Introduction to operations research, 3rd edn. San Francisco, Holden-Day, Inc., 829 p

Kumar DN, Reddy J (2007) Multipurpose reservoir operation using particle swarm optimization. J Water Resour Plann Manag 133(3):192–201

Matsukawa J, Finney AB, Willis R (1992) Conjunctive use planning in Mad River basin, California. J Water Resour Plann Manag 118(2):115–131

Mayer A, Muñoz-Hernandez A (2009) Integrated water resources optimization models: an assessment of a multidisciplinary tool for sustainable water resources management strategies. Geogr Compass 3(3):1176–1195

Modarresizdi M, Asif Vaziri A (2003) Operations research - mathematical programming, vol. 2. Young Publications

Nikkami D, GHafouri A, Mahdian M, Bayat R, Chamheidar H, Arsham A (2012) Optimizing the management of soil erosion and sediment yield in one of the Roodzard representative sub-basins. Final report of the research project, Tehran 136 pp (in Persian)

Playán E, Mateos L (2006) Modernization and optimization of irrigation systems to increase water productivity. Agric Water Manag 80(1–3):100–116

Rao SS (1984) Optimization theory and application, 2nd edn. Wiley, 1247 p

Soltani (1994) Irrigation program for optimal use of water resources in Iran. Water Magazine. Soil Mach 3:10

Zahraei B (2016) Bio Online site (Iran's ranking in water resources management
Zeinali M, Azari A, Heidari MM (2020) Multiobjective optimization for water resource management in low-flow areas based on a coupled surface water-groundwater model. J Water Resour Plann Manag 146(5):04020020

Chapter 3
The Role of Water Information and Data Bases in Water Resources Management

Mahsa Jahanddideh-Tehrani, Omid Bozorg-Haddad, and Ioannis N. Daliakopoulos

Abstract Water information and data bases are used for addressing environmental, physical, social, political, economic, and ecological issues of water supply, consumption, availability, and accessibility. Different disciplines of water resources management, such as flood risk assessment, water supply management, reservoir operation, and water sanitation require incorporation of different types of water information for efficient planning and decision making. Therefore, this chapter aims to investigate the main features of water information, data sources, water data challenges, water data processing, analysis and dissemination which also play a key role in integrated and sustainable water resources management. This chapter begins with an introduction to the importance of data and information in various water disciplines as well as indication of data life cycle. Thereafter, different types of water data and sources (measurements, models, remote sensing, and administrative institutes) are proposed. Additionally, the challenges and limitation of water data, such as poor data quality, lack of integrated water portals, limited funds, and big data problems are discussed. This section is followed by indication of water data processing key points and steps of water data dissemination. Additionally, the World Hydrological Cyclone Observing System (WHYCOS), Global Runoff Data Centre (GRDC), and Bureau of Meteorology (BOM) are introduced as examples of important water data systems which improve development in delivery and use of water data, and evaluation of environmental impacts and risks. Finally, the chapter revealed recommendations

M. Jahanddideh-Tehrani
School of Engineering and Built Environment, Gold Coast Campus, Griffith University, Queensland, Australia
e-mail: mahsa.jahandideh-tehrani@griffithuni.edu.au

O. Bozorg-Haddad (✉)
Department of Irrigation and Reclamation Engineering, Faculty of Agricultural Engineering and Technology, College of Agriculture and Natural Resources, University of Tehran, Karaj, Tehran, Iran
e-mail: OBHaddad@ut.ac.ir

I. N. Daliakopoulos
School of Agriculture, Hellenic Mediterranean University, Heraklion 74140, Greece
e-mail: idaliak@hmu.gr

O. Bozorg-Haddad (ed.), *Essential Tools for Water Resources Analysis, Planning, and Management*, Springer Water, https://doi.org/10.1007/978-981-33-4295-8_3

to improve water data information and portals for the purpose of efficient water resources planning and management.

Keywords Water data · Measurements · Data processing · Data dissemination · Data management · Resolution

3.1 Introduction

Limited access to safe water, sanitation, and increased demand threatens human health, environmental sustainability, and economic growth, while leading to poor water security and supply. The increasing pressure on water resources is the result of global drivers such as rapid population growth, industrialisation, inefficient water management, water pollution, climate change, serious imbalance between supply and demand, but also poor water data (World Bank 2018). Therefore, water security, reliable water supply, as well as environment and ecosystem protection are directly dependant on the level of efficiency, integration, and sustainability of water resources management.

Data and information are denoted as the "life blood" of a research with the purpose of efficient water management and decision making. Hence, data is the foundation for well-informed decisions and sustainability that is required by all decision makers. Water data, or in other words measurements, is a key element in water resources management projects and planning. Water information is generated from water data through a process of analysis, integration, and interpretation. In other words, water information is the processed and synthesized form of water data. The generated water information should be combined with infrastructure and institution in order to reach an efficient water resources planning and management for both the current and uncertain future periods (Cantor et al. 2018).

Different water management disciplines, such as the flood response, drought management, water supply systems, reservoir operations, water sanitation, groundwater management, and irrigation supply require incorporation of different types of water information for better decision making and water planning. Thus, decision makers and water organizations need to access water information with proper spatial and temporal resolution (Cantor et al. 2018). Additionally, provision of sound data with proper quality is essential as it can affect the decisions and management strategies. However, there have been debates among organizations and water sector participants that some provided data do not meet the need of water users. Additionally, some water information is not shared openly. Despite the development of technology and modern measuring devices, there are still many gaps in available water data, and the quality of some water information is also still unacceptable (BOM 2017). Therefore, unmet data needs can interfere the efficient and sustainable water management.

Fig. 3.1 Data life cycle (Adopted from NSF DataONE Project 2017)

Figure 3.1 indicates the data life cycle (NSF DataONE project 2017) and the required steps from the data need evaluation to the data application. The main nine steps of the data life cycle are as follow:

- Data need evaluation and planning: needs of decision makers are fully evaluated;
- Data collection: observations are performed;
- Data quality assurance: checks and inspections are required to analyse the data quality;
- Data description and documentation: data should be accurately described using the metadata standards and guidelines;
- Data preservation and archival: data should be preserved in a specific archive (e.g., data service and data centre);
- Data discovery: both useful and relevant data information should be obtained;
- Data integration: data from different sources should be combined;
- Data analysis: data should be interpreted; and
- Data release planning: data should be described, managed, and made accessible.

Globally, water data and information is essential for sustainable planning, policy, management, and prediction. Additionally, water data provides decision makers with proper information about water demand, weather patterns, hydrologic modelling, infrastructure needs, climate change, and extreme events. This book chapter examines the important role of water data and information in integrated water resources management. It also provides an in-depth understanding of different water data types, sources, challenges, processing, and dissemination, which are essential for sustainable planning, policy, management, prediction, and decision making. Thereafter, examples of two international information systems (WHYCOS and GRDC) and a national water portal (BOM) are proposed to indicate how water data organisation can enhance climate trend analysis, water information products and services, and sustainable protection of water resources systems through regional and international cooperation. Finally, conclusions summarizes focal points of this work, and recommendations are proposed to help stakeholders discover and overcome limitations of current water data.

3.2 Water Data Type

Water data is a set of information that can address issues associated with physical, environmental, social, economic, ecological, biological, chemical, cultural, and political parameters in disciplines such as water use, availability, and accessibility. Additionally, water data can consist of a complex mix of different formats, including spreadsheets, satellite images, geospatial databases, and photographs (Laituri and Sternlieb 2014).

Generally, water data can be classified in two main categories: (1) Primary data and (2) secondary data. Primary data indicates the collection of raw data, including the data collected through measurements by technician and/or automated sensors. Measurements of water quality parameters, streamflow and precipitation are examples of primary data. Secondary data are obtained from models, specific methods, and lab experiments under controlled condition. Typically, models and methods apply primary data from different sources to compile and analyse them for estimating a variable. For instance, precipitation pattern, streamflow condition, irrigated area, and crop data can be combined through a model or method to estimate the agriculture water demand (Ziman 2016). Climatological models, hydrological models, processed remote sensing data or satellite images are some examples of secondary data (Laituri and Sternlieb 2014). In addition to these classes of water data, 12 extensive groups of water data can be developed based on disciplines or sectors where water data is applied (BOM 2017). Detailed classifications of water data with respective examples are shown in Table 3.1.

Table 3.1 Water data types together with data sources and examples

Water data	Data provider	Example
Hydrometric data	Hydrology services	discharge, water level, water quality parameters, flood inundation area, flood level, and water temperature
Meteorological data	Meteorological services	precipitation, humidity, evaporation, and evapotranspiration
Groundwater data	Geological and mining institutes	groundwater level, storage capacity, and permeability capacity
Water storage data	Geological and mining institutes	water storage bathymetry and level, storage volume, and storage inflow and outflow
Water use data	Water and environmental regulation organizations; and water resources institutes	water supply from rivers, groundwater and storages
Water quality data	Environmental protection organizations; and health-related institutes	turbidity, salinity, PH, nutrients, suspended sediment, and phosphorous
Wastewater data	Environmental protection organizations; and health-related institutes	stormwater volume, treated water, and sewage volume
Water pollutant data	Environment protection organizations; and environment and energy institutes	concentrations of fertilizers, bacteria, algae, and industrial waste
Manufactured water	Environmental and conservation organizations; and health-related institutions	water derived from recycling, desalination, and stormwater harvesting
Ecosystem data	Environmental organizations; and water and natural institutes	springs, lakes, ponds, caves, and wetlands
Water rights data	Water share trading institutes; and statistics institutes	water ownership, border rivers water, water transfers, and water license conditions
Administrative data	Water market institutes	water prices, water infrastructure expenses, and water access and sharing rules

All the stated data are essential for efficient decision making in different water disciplines, including the sectorial water management, integrated water sector planning, climate change adaptation, global and regional reporting, as well as operational and emergency management. Additionally, the type of required data for a specific project relies on the type of activity and plans. For instance, a flood inundation project will not require the same data as a peak flood prediction project. Additionally, the temporal and spatial resolution of the required water data should be specified for each project. For instance, the hourly river flow is generally applicable to peak flow prediction, and estimation of flash flood peak flows requires even higher temporal river flow resolution (e.g., minutes), while the daily river streamflow is required for analysis of river flow due to changes in land use pattern. Thus, the researchers and managers are required to consider the required data type and resolution from the beginning of a project.

3.2.1 Water Data Source

Different distinct types of water data are required to determine water quality and quantity to achieve efficient integrated water resources management. Water data is obtained through different sources and methods. Basically, the water data comes from four sources: (1) measurements; (2) models; (3) remote sensing; and (4) administrative institutes (BOM 2017). Each of the stated sources will be discussed in the next four sections.

3.2.1.1 In Situ Measurements

Water resources management constitutes a data-driven discipline, which involves consistent measurements of water quantity and quality. The measurements are mainly concerned with monitoring meteorological and hydrological variables (BOM 2017). In order to measure water variables, including physical, chemical, and biological parameters, specific methods and instruments are applied. For example, water level can be measured using different techniques such as a float and shaft encoder, a radar or an ultrasonic level sensor (NIWA 2019), turbidity by determining the intensity of light scattered by suspended particles in the water column (EHMP 2006–07), and biological contamination using *E. coli* concentration as a marker (Edberg et al. 2000). In addition to the variables which come from direct measurement, some other variables are calculated according to a mathematical transformation of a direct measurement. For instance, river discharge is computed by multiplying the water area in a channel cross section by the water average velocity in that cross section (BOM 2017).

In situ measurements can yield the most reliable water data, provided that the measuring devices are precise and properly calibrated. The restriction of direct measurements is the high cost of purchasing, installing, calibrating, fixing, and maintaining the measuring devices and equipment. Additionally, regular monitoring should be conducted periodically to obtain the proper data length and spatial coverage, and therefore, constant financial support and effort are required to manage the long and intensive monitoring programs (BOM 2017). Due to the discussed high costs, many poor countries suffer from improper and inadequate water data, which can directly affect the water management planning and strategies. This issue necessitates the provision of financial support by international agencies and organizations to improve the water monitoring networks in poor countries dealing with critical water supply and management problems.

3.2.1.2 Remote Sensing

In 1957, the Sputnik 1 was the first artificial earth satellite that placed into orbit. This satellite had significant impact on humanity's perception of space, and announcing a new era if earth observation. After launching the Sputnik 1, several systems (e.g., Television and InfraRed Observation Satellite (TIROS 1), Nimbus 1 etc.) were launched in the following year to weather and climate conditions. Such satellite missions led to dramatic advances in remote observations and measurements (McCabe et al. 2017). Remote sensing refers to the process of inferring surface parameters, which are derived from measurements of the emitted/reflected electromagnetic radiations, from the land surface (Schmugge et al. 2002). The application of remote sensing in hydrological sciences has provided new datasets with high temporal and spatial resolution for continuous water resources observations, which addresses the low resolution and expensive in situ measurements. Figure 3.2 indicates the earth observing system applied in hydrological sciences. The stated system consists of several components, signals of opportunity (e.g., cars, mobile phones), Doppler radar, mobile rovers, smart phone and citizens science (e.g., simple image capturing, plug-in and Bluetooth technologies), cell signals (e.g., antennas), unmanned aerial vehicle (UAV) (e.g., drones), research balloons, airborne vehicles, CubeStas, high definition (HD) videos, satellite missions (e.g., Landsat 1–8), sensors aboard the International Space Station (ISS), and geosynchronous meteorological satellites (McCabe et al. 2017).

There are two types of remote sensing instruments: i) active and ii) passive. These instruments are mounted on satellites, drones, vehicles, aircrafts, and ground-based structures to measure different water data parameters, such as soil moisture, evaporation, water temperature, rainfall rate, sediment concentration, flood inundation area, rainfall distribution, surface topography, wind speed and direction, cloud composition, and air humidity (BOM 2017). Active sensors provide their own source of energy for emitting radiation towards the investigated target, and then measure the time of arrival radiation reflected from the target. The majority of active sensors are able to perform in the microwave portion of the electromagnetic spectrum. Hence, these sensors can potentially penetrate the atmosphere under rough conditions. Laser altimeter, radar, Lidar, Ranging instrument, Scatterometer, and Sounder are examples of active sensors (NASA 2019).

On the other hand, passive sensors detect natural energy (light wavelength) which is emitted or reflected from the observed scene. The common source of radiation measured by passive sensors is the reflected sunlight. Passive sensors include various types of spectrometers and radiometers. These sensors are able to perform in the infrared, visible, thermal infrared, and microwave portions of the electromagnetic spectrum. The Accelerometer, Hyperspectral Radiometer, Imaging Radiometer, Sounder, Spectrometer, and Spectroradiometer are some examples of passives sensors (NASA 2019).

Remotely sensed data provides continuous spatial coverage and are typically presented in gridded form. The most temporally regular remotely sensed data set is obtained from satellites (BOM 2017). Since its advent in the 1960s, satellite remote

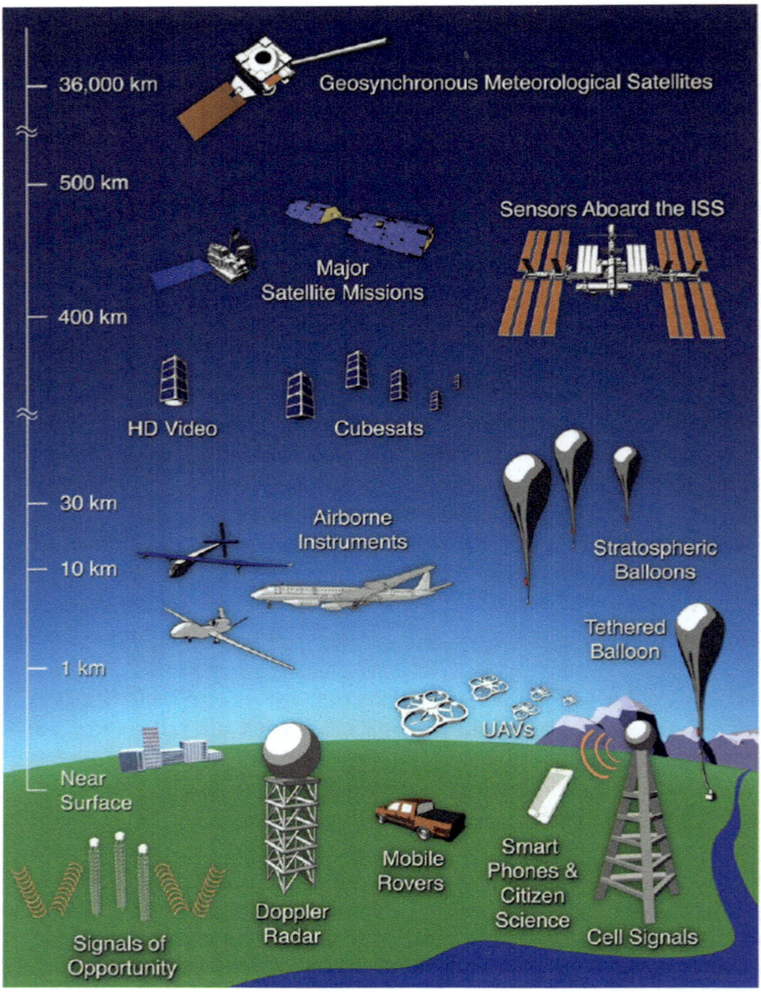

Fig. 3.2 Earth observing system (Adopted from McCabe et al. 2017)

sensing has been widely used as a complete source of information. Satellite sensors can measure almost all components of hydrological cycle, such as precipitation, evaporation, surface water, soil moisture, lake and river levels, surface and subsurface water storage, and snow storage (Sheffield et al. 2018). However, satellites have relatively coarser spatial resolution than in situ measurements due to their large distance from the earth in relation to the resolution of the sensor and the attenuation effect of the atmosphere between the sensor and the observation point. Low earth orbiting satellites present much higher spatial resolution with lower temporal resolution (BOM 2017).

The only issue that should be carefully considered in remote sensing is the requirement of significant information technology infrastructure for data management. Additionally, complex image processing tasks are needed to provide the data with suitable format (BOM 2017). In summary, remote sensing tools play an important role in improving water resources management through provision of large spatial and high temporal resolution.

3.2.1.3 Models

In order to tackle the shortcomings and costs associated with in situ measurements and also to address the poor spatial support of in situ point measurements, different types of models and methods can be applied to generate water data. Modelling approaches are useful tools to address poor spatial support of in situ point measurements. For instance, required spatially distributed water surface elevations for flood magnitude, extent and timing estimations are derived from hydrodynamic models. Therefore, models and methods are applied for different purposes. For example, different rainfall-runoff models have been developed for generating the runoff data in ungauged catchments. Conceptual rainfall-runoff model applications such as Nedbor-Afstromings Model (NAM) (Makungo et al. 2010; Faiz et al. 2018; Zhenlei et al. 2019), Hydrologiska Byråns Vattenbalansavdelning (HBV) (Engeland and Hisdal 2009; Osuch et al. 2019), and The Hydrologic Engineering Center Hydrologic Modeling System (HEC-HMS) (Gumindoga et al. 2017) are examples of model applications, which converts the rainfall to runoff in a defined catchment. In the field of water resources engineering, data-driven approaches have gained popularity for different water modelling purposes (Kim et al. 2013; Jahandideh-Tehrani et al. 2019; 2020a). Artificial neural network (ANN), support vector machine (SVM), and random forest (RF) are few examples of data-driven approaches, which have been extensively applied in the area if water related modellings such as ground water level prediction (Daliakopoulos et al. 2005; Tsanis et al. 2008; Baudron et al. 2013), rainfall-runoff modelling (Daliakopoulos and Tsanis 2016; Berezowski and Chybicki 2018; Jahandideh-Tehrani et al. 2020b), reservoir operation (Niu et al. 2019; Ahmad and Hossain 2019), flood prediction (Zhou et al. 2019; Ahmadalipour and Moradkhani 2019).

Some models are also used to extrapolate to past (e.g., paleoclimate) and future conditions, such as climate models. Future precipitation, evaporation, wind, and temperature can be generated using different climate models, such as Hadley Center Coupled Model (HADCM3), Norwegian Erath System Model (NorESM), and The Fifth Version Model for Interdisciplinary Research on Climate (MIROC5) (Randall et al. 2007). In order to summarize a large amount of information, specific approaches such as system dynamics (SD) are applied to analyse and interpret information. Using flow diagrams and feedback loops facilitate the analysis and conclusion for complex information (Ahmad and Simonovic 2004; Jahandideh Tehrani et al. 2014; Christias et al. 2020). In summary, models and methods are used for tasks, such as making predictions, data assimilation, and interpreting the large amount of complex information. The main limitation of applying models is the uncertainty resulting from models due to prior assumptions and scale issues. In water resources engineering field, analytical techniques (e.g., Mellin transform technique, and Fourier transform technique) and approximation techniques (e.g., first-order variance estimation (FOVE), probabilistic point estimation (PE) methods, and Monte-Carlo simulation) are the two main uncertainty analysis categories that can be applied to address hydrologic, hydraulic, structural, and economic uncertainties (Tung 2011). Many researchers have discussed model uncertainty analysis and proposed methods to reduce the amount of uncertainty. For instance, Wagener and Wheater (2006) investigated the sources of uncertainty in a rainfall-runoff model by focusing on 10 catchments. They identified that the uncertainty is related to local modelling (process of selection and calibration of the local model structure) and regionalization procedures. To increase the accuracy of model results, many studies have focused on analysing uncertainty and reducing the vulnerability. Her et al. (2019) evaluated the associated uncertainties with multi GCMs in runoff and precipitation projections. In terms of adaptation planning for uncertain climate change impacts, evaluating the robust adaptation decisions to climate change uncertainties was studied by Dessai and Hulme (2007). They concluded that water resources are sensitive to regional climate response uncertainties in climate change impacts. Also, Fletcher et al. (2019) developed a new planning framework to evaluate climate uncertainty over time for flexible planning strategies. In summary, despite the wide applications of models in improving water data, research is still required to improve the accuracy of the model results.

3.2.1.4 Administrative Institutes

Two types of data, including the water rights data and administrative data (mentioned in Table 3.1) are not obtained from measurements or models. Data such as water rights, water pricing, water infrastructure inventories, water demand in different economic activities, water ownership, and basin borders are basically recorded by water management agencies as part of their business purposes. Additionally, such data can be collected by conducting household and business surveys (BOM 2017).

Water rights data and administrative data play a critical role in developing, proposing, and evaluating the water management strategies and policies as well as decision making. For instance, water system governance, which refers to the structure and administrative process, should direct and control operations, decision making, legislation, and finance as part of their tasks. Therefore, administrative data is required in order to determine operational standards and water regulations.

A large number of agencies and organizations are involved in the establishment and administration of water rights in scale of regional, national and international scopes. In Europe, the Water Framework Directive (WFD) is the main foundation of European Union (EU) water policy that aims to ensure sustainable water use and protect aquatic systems through applying legislations to all surface waters (rivers, lakes, transitional and coastal waters) as well as groundwater (Carvalho et al. 2019). In Australia, the Department of Natural Resources and Mines is involved in water system administration in all states (Productivity Commission 2003). While different countries/regions/continents have their own administrative institutes to dictate water related policies (e.g., water pricing, water ownerships, river borders, etc.), there are also international foundations, which deal with global water issues. For example, World Water Council is an international multi-stakeholder organization that enhance political action and awareness to deal with critical water issues through developing partnerships between countries (World Water Council 2020). Pacific Institute, Clean Water Action, and Water.org are other examples of global organizations that aim to solve water conflicts, promote legislation for water protection, and deal with water crisis.

3.3 Key Characteristics of Water Data Type

Different types of water data have their own characteristics and use. In other words, each type of water data should be applied for a specific purpose of water resources management, such as flood management, water quality management, sediment control, and water security management. The key characteristics of water data together with examples are listed below (BOM 2017):

- Parameter type (e.g., river flow data for reservoir operation; water demand data for irrigation management)
- Measurement location (e.g., water quality data for upstream of water offtakes for drinking water supply; groundwater level data for heavily utilized aquifers)
- Spatial coverage (e.g., flood level data for flood prone areas; national scale water assessment)
- Spatial density (e.g., rainfall distribution for rainfall pattern assessment)
- Temporal frequency (e.g., daily or monthly river data for regional water security assessment; high frequency river and rainfall data for flood forecasting)
- Longevity of measurements (multi-decadal data of continuous rainfall for rainfall IDF analysis; historical river flow data for flood frequency and magnitude analysis)
- Latency (e.g., basin time concentration for rainfall-runoff modelling; real-time rainfall and river level data for flood forecasting)
- Precision (e.g., water quality data should be precise in order to confirm safe drinking water supply; water pricing should be assessed carefully to prevent probable issues between stakeholders).

3.4 Challenges of Water Data

In most countries, the nation's water data arrangements cannot satisfy the need of water sector participants, and the processes of data collection are limited. There is an increasing need for systematic data collection in order to improve monitoring and ensure informed decision making (IWRM 2019). Over the next three sections, basic challenges of accessing proper and accurate water data will be discussed.

3.4.1 *Poor Quality of Water Data*

There are many gaps in the available water dataset, and the quality of some water data is poor. Generally, the available water datasets are heterogeneous, dispersed, and poor in quality (BOM 2017). Dispersion and gaps in water data are mainly the result of poorly calibrated measuring devices as well as inefficient equipment maintenance. Apparently, water management will be challenging without good water data. Therefore, policy makers and managers are unable to make sound decisions as the provided water data are unreliable and contradictory (BOM 2017). Additionally, many poor countries, which are facing financial problems, are not equipped with modern and efficient measuring devices and techniques, which leads to poor measurements, and subsequently poor quality datasets (INBO 2018).

Lack of homogeneity in water data causes poor data management as the data is useless, and massive investments on water infrastructure and data collection is wasted. Heterogeneous data mainly occurs when private organizations follow their own specific guidelines and procedures for data collection, and there is no general adopted rules for all data collector organizations (BOM 2017).

3.4.2 *Lack of Integrated Water Portal Systems*

One of the main water data challenges in many countries is the lack of information about the available water data, and also not openly shared datasets. Basically, there is no single established organization/tool/website to provide potential users with information about the different available water types, and characteristics of the water date sets. Detailed metadata, i.e. information on "how the data is produced", "how the data can be accessed", "how the data should be used", "who to contact", and "what the licence of the data is" are not provided by institutions in many countries. Thus, researchers and water data users require considerable time, effort, and investment to identify stated information.

Therefore, data users and researchers need to spend time on basic data gathering and transformation rather than scientific analysis. This problem is noticeable when data is collected by multiple agencies for a specific task purpose. For instance, the required meteorological data models and formats for hydrological purposes (e.g., groundwater and catchment modelling) are different from the meteorological data formats for atmospheric science analysis. As a result of this data difference, a hydrologist should spend significant amount of time on learning the file format and visualization tool used by atmospheric science community. To address this problem, data should be communicated between scientific sub-disciplines systems through standard protocol/web services (Goodall et al. 2008). In this context, organizations such as European Flood Awareness System (EFAS), Global Flood Awareness System (GLOFAS), European drought observatory (EDO), Watershed Index Online (WSIO), United States Geological Survey (USGS) Water Data, BOM climate data online, European Space Agency (ESA), Copernicus Space Component (CSC) and National Aeronautics and Space Administration (NASA) have made efforts to provide users with documented, homogeneous, and well-organized data set, which can be accessed through web services, and indicate datasets in the form of maps, tables, and graphs.

3.4.3 Limited Access to Data

According to the Integrated Water Resources Management (IWRM) principle, the public should participate in decision making processes. In order to pave this way, water institutions and organization should make the water data and information available and accessible to the public. However, much water data is not accessible by the public, or the water data type is not understandable by people with no expert knowledge. Thus, to ensure the participations of the public, water data should be disseminated in an informative and comprehensible way (INBO 2018). In addition to the public, openly shared and freely access to most water data is limited for water data users, particularly researchers. Regarding the fact that in many countries, funds for data collections come from private sectors, institutions are mainly reluctant to share the data freely. Additionally, many national organizations have no tends to share the collected data with the neighbouring countries or other international organizations. This situation is observed particularly when the countries are facing critical water scarcity, and the countries under such conflicts deny publishing information due to security risks (INBO 2018).

Additionally, poor countries, which are facing financial problems, are unable to build efficient water information systems, manage water data, and develop efficient web portals and tools for data sharing. As a result, international organizations and agencies play an important role in providing financial support as well as remote monitoring to deal with the water problem in poor countries (INBO 2018).

3.4.4 Big Data Problems

Regarding the increased amount of water data, several challenges can appear for big water data management. Big water data is associated with increased data volume, variety, and velocity. The amount of measured and processed data is growing rapidly. Additionally, the growing formats of data can make data management challenging. Considering the increased automation in data measurement process, the biases, abnormalities, and noise are also increasing in raw water data. As a result, a suitable and massive validation process is required to treat raw water data. Citizen science data is another aspect of big data complexity. Citizen science refers to the participation of the general public in the process of research design and data collection, which leads to the collection of large volumes of data (Buytaert et al. 2014). In the field of climate modelling, the spatial and temporal dimensions of data are intensive as such data indicates complex process of atmosphere and ocean circulation. Therefore, such climate data requires high storage space and large computer processing power (Cuntz et al. 2007). High volume data with long time periods can lead to high potential of data loss or intrusion. Generally, with high increase in the amount of data, many challenges are coupled with the process of big water data management.

3.5 Data Processing

Data processing refers to the manipulation of data obtained from in situ measurements or remote sensing, before they can be efficiently stored or made available (Hughes 1998). Raw data should be processed and transformed into formats to be identified and applied by researchers, data users, and decision makers (INBO 2018). Regarding the cost, effort, and high volume involved in data collection, proper data processing tools and approaches should be adopted (Hughes 1998). Many tools are available for data processing, including spreadsheets and databases, database management tools, geographic information systems (GIS), Extract-Transform-Load (ETL) tools, modelling tools, and statistical data analysis tools. However, the selected tools and approaches should meet the requirements of the data users and researchers: What data format is required? What is the purpose of using the data? What kind of research will be conducted? And which organization will use the data? (INBO 2018).

3.5.1 Water Data Processing and Analysis

The main purpose of data processing is to convert raw data into information, which meets the need of researchers and decision makers. Key points of water data processing are (Hughes 1998):

- Data encoding and digitizing;
- Quality control, consistency checking, and error checking;
- Data storage resolution;
- Data calibration and stability;
- Data storage and retrieval system.

The stated key points will be discussed below:

Data encoding and digitizing: Converting the measurements to digital is the first step of data processing. Hand-written field sheets or automatically penned charts mounted on a rotating drum are simple forms of data collection, which is still applied in many areas, particularly for collecting rainfall, evaporation, and other meteorological variables. Next, this data should be transferred to a computer database. Lost or damaged data is the main drawback of this manual data recoding. Therefore, modern telemetry, remote sensing, and other modern instruments have been developed to record the data automatically, particularly collecting the "real time" data (Hughes 1998). The vast quantities of recorded data (in situ measurements and remote sensing data) should be transformed into the cloud services in order to be stored and classified and then be accessible through portals and web services (McCabe et al. 2017).

Data quality control: Data collection corresponds with issues such as missing data, recorded data over the incorrect time, accumulated data, instrument failure, etc. Additionally, most databases are not able to identify poor and incorrect data and inform data users of data inconsistency (Hughes 1998). Missing data processing, logical error detection, repeated data processing, abnormal data detection, and inconsistent data processing are main aspects of hydrological data quality control. Prediction methods (e.g., recurrence neural network (RNN), and support vector machine (SVM)), optimization methods (e.g., particle swarm optimization (PSO)), and statistical methods (e.g., adaptive boosting, and statistical control) are some examples of data driven approaches for data quality control (Zhao et al. 2018).

Data storage resolution: Time resolution is of great importance in water data, particularly, the data used in hydrological analysis. For example, required time step for analysing the rainfall intensity in temperate climate is different from the arid and semiarid climates. Normally, the rainfall data with a resolution of one hour should be proper in temperate climate, while short-term variation of intensity is more important in arid and semiarid areas, and hence, minute should be the proper time step in arid areas (Hughes 1998). Therefore, disaggregation methods, such as Hyetos based on Bartlett-Lewis process and Marcov chain model should be applied in order to break down the data into shorter time steps (Koutsoyiannis 2003), which is considered as an important task of data processing.

Data value accuracy: Accuracy of observations is a numeric value that quantifies the data measurements accuracy based on the closeness to the standard value. The stated value accuracy measures the associated uncertainty with data measurements, which is the result of errors in bias and precision. The value accuracy can be estimated using the knowledge of instrument accuracy and measurement method as well as statistical analysis of the records from repeated measurements (Horsburgh et al. 2008).

Additionally, some instrument calibration process (either manual calibration or using table and equations) is required to ensure accurate measurements. For instance, the stage-discharge relationship is commonly used to convert water level measurements of a flume to flow rate or discharge. In the case of a flooding event in a river, calibration process is necessary as riverbank erosion and massive sediment transport occur over flood events (Hughes 1998).

Data storage and retrieval system: As describe in Sect. 4.4, big data requires massive amounts of data to be stored. Metadata catalogues and data storage systems help data users to identify the available datasets and check their characteristics (Hughes 1998). Within this context, recorded observations should be stored with sufficient supporting information (e.g., the location, date and time and type of variable that was observed) about the observations to enable data interpretation and cross-dimension data retrieval and analysis (Horsburgh et al. 2008). The databases should provide users with helpful information on the type measurements, data length, instrument details, the latitude and longitude of measurements, data quality, missing data, time zone, etc. Additionally, an efficient database is able to allow user to access and download the data in different formats, such shape files, text file, and excel files. It is also necessary to design a frequent backup system for the databases in order to prevent data lost in a case of system failure (Hughes 1998). In the scope of data storage and retrieval, Internet of Things (IoT) technology, which has been developed rapidly in recent years, is considered as a key approach in big data processing. Within this context, cloud computing is a new generation of computing architecture that supports big data applications and growing data processing (Fan et al. 2018). As an example of such applications, the cloud-computing services have been applied to watershed management by enabling watershed partnerships for focusing primarily on decision-making activities (Sun 2013). Bürger et al. (2012) also developed an intuitive Web interface to create an integrated hydrologic simulation platform that enables users to access scientific software through a web browser.

3.5.2 *Water Data Visualization*

After the data process step and data preparation, the next step is to present the data in an understandable and efficient way. Regarding the recent improvements of digital culture and information technologies, large datasets become complex to understand. Therefore, visual presentation of data has been developed, which presents data in the form of a graph, map, video, animation, hierarchy, etc. Data visualization can also improve understanding of existing trends, patterns and correlation in the datasets that might not detected in text and number-based data. As a result, stakeholders and decision makers can easily identify the important issues of water resources management to focus on as well as adapting new policies. In the scope of data visualization, the quality of data presentation is important as a poor or incorrect type of data visualization may cause misleading and hinder smart decision making. The visualization tool should be selected based on specific criteria, including i) how data viewers interact with data, ii) the information that should be obtained by the viewers, iii) the function of the data, and iv) the data composition. The three main types of water data presentation are as below (INBO 2018):

- **Maps**: They indicate the geographical distribution of different parameters, such as evaporation, rainfall, temperature. Maps can be produced using GIS software.
- **Key figures**: They are generated from data processing and document analysis. Key figures are understandable by public.
- **Factsheet**: They show a regular summary of data analysis, information, and graphs on different topics such as global warming, climate change, and drought trend.

3.5.3 Water Data Dissemination

After collecting, verifying, processing, and storing the water data, the next important task is to release and disseminate the data according to the target group (e.g., the general public, researchers, and decision makers) (e.g., before and after a flooding event). Data dissemination consists of four main steps (SEQwater 2014):

(1) **Data value identification**: The first step is to evaluate the data value and information asset to decide on the price of the water data. The data value is conducted by surveying business system, water companies, and stakeholders about their needs and interests. Some data can be accessible free of charge considering the expenses of data preparation process and company's needs and interests.
(2) **Data assessment**: Over the second step, the data suitability for release should be assessed based on relevant guidelines, policies, and legislations. Additionally, depending on the target group (e.g., scientific, non-expert purposes), data visualizations, data use instructions, and data use policies should be defined.
(3) **Data publication**: Over the third step, data should be published based on priority factors such as public interest, stakeholder feedback, company and institutes need, and economic values. Additionally, given the type of requirements of different target groups (e.g., the general public, researchers, and decision makers) and required time of hydrological events (e.g., before and after a flood event, before and after a dam construction), data dissemination should be specified.
(4) **Data management**: Over the last step, data quality should be improved with regard to data user's feedback.

In order to release the water data, different strategies and digital tool are available. Below is the list of different water data dissemination plans (INBO 2018):

- **Web portals and websites**: data resulted from monitoring programs is delivered through portals and website. Basically, portals provide organized data, and water data is classified based on different criteria such as collected data from open stations, water data type (e.g., water level, streamflow, and rainfall), collected data from closed station, river type, time periods, water data unit (e.g., mega litre and cubic meters), and measured interval (e.g., hourly, daily, and monthly).

- **Factsheets**: guidelines, water issues (e.g., global warming and climate change), water data change trend analysis (e.g., temperature trend, major flood event trend and evaporation change trend), monitoring programs (e.g., coastal erosion monitoring), management programs (e.g., integrated sub-basin management plans) can be published through digital books.
- **Social media networks**: In addition to portals, websites, and digital books, social media networks also play an important role in water data dissemination. Social media networks increase people's awareness of available water data and establish contact with internet and data users. Additionally, social media are effective way of transferring news and water data updates to users.
- **Smart apps**: smart and free apps are favourable, particularly among the public. Smart applications can potentially target different water users and improve interaction. For instance, "Ma Cons'eau" is a useful free application, which can estimate the people's water consumption through providing a set of simple questions. The amount of water consumption is calculated based on the water price of resident locations.

3.6 Practical Examples

Many organizations, systems and programs have been developed to provide high quality data and services of weather, climate, hydrological, and environmental fields for the purpose of integrated water resources management. Therefore, multiple water data, information, and knowledge should be shared to build cross-functionality between stakeholders.

One example of the water data system is the World Hydrological Cycle Observing System (WHYCOS), launched by the World Meteorological Organization (WMO) in 1993 to improve sustainable development in delivery and use of hydrological data as well as promoting regional and international cooperation. WHYCOS aims to enhance the sustainable protection, use and management of water resources systems as well as support decision makers through reliable and data provision. In terms of regional water resources management, HYCOS promotes data and information products, such as flood forecasts and warning data. At global scale, the WHYCOS International Advisory Group (WIAG) was introduced to provide support for policy guidelines and future development of the projects (WMO 2019).

Another example of global water data organization is the Global Runoff Data Centre (GRDC), which was established three decades ago. GRDC is an international archive of streamflow data, which aims to enhance analysis of global climate trends and evaluate environmental impacts and risks. The Global Runoff Data Base (GRDB) at GRDC was built on an initial dataset collected in the early 1980s. GRDB is a unique tool for obtaining river streamflow data at daily and monthly time step from over 9,500 gauging stations in 161 countries.

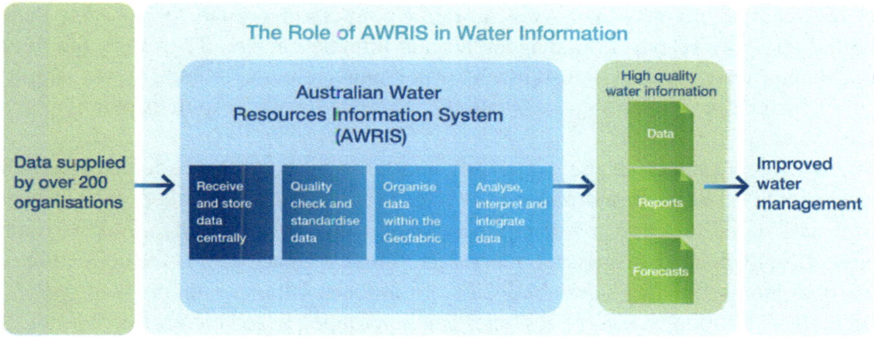

Fig. 3.3 The role of AWRIS in water information in Australia (Adopted from AWRIS Information Sheets 2019)

In terms of national scale, the Bureau of Meteorology (BOM) in Australia has developed the Australian Water Resources Information System (AWRIS). As shown in Fig. 3.3, the purpose of AWRIS is to receive and manage water data and information as well as supporting and disseminating different water information products and services.

AWRIS is able to store and manage information about groundwater levels, river flow, the quality of water in rivers and aquifers, water trades, water use and restrictions, and water volume in storages in a central database. Next, the stored data is checked for quality and standards to ensure that the stored data is consistent with the current available data. Over the next step, the Australian Hydrological Geospatial Fabric cast the stored water information in a spatial context through encoding the spatial connections between Australia's hydrological features (e.g., rivers, lakes, aquifers, dams, channels etc.). As indicated in Fig. 3.3, the final step includes analysing, interpreting, and integrating the water data. AWRIS delivers variety of water data, reports, and forecasts, such as real-time water reports, regular national water resources assessments, improved flood warning systems, and seasonal streamflow forecasts. The main advantage of AWRIS is that water users are able (i) to identify water resources condition in Australia through integrated viewpoint, (ii) to prepare benchmark reports using quality and transparent data, (iii) to check the details of standards, (iv) to control the quality of applied water data, and (v) to easily use the data for planning, analysis, reporting, and modelling (AWRIS information sheets 2019).

3.7 Summary

Limited water resources and increased demand for the resources leads to growing challenges facing water resources systems. This growing pressure on water resources is the result of many issues, such as rapid population growth, poor water data, inefficient water resources management, and climate change. Therefore, the need to enhance water security, sustainable water management, reliable water supply, and environment protection necessitate the integrated water resources planning and management.

The chapter underlines that water data and information are considered as the key components of water-based research, which provide decision makers with proper information in different water disciplines. Water data can be classified into two main categories, primary data and secondary data, where primary data is the collection of raw data, and secondary data obtained from models, lab experiments, and specific methods. Additionally, water data types can be grouped based on disciplines/sectors that water data is used. For instance, hydrometric data, meteorological data, and groundwater data are applied in hydrology services, meteorological services, and geological and mining institutes, respectively. Different distinct water data is obtained from four main sources: (1) measurements; (2) models; (3) remote sensing; and (4) administrative institutes. The chapter also discuss the limitation of water data, including poor quality, lack of integrated water portal systems, limited funds for data provision, and big data problems. In order to efficiently store and apply data, water data should be processed and analysed through key steps: (1) data encoding and digitizing; (2) quality control, consistency checking, and error checking; (3) data storage resolution analysis; (4) data calibration and stability; and (5) data storage and retrieval systems. After the data processing steps, water data should be visualized and presented in an understandable way to find our existing climate trend, pattern and correlation for the purpose of adapting new policies and management strategies in the realm of water resources. Thereafter, water data should be released and disseminated through web portals and websites, E-books, media and social networks, and smart apps to be used by decision makers, researchers, and policy makers for efficient water management planning.

In general, the outcomes of the present chapter will benefit different water disciplines sectors through provision of discussion on the role, current condition, and importance of water data and information in water resources management.

3.8 Recommendations

Regarding the important role of data provision in integrated water resources management, it is recommended:

(1) **To provide public easy access to water data with the purpose to serve many different contexts**: In order to improve decision making strategies, enhance national and international research studies, and ensure efficient water resources management, different water data sources and types should be highly accessible. Additionally, free provision of some data sources can potentially increase people's awareness about the responsibility and importance of different water

sectors. As a results of this increased awareness, public tends to have interactions and collaborations with experts, which leads to efficient water supply and management.

(2) **To improve collaboration and engagement between water data system or portal developers and users**: Identification of the intended purpose of data provision is a fundamental need for achieving efficient data use. Therefore, before providing the data, sound understanding of the need of stakeholders and data users is essential to ensure the relevant data collection.

(3) **To develop a single global data system**: Despite the availability of many different sources, portals and websites for data users, some researchers have challenge to identify the most accurate and efficient source of data for their studies. In order to address this issue, multiple sources of water data at different spatial and temporal resolutions should be consolidated in a single global data system.

(4) **To provide data in different formats and resolutions to target different decisions making strategies**: Based on the type of decision-making strategy, research project, and water management strategy, different temporal and spatial resolutions of data should be provided. Hence, an efficient water data system should provide resolution flexibility in data as well as ensuring the quality and integrity of data.

(5) **To propose and build a novel water data system for water data format conversion**: Many researchers and water data users should convert the format of the collected water data into the required input data format of different water software packages. The process of data format conversion is time and money consuming. Therefore, the generation and provision of different water data formats help users to save time in running water software and models.

(6) **To develop efficient water data systems, which are able to generate and update information from water data automatically**: Water data systems should be improved in order to automatically generate some illustrations (maps, figures, and tables) from the input data. Additionally, the new collected data should automatically update the water data system by providing updated illustrations and information.

(7) **To develop more accurate data gap filling methods**: Some data, such as future meteorological data under climate change, should be estimated. There are also gaps in some data due to different reasons, such as defective measuring device. In order to inform decision making and researchers with proper data, the stated limitations and gaps should be addressed.

(8) **To integrate different data sources to ensure quality data**: Various organizations provide water data based on different standards of data measurements, processing, and collection. Water organizations and sectors should integrate water data system by satisfying specific level of standards in terms of data quality, processing, measurements, and documentations.

These recommendations will help water data providers, modellers, and researchers to detect the limitations of current water data and to identify room for improvements.

Acknowledgements The authors thank Iran's National Science Foundation (INSF) for the support of this research.

References

Ahmad SK, Hossain F (2019) A generic data-driven technique for forecasting of reservoir inflow: application for hydropower maximization. Environ Model Softw 119:147–165. https://doi.org/10.1016/j.envsoft.2019.06.008

Ahmadalipour A, Moradkhani H (2019) A data-driven analysis of flash flood hazard, fatalities, and damages over the CONUS during 1996–2017. J Hydrol 578. https://doi.org/10.1016/j.jhydrol.2019.124106

Ahmad S, Simonovic SP (2004) Spatial system dynamics: new approach for simulation of water resources systems. J Comput Civil Eng 18(4)

AWRIS Information Sheets (2019) Australian Water Resources Information System, Bureau of Meterology. http://www.bom.gov.au/water/about/wip/awris.shtml

BOM (2017) Good practice guidelines for water data management, policy, World Water Data Initiative, Bureau of Meteorology, Melbourne, Australia

Baudron P, Alonso-Sarría F, García-Aróstegui JL, Cánovas-García F, Martínez-Vicente D, Moreno-Brotóns J (2013) Identifying the origin of groundwater samples in a multi-layer aquifer system with random forest classification. J Hydrol 499:303–315

Berezowski T, Chybicki A (2018) High-resolution discharge forecasting for snowmelt and rainfall mixed events. Water 10(1). https://doi.org/10.3390/w10010056

Bürger CM, Kollet S, Schumacher J, Bosel D (2012) Introduction of a web service for cloud computing with the integrated hydrologic simulation platform ParFlow. Comput Geosci 48:334–336

Buytaert W, Zulkafi Z, Grainger S, Acosta L, Alemie TC, Bastiaensen J, De Bievre B, Bhusal J, Clark J, Dewulf A, Foggin M, Hannah DM, Hergarten C, Isaeva A, Karpouzoglou T, Pandeya B, Paudel D, Sharma K, Steenhuis T, Tilahun S, Hecken GV, Zhumanova M (2014) Citizen science in hydrology and water resources: opportunities for knowledge generation, ecosystem service management, and sustainable development. Front Earth Sci. https://doi.org/10.3389/feart.2014.00026

Christias P, Daliakopoulos IN, Manios T, Mocanu M (2020) Comparison of three computational approaches for tree crop irrigation decision support. Mathematics 8(5). https://doi.org/10.3390/math8050717

Cuntz N, Leidl M, Darmstadt T, Kolb G, Salama C, Bottinger M, Klimarechenzentrum D, Hamburg G (2007) GPU-based dynamic flow visualization for climate research applications. In: Proceedings of the simulation und Visualisierung, SCS Publishing House, pp 371–384

Cantor A, Kiparsky M, Kennedy R, Hubbard S, Bales R, Pecharroman LC, Guivetchi K, McCready C, Darling G (2018) Data for water decision making: informing the implementation of California's open and transparent water data act through research and engagement, Centre for Law, Energy & the Environment, UC, Berkeley School of Law, CA. https://www.law.berkeley.edu/research/clee/research/wheeler/data/

Carvalho L, Mackay EB, Cardoso AC, Baattrup-Pedersen A, Birk S, Blackstock KL, Borics G, Borja A, Feld CK, Ferreira MT, Golbevnik L, Grizzetti B, Hendry S, Hering D, Kelly M, Langaas S, Meissner K, Panagopoulos Y, Solheim AL (2019) Protecting and restoring Europe's waters: An analysis of the future development needs of the Water Framework Directive. Sci Total Environ 658:1228–1238

Daliakopoulos IN, Coulibaly P, Tsanis IK (2005) Groundwater level forecasting using artificial neural networks. J Hydrol 309(1–4):229–240

Dessai S, Hulme M (2007) Assessing the robustness of adaptation decisions to climate change uncertainties: a case study on water resources management in the East of England. Glob Environ Change 17(1):59–72

Daliakopoulos IN, Tsanis IK (2016) Comparison of an artificial neural network and a conceptual rainfall–runoff model in the simulation of ephemeral streamflow. Hydrol Sci J 61(15). https://doi.org/10.1080/02626667.2016.1154151

EHMP (2006) Ecosystem health monitoring program annual technical report, South East Queensland Healthy Waterways Partnership, Brisbane, 2006-07. https://hlw.org.au/download/ehmp_2006_07_technical_report/

Engeland K, Hisdal H (2009) A comparison of low flow estimates in ungauged catchments using regional regression and the HBV-model. Water Resour Manag 23(12):2567–2586

Edberg SC, Rice, EW, Karlin RJ, Allen MJ (2000) Escherichia coli: the best biological drinking water indicator for public health protection. J Appl Microbiol 88 (S1). https://doi.org/10.1111/j.1365-2672.2000.tb05338.x

Faiz MA, Liu D, Fu Q, Li M, Baig F, Tahir AA, Khan MI, Li T, Cui S (2018) Performance evaluation of hydrological models using ensemble of general circulation models in the northeastern China. J Hydrol 565:599–613

Fletcher S, Lickley M, Strzepek K (2019) Learning about climate change uncertainty enables flexible water infrastructure planning. Nat Commun 10. https://doi.org/10.1038/s41467-019-09677-x

Fan Y, Zhu Q, Liu Y (2018) Cloud/fog computing system architecture and key technologies for South-North water transfer project safety. Wireless Commun Mobile Comput. https://doi.org/10.1155/2018/7172045

Goodall JL, Horsburgh JS, Whiteaker TL, Maidment DR, Zaslavsky I (2008) A first approach to web services for the National Water Information System. Environ Model Softw 23(4):404–411

Gumindoga W, Rwasoka T, Nhapi, I., and Dube, T. D (2017) Ungauged runoff simulation in Upper Manyame Catchment, Zimbabwe: application of the HEC-HMS model. Phys Chem Earth Parts A/B/C 100:371–382

Hughes DA (1998) Data processing in hydrology. In: Encyclopedia of Hydrology and Lakes, Encyclopedia of Earth Science. Springer, Dordrecht

Horsburgh JS, Tarboton DG, Maidment DR, Zaslavsky I (2008) A relational model for environmental and water resources data. Water Resour Res 44(5). https://doi.org/10.1029/2007WR006392

Her Y, Yoo SH, Cho J, Hwang S, Jeong J, Seong C (2019) Uncertainty in hydrological analysis of climate change: multi-parameter vs. multi-GCM ensemble predictions. Sci Rep 9. https://doi.org/10.1038/s41598-019-41334-7

INBO (2018) The handbook on water information systems: Administration, processing and exploitation of water-related data, International Network of Basin Organization, Paris, France, 2018. https://www.riob.org/pub/HandBook-SIE-en/

IWRM (2019) Integrated Water Resources Management Data Portal. http://iwrmdataportal.unepdhi.org/

Jahandideh-Tehrani M, Bozorg-Haddad O, Mariño MA (2014) Power generation simulation of a hydropower reservoir system using system dynamics approach: case study of Karoon reservoirs system. J Energy Eng 140(4):04014003. https://doi.org/10.1061/(ASCE)EY.1943-7897.0000179

Jahandideh-Tehrani M, Bozorg-Haddad, Loáiciga HA (2019). Application of non-animal–inspired evolutionary algorithms to reservoir operation: an overview. Environ Monitor Assess 191. https://doi.org/10.1007/s10661-019-7581-2

Jahandideh-Tehrani M, Bozorg-Haddad, Loáiciga HA (2020a) Application of particle swarm optimization to water management: an introduction and overview. Environ Monitor Assess 192. https://doi.org/10.1007/s10661-020-8228-z

Jahandideh-Tehrani M, Jenkins G, Helfer F (2020b) A comparison of particle swarm optimization and genetic algorithm for daily rainfall-runoff modelling: a case study for Southeast Queensland, Australia. Optim Eng. https://doi.org/10.1007/s11081-020-09538-3

Koutsoyiannis D (2003) Rainfall disaggregation methods: theory and applications. In: Proceedings, workshop on statistical and mathematical methods for hydrological analysis, Rome, Universita' degli Studi di Roma "La Sapienza, pp 1–23

Kim S, Shiri J, Kisi O, Singh VP (2013) estimating daily pan evaporation using different data-driven methods and lag-time patterns. Water Resour Manage 27:2267–2286

Laituri M, Sternlieb F (2014) Water data systems: science, practice, and policy. J Contemp Water Res Educ Banner 153(1):1–3

Makungo R, Odiyo JO, Ndiritu JG, Mwaka B (2010) Rainfall–runoff modelling approach for ungauged catchments: a case study of Nzhelele River sub-quaternary catchment. Phys Chem Earth, Parts A/B/C 35(13–14):596–607

McCabe MF, Rodell M, Alsdorf DE, Miralles DG, Uijlenhoet R, Wagner W, Lucieer A, Houborg R, Verhoest NEC, Franz TE, Shi J, Gao H, Wood EF (2017) The future of Earth observation in hydrology. Hydrol Earth Syst Sci 21(7):3879–3914

NASA (2019) https://earthdata.nasa.gov/user-resources/remote-sensors

NIWA (2019) https://www.niwa.co.nz/

NSF DataONE (2017). Data life cycle. https://www.dataone.org/data-life-cycle

Niu WJ, Feng ZK, Feng BF, Min YW, Cheng CT, Zhou JZ (2019) Comparison of multiple linear regression, artificial neural network, extreme learning machine, and support vector machine in deriving operation rule of hydropower reservoir. Water 11(1). https://doi.org/10.3390/w11010088

Osuch M, Wawrzyniak T, Nawrot A (2019) Diagnosis of the hydrology of a small Arctic permafrost catchment using HBV conceptual rainfall-runoff model. Hydrol Res 50(2):459–478

Productivity Commission (2003) Water rights arrangements in australia and overseas, commission research paper, Productivity Commission, Melbourne

Randall DA, Wood RA, Bony S, Colman R, Fichefet T, Fyfe J, Kattsov V, Pitman A, Shukla J, Srinivasan J, Stouffer RJ, Sumi A, Taylor KE (2007) Cilmate models and their evaluation. In: Solomon S, Qin D, Manning M, Chen Z, Marquis M, Averyt KB, Tignor M, Miller HL (eds) Climate change 2007: the physical science basis. contribution of working group I to the fourth assessment report of the intergovernmental panel on climate change. Cambridge University Press, Cambridge, United Kingdom and New York, NY, USA

Sun A (2013) Enabling collaborative decision-making in watershed management using cloud-computing services. Environ Model Softw 41:93–97

Schmugge TJ, Kustas WP, Ritchie JC, Jackson TJ, Rango A (2002) Remote sensing in hydrology. Adv Water Resour 25(8–12):1367–1385

Sheffield J, Wood EF, Pan M, Beck H, Coccia G, Serrat-Capdevila A, Verbist K (2018) Satellite remote sensing for water resources management: potential for supporting sustainable development in data-poor regions. Water Resour Res 54(12):9724–9758

SEQwater (2014) Seqwater Open Data Strategy 2013–2017, Southeast Queensalnd Water. https://www.seqwater.com.au

Tung YK (2011) Uncertainty and reliability analysis in water resources engineering. J Contemp Water Res Educ 103(1):13–21

Tsanis IK, Coulibaly P, Daliakopoulos IN (2008) Improving groundwater level forecasting with a feedforward neural network and linearly regressed projected precipitation. Hydroinformatics 10(4):317–330

Wagener T, Wheater H (2006) Parameter estimation and regionalization for continuous rainfall-runoff models including uncertainty. J Hydrol 320(1–2):132–154

World Bank (2018) World Bank Annual Report 2018. World Bank, Washington, DC. doi:10.1596/978- 1-4648-1296-5. License: Creative Commons Attribution–NonCommercial–NoDerivatives 3.0 IGO (CC BY-NC-ND 3.0 IGO)

WMO (2019) World Hydrological Cycle Observing System (WHYCOS), World Meteorological Organization. https://hydrohub.wmo.int/en/world-hydrological-cycle-observing-system-whycos

World Water Council (2020) World Water Council's Brochure. https://www.worldwatercouncil.org/en

Ziman M (2016) Data intelligence for improved water resource management. Master thesis, Nicholas School of the Environment, Duke University, 2016
Zhou Y, Guo S, Chang FJ (2019) Explore an evolutionary recurrent ANFIS for modelling multi-step-ahead flood forecasts. J Hydrol 570:343–355. https://doi.org/10.1016/j.jhydrol.2018.12.040
Zhen-lei W, Hong-yue S, Hao-di X, Gang W, Wei X (2019) The effects of rainfall regimes and rainfall characteristics on peak discharge in a small debris flow-prone catchment. J Mountain Sci 16. https://doi.org/10.1007/s11629-018-5260-3
Zhao Q, Zhu Y, Wan D, Yu Y, Cheng X (2018) Research on the data-driven quality control method of hydrological time series data. Water 10(12). https://doi.org/10.3390/w10121712

Chapter 4
The Role of Data Mining in Water Resources Management

Ali Arefinia, Omid Bozorg-Haddad, and Heejun Chang

Abstract Managing water resources from the past has been a concern of managers and decision-makers in different societies. With the development of various sectors of industry and agriculture and the increasing population that has increased the need for water resources, traditional management of water resources has been challenged. Managers and decision-makers in this sector are faced with a large amount of data with a variety of characteristics and complex relationships among them, whose analysis and management are difficult in some basins through traditional analyses. Data mining is a powerful technology in the management and organization of high volume information. In this chapter, while introducing various data mining methods and their related terms and concepts, and describing the importance and necessity of using data mining methods in water resources, its application in predicting water quality parameters and surface runoff forecasting in hydrology, so we try to better understand this issue.

Keywords Data mining · Water resource management · Backup vector machine · Artificial neural network · Decision tree

A. Arefinia · O. Bozorg-Haddad (✉)
Department of Irrigation and Reclamation Engineering, Faculty of Agricultural Engineering and Technology, College of Agriculture and Natural Resources, University of Tehran, Karaj, Tehran 31587-77871, Iran
e-mail: OBHaddad@ut.ac.ir

A. Arefinia
e-mail: Ali.Arefinia@ut.ac.ir

H. Chang
Department of Geography, Portland State University, Portland, OR, USA
e-mail: changh@pdx.edu

O. Bozorg-Haddad (ed.), *Essential Tools for Water Resources Analysis, Planning, and Management*, Springer Water, https://doi.org/10.1007/978-981-33-4295-8_4

4.1 Introduction

Data mining is a repeatable process where progress is made by exploring through automated or manual methods. Data mining is the most beneficial exploratory analysis scenario in which there is no pre-determined conception of a significant outcome (Kantardzic 2002). In another definition, the data collection process, identification, and modeling based on large amounts of data are to discover an order or relationship that is initially unknown, to obtain useful and transparent results for the database owner (Giudici 2003). The first research on the decision tree algorithms can be found in studies by Buntine (1993), Moret (1982), Murthy (1998), and Safavian et al. (1998). The current view of neural networks began in the 1940s with the research of McCulloch and Pitts (1943), which showed that artificial neuron networks can calculate any definite and arithmetic function and is often referred to as the basis of the other neural network researches. Genetic Programming (GP) is the solution provided by Koza (1992), which enables computers to solve problems without their explicit programming. This tool includes the use of genetic algorithms (GAs), introduced by Holand (1992), which serves the production of automated computer programs. The first GP research activity was reported in the early 1980s. For example, Smith (1980) reported a GA-based learning system. The vector support algorithm is based on the theory of statistical learning that has been developed over the past three decades by Vapnik and Chervonenkis (1964) and Vapnik (1982, 1995).

4.2 Definitions and Terms

In this section, we introduce the most important terminology in data mining:

- Root node: This node contains all existing instances, and the next level is the first division of the original set into two more homogeneous sets.
- Decision node: When a node is subdivided into subgraphs, it is called a decision node.
- Leaf/Terminal Node: The node that does not divide or, in the sense of being divided sequentially through them, is called Leaf or Terminal Node.
- Pruning: When we remove a node from a node, it's called Pruning. In fact, this is an antidote to divide.
- Split/sub-tree: A part of the whole tree is called a branch or under the tree.
- Father and child nodes: A node subdivided into several sub-nodes is called the parent node or parent node for its sub-nodes. While under the nodes of the parent, they are known as child nodes.
- Dendrites: Dendrites are similar to trees that have many branches that, by joining them, ultimately form the string of neurons. Dendrites act as the entry point for the main body of the neuron. On average, there are about 103 to 104 dendrites per neon, which is a surface area for input signals for the neuron.

- Synapse: This is the storage space of past experiences (the basis of knowledge). Synapse provides long-term memory for collecting past experiences, receiving information from other neurons, and providing output for Axon.
- Soma: The cell body of the neuron is called soma. This part is the largest part near the center of the neon and receives synaptic information and provides more information processing. Often, all the logical functions of the neuron are stored in the soma.
- Axon: The Neon Exit Line is called Axon. This output is, in fact, an action and operation potential that is transmitted to other neurons for processing more.
- Input layer: The layer through which the vector of independent variables (inputs) is received. In fact, the input layer acts as a processor to provide the network after receiving input data. It should be noted that the input layer is not computable because it lacks input weights and stimulus functions.
- Hidden layers: Layers that take information from the previous layer and with the help of processing elements, which are in fact, the same neurons, perform analytical operations on them.
- Output layer: This layer, which is the last layer of the ANN, shows the network output in response to a specific input vector.
- Actuator Function (Activation Function): The actuator function is a nonlinear filter that is used to change the output of the neuron.

4.3 Basics and Logic of the Subject

Due to the widespread application of data mining in various scientific disciplines, various methods have been proposed for multiple uses. The predictive methods of continuous values in data mining are decision tree, ANN, ANFIS, GP, BN, and SVM, which we will focus on:

4.3.1 Decision Tree

The decision tree is a tree-like structure similar to the trendline. The decision tree consists of several nodes and branches. Knots, also known as decision nodes, specify the questions for one or more variables. As far as this question is concerned, there are several possible answers to be solved in the form of a branch of each decision node. Thus, tree-based models divide the space of variables into a series of parts and then assign a constant value to each part. Each part that has a certain amount is called a leaf node. A decision tree is one of the most useful methods in categorical issues. This method is used in the training process to classify a tree and then for unclassified data. Using the decision tree because of the ease of use and understanding, accuracy, and speed is a common practice in solving data problems. The decision tree applies to issues that can be presented in such a way as to give a single answer as the name of

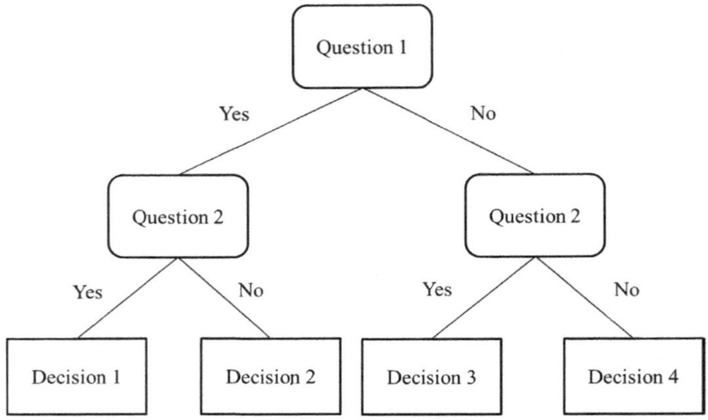

Fig. 4.1 Decision tree structure

a class or class. This means that successive nodes of a tree are like asking consecutive questions. If there is a unique and suitable answer to these sequential questions, then this unique answer will be the proper cluster/cluster for the clustering process. For this purpose, various algorithms have been introduced to construct decision trees, including CART, CHAID, KStar, M5Rules, REPTree, C5.0, and QUEST. The general structure of a simple decision tree is depicted in Fig. 4.1.

4.3.2 Business Network (BN)

BNs, also known as believable networks, are non-directional directed graphs that represent a set of random variables and their independent communication. These graphical structures are used to represent information in an area of uncertainty. Specifically, each node in the graph represents a random variable, and connections between nodes represent probabilistic dependencies between variables. In some cases, in data analysis, data itself is the only source of available information about a phenomenon. In these cases, conditional affinities are often identified by the statistical and probabilistic methods that are derived from the probability hypothesis. The probabilistic hypothesis suggests a way to link previous information about a phenomenon in a way with the process of data analysis. The business network training process begins with the distribution of the initial probability for the network nodes, either because of the lack of information hypothetical or due to past analyzes. The previous distribution is called because this probability distribution is not prepared before any thinking about data or related to past analyzes. By introducing new data sets into the network, the distribution of probabilities in each node is updated. This update is such that the probability distribution in each node varies from the previous distribution to the later distribution by using the probability hypothesis.

In general, BN combines the principles of graph theory, probability theory, computer science, and statistics for modeling together and uses these principles to distinguish, predict, and classify. BN in this unit can only predict a continuous or discrete variable. BN also does not work on issues that need to predict or estimate more than one output.

4.3.3 Genetic Programming (GP)

Genetic programming was introduced by John Koza (1992) and gradually developed in the early 1990s. Genetic programming is one of the developmental branches that, using the genetic algorithms and the concepts of decomposition trees for specific applications, instead of writing the code of the program, allows the computer to know only the general concept of the work, the program Get ready for us.

Koza (1992) was one of the main advocates of GP ideas. Based on extensive theoretical background and test results on many different domains such as in modeling of geotechnical engineering systems (Shahin 2015), application of genetic programming in hydrology (Fallah-Mehdipour and Bozorg-Haddad 2015), application of gene-expression programming in hydraulic engineering (Zahiri et al. 2015), and genetic programming applications in chemical sciences and engineering (Vyas 2015). He showed the ability of GP as an automated invention which produced new and outstanding results for a variety of issues. In addition to these engineering approaches, there was an increasing interest in understanding how GP works. Although GP was successfully used to solve problems, the development of GP theory was relatively difficult due to computational constraints in the 1990s. Since the early 2000s, a new GP theory has emerged, which has since been rapidly developed in GPs (Langdon and Poli 2002).

This tool works by applying genetic algorithms (GAs) to generate automated computer programs. GAs was inspired by natural evolution and their application in computers. Natural evolution is the creation of complex organisms (such as plants and animals) from simpler forms of single-cell life. GAs are simple models of the concepts of natural evolution and inheritance. The growth of plants and animals from seeds or eggs is controlled by genes inherited from their parents. The genes are stored on a DNA stream or in more strands. The child's DNA involves mixing or changing the parent DNA. Natural evolution occurs only when the most elegant people survive for generation and transfer of DNA to the subsequent generation. In the GP, the population is the same computer program. In order to simplify the production process of new programs from two-parent programs, programs are written as trees. New programs are created by removing branches from a tree and embedding into another tree. This simple process ensures that the new program is still a tree and is structurally valid. The ability of the algorithm to search and exploit simultaneously, the growing theoretical and growing theoretical and practical foundations of the real-world affairs, reinforces the conclusion that genetic algorithms are a very strong and coherent way to optimize.

One of the drawbacks of the GP method is that in the GP there is a less practical use of the relationships due to the possibility of creating diversity in the provision of mathematical relationships with the tree structure, as well as the prolongation of some of the resulting relationships. On the other hand, due to the consideration of the tree structure for the extraction of mathematical relations, the GP only can provide one relation between input and output sets. While in such systems with more than one output, there is a need to extract more than one relation. Another disadvantage of the GP is that this trunk technique can estimate a variable and can not be used for problems requiring multi-variable simultaneous estimation. GP can be applicable in many water resources related problems such as a genetic programming approach to rainfall-runoff modeling (Savic et al. 1999), Real-time operation of reservoir system by genetic programming (Fallah-Mehdipour et al. 2012), Seawater level forecasting using genetic programming (Ghorbani et al. 2010), Forecasting monthly urban water demand using extended Kalman filter and genetic programming (Nasseri et al. 2011), and Daily pan evaporation modeling using linear genetic programming technique (Guven and Kişi 2011).

4.3.4 Artificial Neural Network (ANN)

The ANNs are mathematical and flexible methods that inspire the human brain and the nervous system. An ANN is a nonlinear function that establishes the relationship between input variables and network outputs like the biological neural network system. ANNs can be used in modeling complex systems. Also, these networks can provide a nonlinear mapping between inputs and outputs by selecting an appropriate number of layers and neurons. The overall ability of ANN is to learn nonlinear communication between data and generalize the results for other data.

4.3.4.1 ANN Operation Mode

In many water system issues, there is an intuitive look at the process of a natural process, but the details of the process are not there. In such cases, the so-called "black box" is used. These models have been created with a pattern of nature. The reason for the word black box is that relationships and operators within that model are unknown for most people. One of the most famous and most used is ANN. ANNs are part of intelligent dynamic systems that are based on empirical data and, by processing these data, transfer knowledge or law beyond the data to the network structure. The ANNs are used to categorize and predict data with different sizes. In summary, ANNs have found the most applications in solving three groups of problems: (1) issues that are not algorithmic solutions, (2) problems with a very complex algorithm solution, and (3) Problems that humans are more successful in solving their problems.

4.3.4.2 ANN Types

There are now a large number of different types of ANNs, such as single-stranded perceptron networks (SLP), multi-layer perceptron (MLP) networks, recursive networks, competitive networks, etc., which also have algorithms various methods, such as BP, are trained.

4.3.4.3 ANN Training Algorithms

Although the idea of neural networks was established in the 1950s and then its mathematical foundations were founded, its development was not without the problem of the lack of high processing and theoretical defect in the network learning algorithm faced by the years left unaltered. In the years that followed, researchers proposed different methods for ANN teaching. Hebb supervised learning algorithm (Hebb 1949) and the Widrow-Hoff learning algorithm (Widrow and Stearns 1985) can be mentioned. Each of the proposed algorithms had problems during the 1980s, including the instability of algorithm convergence, low speed, or limited application in some ANNs. Training in natural systems occurs comparatively, which means that changes occur in synapses as a result of training. Adequate training on ANNs is also true. In these networks, training is done through a sample, which means that often (but not always) a set of dependent and independent variables is assigned to the ANN, and ANN uses these samples to adjust weights so that better responses can be generated if new weights are applied. In fact, ANN training is stored in its communication weights.

4.3.4.4 Error Back-Propagation Algorithm (BP)

This algorithm is a training ground used to train MLP networks. The BP algorithm consists of two main paths, including commuting routes. On the path, the input vector is applied to the MLP network and its effects are disseminated through the hidden layer to the output layer, and the output vector represented in the output layer forms the true MLP network response. In this way, network parameters are considered constant and unchanged. On the return path, unlike the path, the parameters of the MLP network are changed and adjusted, which are set according to Delta's rule, and the error vector is formed in the network output layer. The error vector is equal to the difference between the observational response and the computational response of the network. The error value after the calculation is distributed in the return path of the output layer and through the network layers throughout the network. Because this distribution is contrary to the communication path Synaptic weights, BP expression is chosen for this algorithm. In the sequel, the algorithm changes the network parameters so that observation values and computational values are closer.

4.3.5 Fuzzy Neural Network Inference (ANFIS)

ANFIS consists of the combination of two ANN structures and fuzzy logic, in which the advantages of both ANN and fuzzy logic are utilized. This means that ANFIS has been used as an educational feature of ANN as well as the ability to model a fuzzy inference system, which increases the decision-making power in conditions of uncertainty and as a result of increasing accuracy. Fuzzy concepts emanate from fuzzy phenomena occurring in nature. A fuzzy phenomenon has a wide range of variations and is difficult to describe. The rain phenomenon can be mentioned, for example, because the intensity of the rain varies from light to heavy rain (Kantardzic 2002). Most of the concepts formed in the human mind to understand and categorize natural phenomena are fuzzy, and the boundaries of these categories are important. These boundaries depend on the personality of each person in that field. For example, rain phenomena can be classified into rainstorms, mild rainy rain, and heavy rain, which makes it hard to define the boundaries of these categories. Contrary to traditional classifications, a fuzzy classification includes unambiguous and clear categories that are a gradual transformation from belonging to a certain set to the non-belonging to that set. This transformation is described by membership functions. Not be A fuzzy set is completely described by its membership function, and the membership function is a relation that determines the degree of belonging of value to a category. This function can take values between 0 and 1, indicating the degree of membership of value to a specific category. Usually, for the simplicity of the function, the membership function is expressed as a mathematical relation, most notably triangular, trapezoidal, and Gaussian membership functions.

ANFIS has a good ability to train and classify, and it also has the advantage of constructing fuzzy rules using numerical information or knowledge of the user. The ANFIS architecture is equivalent to a five-layer forward-looking grid that employs neural network learning algorithms combined with the fuzzy argument to the mapping of the input space to the output space.

4.3.6 Support Vector Machine (SVM)

SVM is one of the most popular supervised learning models used for categorization. SVM is a data-mining method introduced by Vapnik (1995) that is based on statistical learning theory (Kashif Gill et al. 2007). This model is among the relatively new models for categorization, which in recent years has shown good performance over older methods, including MLP networks. The SVM is a linear data categorization that attempts to divide the data into a line that has a greater margin of confidence for categorization, and the goal is to find the optimal line for data categorization. For example, Fig. 4.2 shows a set of two-dimensional data that is dual.

Where each point represents a given data, and each data has two variables. As shown in Fig. 4.2, a large number of data lines can be categorized, and the SVM

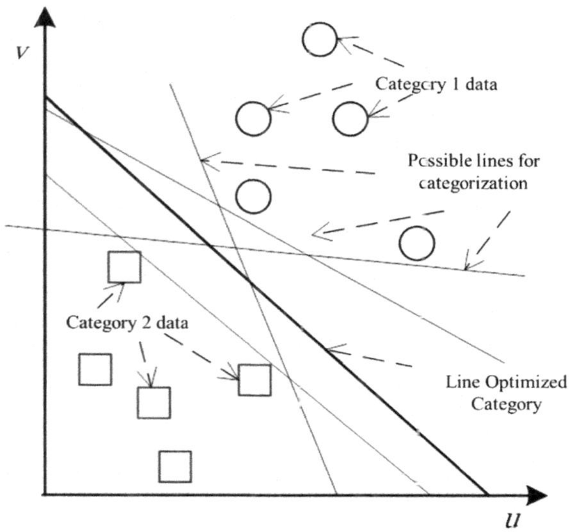

Fig. 4.2 Collections of two-dimensional data belong to the same category

internal algorithm provides a pathway for finding this optimal line. This algorithm states that the line that passes near the margin points of each category is not appropriate and does not properly divide the categories. That is why the goal is to identify a line that has the maximum distance from the marginal points of the categories. Finally, as shown in Fig. 4.3, this problem leads to finding vectors in the boundaries of the groups called the support vectors and adapted to the boundary points of the classes. The optimal line passes through the middle of the support vectors for maximum separation of the groups.

It should be noted that the SVM method can only classify a dependent variable for multiple independent variables. This means that SVM places them in categories by identifying the support vectors among the input variables, which in these categories is estimated/predicted by a dependent variable for the corresponding independent variables.

4.4 Importance and Necessity of the Subject

Rapid advances in storage technology and data gathering enabled organizations to accumulate huge amounts of data. Extracting useful data from a dataset is very challenging. Often, traditional data analysis tools and tools cannot be used because of the large size of a data set. Also, sometimes updated analyzes of events cannot be solved with the help of existing and traditional approaches, so new methods should be created. Data mining is a technology that processes complex volumes of data

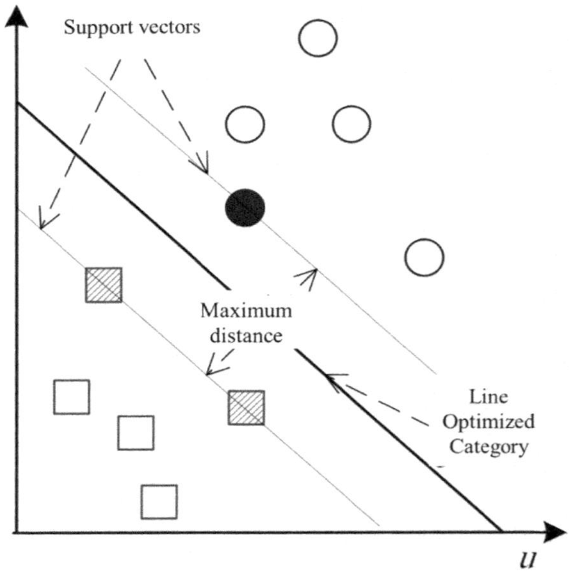

Fig. 4.3 Support vectors and optimal line for data sorting

with complex algorithms and has expanded the opportunities available to explore and analyze data types in new ways.

Data mining is commonly used for 6 main purposes (Fayyad et al. 1996):

1. Finding abnormalities: Data mining can help find abnormal data or data that is heterogeneous which is different in terms of uniformity in a huge amount of data
2. Analysis of Cohort Rules: Using data mining, it can be clearly identified among the database as to how many items are grouped in a group. Alternatively, they can also introduce dependency modeling analysis.
3. Clustering: Detects the same structures and categories lying between the data. When the mass of data is available and the goal is to identify clusters among them (which have the maximum difference within themselves with the maximum similarity and between clusters), this method is used.
4. Categorization: This method is used when data categories are predefined and targeting new data in the best category.
5. Regression: Its purpose is to find the function of the data with the minimum error. This function matches the dependent variables by combining independent variables either linearly or non-linearly. The difference in regression with the analysis of cohort rules is that it expresses regression of the function with logical and mathematical relations, but in associative rules, the dependence between variables is not always known by mathematical relations.
6. Summary: Methods that provide a more compact model of the raw data set, including image reporting and production. This is especially important in large

data warehouses, which need to be summarized in order to show the overall data to the public or decision-makers.

Nowadays, the use of data mining tools in different stages of modeling, simulation, optimization also has a special place in the process of planning and management of water resources. Since water resources data shows natural processes, and most of these phenomena have a lot of uncertainty, working with them is often challenging, and this issue has always been a concern of water science experts around the world. Data mining can be used to simulate surface water, groundwater, qualitative simulation of aquifers, estimate evaporation, etc. Most conceptual models require much input for simulation, which is considered to be a problem with such models. In this case, it is necessary to use models that can be used simplicity and time savings with an intelligent mathematical structure to understand the relationships between inputs and outputs as a precision instrument for simulation. In many engineering optimization problems, such as optimization issues related to water resources systems, simulators need to be used to calculate the target function. The low speed of the implementation of some simulators to use in optimization problems causes the problem of prolonging the optimization process. In this case, the use of data mining tools will be very effective in reducing the implementation time of the simulation process.

4.5 Method of Work and Process

Data mining is one of the stages of Knowledge Discovery in Databases (KDD) (Fayyad et al. 1996). Data mining is, therefore considered as one of the most important stages in the discovery of knowledge. The KDD process that data mining is one of its subsets consists of the following steps:

1. Data selection: This means the selection of data from the data in the database.
2. Data deletion: The deletion of data that is naturally difficult should be removed from the original data.
3. Uniform data: Since data may be received from different sources and sources, they may be uneven. The unevenness of the data makes it difficult to analyze them. Therefore, at this stage, the data must be uniform in terms of different criteria. Scales and other statistical characteristics of these criteria. After this step, the data is uniformly generated.
4. Data conversion: This step involves converting selected data into suitable formats for use in various data mining techniques.
5. Data mining: An essential step in the KDD process, in which various statistical methods and machine learning are used to extract patterns, this step involves the following steps:

A: Selection of data mining objectives (classification, clustering, prediction, determination of dependency).

B: Select the appropriate data mining method (decision tree, ANN, ANFIS, GP, BN, and SVM).

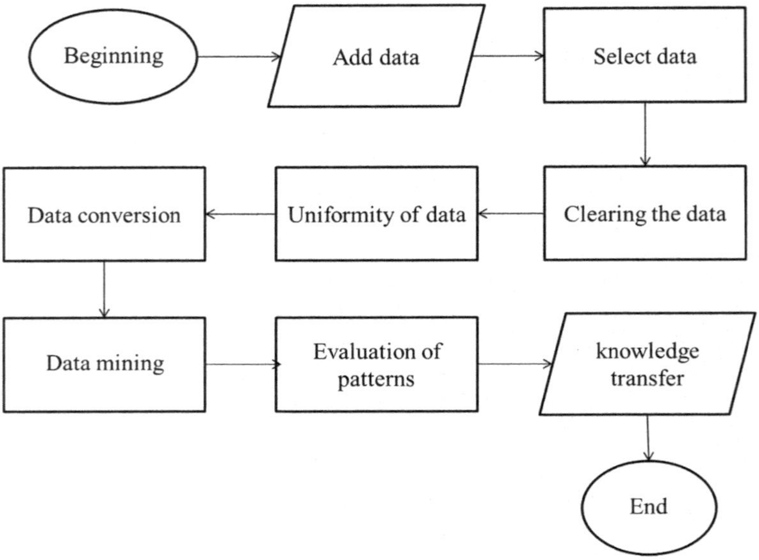

Fig. 4.4 Flowchart of KDD Processes

C: Perform a search query and search for appropriate patterns.

6. Model assessment: Identify appropriate patterns among patterns derived from the data mining process based on data mining objectives.

7. Knowledge provision: Provide extracted knowledge (template) using information display methods (including equation, image, diagram, etc.). Figure 4.4 illustrates the different stages of the KDD process that data mining is one of its subsets.

The data mining process involves several steps. These multiple stages begin with understanding the purpose of the project and what data is available and ending with the final analysis. Key data mining processes can be called model learning and understanding processes, modeling and evaluating the model, and finally using the model. These steps will become clearer once the data is organized regularly.

4.6 Practical Examples

Data mining has many applications in various sciences, including hydrology and water resources. Two examples of its application in predicting water quality parameters and surface runoff forecasting in hydrology are mention below.

Monitoring and evaluating the quality of water resources is a very costly and time-consuming process. Therefore, it is important and necessary to choose a method in which with relatively low hydrochemical parameters it is possible to have a relatively accurate prediction of water quality parameters, and data mining is a method that is widely used today in predicting water quality parameters.

In a 2017 study, Solgi et al. used predictive model vector regression (SVR) and flexible fuzzy neural system (ANFIS) models to predict the biochemical oxygen demand (BOD) in the Karun River in western Iran. Combinations also used waveguide conversion, and after analyzing the parameters with wavelet conversion, the principal component analysis (PCA) was used to identify important components. For this modeling, the monthly time series of the BOD index was used. The results showed that the SVR model with RMSE = 0.033 (mg/l) and $R^2 = 0.843$ performs better than the ANFIS model with $R^2 = 0.828$. Also using wavelet conversion on SVR input data the model has improved the results to $R^2 = 0.937$ and RMSE = 0.021 (mg/l). Therefore, combining SVR with wave conversion (WSVR) was a good idea to improve the prediction of BOD values in the Karun River. Finally, this combination was recognized as a suitable model. In this example, using a data mining tool and without the need to spend money and time and laboratory operations, the amount of BOD in the next six months was predicted.

Runoff prediction is an important and necessary thing in hydrological research. Sarzaeim et al. (2017) predicted runoff in the Idooghmush basin of Iran, under the conditions of climate change by genetic programming (GP) data mining methods, artificial neural network (ANN), and provided a support vector tool (SVM). The input of data mining methods is the monthly precipitation and temperature, which are obtained using (HadCM3) and the general ocean circulation model (OOGCM), then using them, the runoff is predicted by the aforementioned data mining methods. Their predictive skills were compared using several standard models of model performance, and the results showed that SVM performed better than GP and ANN by 7% and 5%, respectively.

4.7 Summary

Today, due to the advancement of data mining models and their effective efficiency in engineering, these models are more likely to be considered by engineers and water resource managers. The simplicity of working with these models is considered to be their virtues, which has led to their popularity in various sciences. Through the learning process, the intelligence system can find the complex rules contained within the parameters, and by obtaining a general mathematical rule and obtaining new inputs to simulate and predict the parameter (S) to target. In general, the use of data mining in water resources is divided into surface waters and underground waters, such as rainfall-runoff simulation, prediction of evaporation from the surface of aquifers, simulation of the transfer process and the spread of pollution in groundwater, the qualitative simulation of reservoirs and other aquatic structures. In this chapter, introducing various data mining methods and their related terms and concepts, the importance and necessity of using data mining methods in water resources are described, and in the following, with some examples in this field, we try to better understand this issue.

Acknowledgements The authors thank Iran's National Science Foundation (INSF) for the support of this research.

References

Buntine W (1993) Learning classification trees. Artificial intelligence frontiers in statistics. Chapman & Hall, London, pp 182–201

Fallah-Mehdipour E, Bozorg-Haddad O (2015) Application of genetic programming in hydrology." In: Handbook of genetic programming applications. Springer, Cham, pp 59–70

Fallah-Mehdipour E, Bozorg-Haddad O, Mariño MA (2012) Real-time operation of reservoir system by genetic programming. Water Resour Manag 26(14):4091–4103

Fayyad U, Piatetsky-Shapiro G, Smyth P (1996) From data mining to knowledge discovery in databases. AI Mag 17(3):37–54

Ghorbani MA, Khatibi R, Aytek A, Makarynskyy O, Shiri J (2010) Seawater level forecasting using genetic programming and comparing the performance with artificial neural networks. Comput Geosci 36(5):620–627

Giudici P (2003) Applied data mining: statistical methods for business and industry. Wiley, London, UK, p 364

Guven A, Kişi Ö (2011) Daily pan evaporation modeling using linear genetic programming technique. Irrig Sci 29(2):135–145

Hebb, DO (1949) Organization of behavior. Wiley

Kantardzic M (2002) Data mining: concepts, models, methods, and algorithms. Wiley, New York, p 360

Kashif Gill M, Asefa T, Kaheil Y, Mckee M (2007) Effects of missing data on the performance of learning algorithms for hydrologic predictions: implications to an imputation technique. J Water Resour Res 43(7):W07416

Koza JR (1992) Genetic programming: on the programming of computers by natural selection. MIT Press, Cambridge, MA

Langdon WB, Poli R (2002) Foundations of genetic programming. Springer Verlag, Berlin Heidelberg New York

McCulloch W, Pitts W (1943) A logical calculus of the ideas immanent in nervous activity. Bull Math Biophys 5:115–133

Moret BME (1982) Decision trees and diagrams. Comput Surv 14(4):593–623

Murthy SK (1998) Automatic construction of decision trees from data: a multi-disciplinary survey. Data Min Knowl Disc 2(4):345–389

Nasseri M, Moeini A, Tabesh M (2011) Forecasting monthly urban water demand using extended Kalman filter and genetic programming. Expert Syst Appl 38(6):7387–7395

Sarzaeim P, Bozorg-Haddad O, Bozorgi A, Loáiciga HA (2017) Runoff projection under climate change conditions with data-mining methods. J Irrig Drainage Eng 143(8):04017026

Savic DA, Walters GA, Davidson JW (1999) A genetic programming approach to rainfall-runoff modeling. Water Resour Manage 13(3):219–231

Shahin MA (2015) Genetic programming for modeling of geotechnical engineering systems. In: Handbook of genetic programming applications. Springer, Cham, pp 37–57

Smith SF (1980) A Learning System Based on Genetic Adaptive Algorithms. Ph.D. thesis, University of Pittsburgh

Solgi A, Pourhaghi A, Bahmani R, Zarei H (2017) Improving SVR and ANFIS performance using wavelet transform and PCA algorithm for modeling and predicting biochemical oxygen demand (BOD). Ecohydrol Hydrobiol 17(2):164–175
Vapnik V (1995) The nature of statistical learning theory. Springer, New York
Vapnik V, Chervonenkis A (1964) A note on one class of perceptions. Autom Remote Control 25
Vapnik VN (1982) Estimation of dependences based on empirical data. Springer, Berlin
Widrow B, Stearns SD (1985) Adaptive signal processing. Prentice-Hall, Englewood Cliffs, NJ
Zahiri A, Dehghani AA, Azamathulla HM (2015) Application of gene-expression programming in hydraulic engineering. In: Handbook of genetic programming applications. Springer, Cham, pp 71–97

Chapter 5
Remote Sensing Application in Water Resources Planning

Hossein Rezaei, Omid Bozorg-Haddad, and Mehmet Cüneyd Demirel

Abstract Remote sensing-based satellite data and processing tools as part of geographic information system (GIS) have been utilized in many disciplines in earth sciences, water resources management in particular. Retrieving spatial data and processing technology require a systematic knowledge and experience for an effective use that is the main motivation of this chapter. Field work in large areas and installing gauges have been painstaking and costly. Even the developed countries began to remove gauges measuring eddy-covariance and other meteorological variables due to the high maintaining cost. Instead, they invest on satellite technologies and radar systems. Today, most field works and conventional point data collection have given its place to processing synoptic satellite images using open source GIS tools. The use of spectral indicators and remote sensing technologies to control and monitor the water quality and quantity of rivers, reservoirs and groundwater has been very cost-effective. Different variables that can be remotely measured in water quality are salinity, suspended sediment, water color, extent of oil spill and eutrophication, growing phytoplankton and algal bloom. Also, estimation of land cover and land use, actual evapotranspiration, land surface temperature, runoff, preparation of flood maps, determination of snow cover and depth changes may benefit from remote sensing-based satellite data and GIS technology. To do operations in physical sciences, basic knowledge in remote sensing and geographic information systems (GIS) is necessary as they depend each other. First, basic data is collected and then processed by sensors of remote sensing satellites using different color spectra or recorded thermal properties. This leads to the creation of raw databases that are processed in GIS to enhance data and utilize information management and

H. Rezaei · O. Bozorg-Haddad (✉)
Department of Irrigation and Reclamation Engineering, Faculty of Agricultural Engineering and Technology, College of Agriculture and Natural Resources, University of Tehran, Karaj, Tehran 31587-77871, Iran
e-mail: OBHaddad@ut.ac.ir

H. Rezaei
e-mail: HosseinRezaie18@ut.ac.ir

M. C. Demirel
Department of Civil Engineering, Istanbul Technical University, Istanbul, Turkey
e-mail: demirelmc@itu.edu.tr

O. Bozorg-Haddad (ed.), *Essential Tools for Water Resources Analysis, Planning, and Management*, Springer Water, https://doi.org/10.1007/978-981-33-4295-8_5

store layer composition. Modeling, production of output maps and spatial analysis are very fast and accurate with GIS tools such as gdal, pyproj, pymodis libraries in Python language and 3D data storage in common data format (netCDF). GIS is a very powerful management tool for planners and designers to adopt appropriate land and water management strategies. Since remote sensing and GIS are deep and extensive source, studying the principles and methods require a structured summary of basics and relevant applications in the field of water management and engineering.

Keywords Satellite images · Radar systems · GIS · Spectral indicators · Spatial analysis

5.1 Introduction

Human eyes have always been one of the most important tool to describe, identify environmental issues. When a person uses their sense of sight to observe the environment, he or she receive color and spatial information in an instant with the help of the joint function of the eye and the brain and then It is sent to different parts of the brain. So that the types of colors and spatial information felt simultaneously coincide. Although the brain and eye perception systems have achieved the ultimate evolution, there is still a great deal of information on the Earth's surface that the eyes of humans cannot observe, understand, or process. While human rational perception is able to make and use many advanced tools and equipment that can receive more useful information from Earth's surface objects in addition to the sun's visible light, using other spectral ranges of electromagnetic waves. Remote sensing (RS) is a method of data collection in which there is no direct physical contact with the measured objects. There are many definitions of remote sensing in the literature. We will list only some of the most common ones as below.

Based on Colwell et al. (1983), remote sensing science and technology is the acquisition of information about an object, region, or phenomenon by processing and analyzing data obtained by a device (without direct contact with the object, region, or phenomenon being studied). Sabins (1997) defines remote sensing as the science of processing and interpreting images that are the result of the interaction of electromagnetic energy and objects. Chandra and Ghosh (2002) defines remote sensing as the process or method of obtaining information about an object, area, or phenomenon through data obtained by a device without direct contact with them.

In addition to usual photography, this industry's potential applications were also noted, and the first image of a balloon was taken in 1859 near Paris. The most important evolution in remote sensing was created in 1909 by the invention of the airplane by the Wright brothers. The aircraft was more stable than the balloon and also had more controllability, so balloon imaging quickly gave way to the airplane. The first extensive use of aerial photographs was in World War I (1918) when about 56,000 aerial photographs were taken to identify enemies' locations and equipment. The period between the two world wars was an opportunity to develop and growth the applications of aerial photography in civilian purpose including geology, forest and rangeland surveying, and map preparation. World War II not only led to develop and interpret aerial photography, but also the emergence of special systems for detecting objects such as infrared and radar at the same time.

This continued in Iraq and Afghanistan wars to identify crucial targets using high resolution and not publicly available satellite data from IKONOS. Figure 5.1 illustrates the oil facility of Rumalia, Iraq Retrieved from IKONOS satellite.

Infrared photography, which was known to be useful for identification operations, now used to study a variety of plant coverings, and multi-band imaging has gradually established itself among experts in the field.

With the production of computer-generated maps, many new tools for spatial data and maps were developed. The data processing operation requires a set of powerful tools for collecting, storing, retrieving, converting, and displaying real-world spatial data for a variety of purposes. There are several definitions of GIS as follows:

Burrough (2000) describe geographic information as a powerful tool for collecting, storing, retrieving, and displaying real-world locations. Clarke (2001)

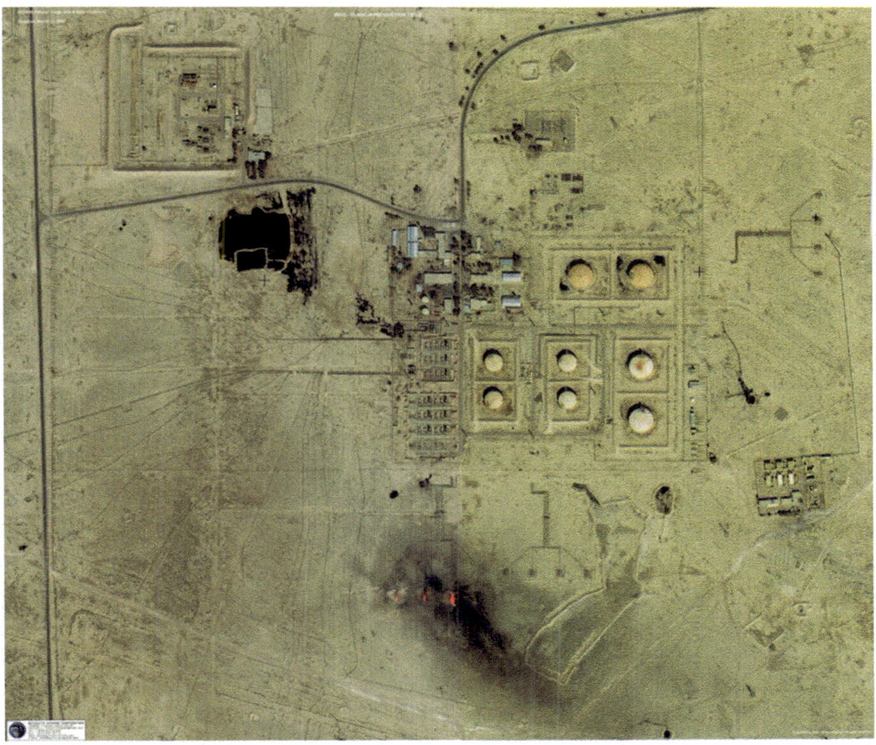

Fig. 5.1 The map of Rumalia oil facility in Iraq. (https://www.satimagingcorp.com/gallery/ikonos/ikonos-rumaliafield-lg/)

defines geographic information as a computer-aided system for controlling, storing, retrieving, analyzing, and displaying locations in a specific organization.

The 1960s were crucial in the history of remote sensing and GIS, as human endeavors to access space came to fruition, the culminating of that is travel humans to the moon. After that, for the first time in the United States, work began on the GIS started. Some of the first satellites launched into space presented as follows.

In 1957, the world's first satellite was launched by the former Soviet Union, Sputnik-1, surprising the world. In 1964, the first meteorological maps of the Earth's surface were prepared by the Nimbus satellite, which had a meteorological application in war.

In 1972, the first satellite of natural resources was launched, which was first called ERTS-A, then renamed Landsat, and this year rapid advances in digital image processing was saw. In 1977, the first European meteorological satellite, Meteosat, was launched with the aim of providing visible and infrared images day and night.

Due to the high efficiency of remote sensing techniques, GIS and their interdisciplinary nature (defense, reclamation and earth sciences), their widespread applications are observed in various sciences e.g. illegal housing, drugs, mining, and deforestation. Applications of remote sensing and GIS in water resources science including estimation of potential and true evaporation and transpiration, determination of plant coefficient, determination of flood ranger of the river, estimation of evaporation from free water level, estimation of precipitation in a zone, determination of salinity and humidity. There are numerous articles and research on estimating soil moisture, determining the drought index, estimating the volume of snow and its equivalent water, analysis of erosion and sedimentation rates, groundwater analysis (Ganapuram et al. 2009; Guzinski et al. 2013; Nijzink et al. 2018; Stisen et al. 2011). The literature review on the use of remote sensing in water resources engineering and management is summarized below.

Ahmad et al. (2005) investigated a new technique for estimating net groundwater utilization within irrigated lands by combining remote sensing approaches. This case study was conducted in Pakistan. This technique is a combination of remote sensing methods and water balance concepts. They concluded that the harvest values were 0.8 to 1.6 mm per day, which averaged 82 mm per year.

Alparslan et al. (2007) investigated water quality in the reservoir of Ömerli Dam using remote sensing techniques. This case study was conducted in Turkey. They concluded that remote sensing methods were reliable and inexpensive, and that satellite imagery could be used to detect changes in water quality, sources of pollution and its effects. Stisen et al. (2008) investigated the potential of applying remote sensing-based input data in hydrological model for Senegal River basin in West Africa. Yilmaz et al. (2008) test the relationship between vegetation water content (VWC) and equivalent water thickness (EWT) for three different vegetation types. EWT and VWC estimated by data from SMEX05 and SMEX02.

Ghazanfari Moghaddam et al. (2011) compared the precipitation data obtained from the PERSIANN model, co-kriging interpolation method and inverse distance weighting. This case study was conducted in Iran. The results of this study illustrate that the use of precipitation estimation models, based on satellite images that have high temporal and spatial resolution, and can be a good alternative to interpolation methods. Vinukollu et al. (2011) developed and inter-compare three process-based evapotranspiration products over lands, based on sensors on the NASA Aqua satellite platform. Stisen et al. (2011) analyzed the spatial pattern of remotely sensed variations in surface temperature to improve a physical based hydrological model. Guzinski et al. (2013) presented modifications for the dual temperature difference model that enable applications using thermal observations from polar orbiting satellites.

McCabe et al. (2017) explored innovative and smart observation approaches and highlight new remote sensing platforms including small cube satellites, drones and LIDAR.

Demirel et al. (2018) determined the appropriate spatial model parametrizations and objective functions to evaluate distributed hydrologic model outputs. In addition, they incorporated spatial pattern of satellite based on actual evapotranspiration data in the model calibration and validation.

Miao et al. (2019) investigated the water quality of urban rivers in China using remote sensing. The study used Landsat-8 satellite data, the vegetation index (NDVI) and the Canadian water quality index. They concluded that the results of the collected ground samples and the data obtained from the remote sensing have a good fit showing the accuracy and proper performance of this approach.

Kalhor and Emaminejad (2019) investigated the relationship between groundwater level and urbanization using remote sensing data to sustainably develop cities. The case study was conducted in the Atlanta metro area of Georgia, USA. The study used remote sensing data from GRACE, MODIS, TRMM, IRIS and GLDAS. They concluded that limited urban development has taken place in the region since 2001–2013, but is projected to increase by 2050, which requires careful and comprehensive planning for sustainable urban development in the region.

Lyazidi et al. (2019) investigated the management of water resources in semi-arid regions by GIS. This case-by-case study of the Gareb-Bouaregin Morocco. In this research, MODFLOW and Arc GIS software have been used. They concluded that the development of this model is very useful for controlling, planning and managing groundwater levels when the water level rises and causes significant damage to agricultural fields.

Ahmadi et al. (2019) investigated the effectiveness of water consumption and improving land drought by remote sensing. This case study was conducted across the United States. In this study, MODIS measurement data and quality flags were used. Dembélé et al. (2020) used spatial pattern information from multiple sources of satellite remote sensing (SRS) data to constrain the model and improve the spatial representation of hydrological processes.

5.2 Definition and Terms

In this section, a brief description of the definitions and terms required to learn remote sensing science and the GIS is presented.

5.2.1 *Remote Sensing*

Remote sensing is the science of extracting information from objects indirectly and by using of a sensor. In other words, remote sensing is the science and art of obtaining information from an object under study by a tool which it is not physically in contact.

5.2.2 *Geographic Information System (GIS)*

GIS is an information system that produces, processes, analyzes, and manages geographic information, and also it is able to collect, store, analyze, and display geographic information. Processing includes changing raster pixel size via upscaling and downscaling methods (resampling), merging image tiles (mosaic), trimming the borders of the domain. The ultimate goal of a GIS is to support decisions based on geographic data, and its basic function is to obtain information that obtained by combining different layers of data in different ways and with different perspectives. The most commonly used GIS tools are ArcGIS, Quantum GIS (QGIS), SAGA, GDAL, PostGIS and GRASS.

5.2.3 *Electromagnetic Spectrum*

The frequency range of electromagnetic radiation is called the electromagnetic spectrum.

5.2.4 *Absorption and Transfer*

Part of the electromagnetic energy is absorbed by various molecules, such as ozone, water vapor, and carbon dioxide, as it passes through the atmosphere. Therefore, during different wavelengths, the amount of absorption by different molecules is different, which causes the identification and selection of different bands to perform different tasks.

5.2.5 Atmospheric Distribution

Particles or gas molecules in the atmosphere, cause the distribution of electromagnetic energy from their initial path when electromagnetic energy collides with them. Atmospheric distribution depends on different factors such as wavelength of the radiation, number of particles and radiated distance.

5.2.6 Pixel and Pixel Data (Raster)

Images in remote sensing are of a raster nature and consist of a matrix of components called pixels (grids). The dimensions of each pixel are the smallest unit identified by the meter or degree-decimal based on the earth projection that shows the ground level. 1 degree at Earth's equator is approximately 111 km. A second of arc, 1/60 of a sea mile (1,852 meters), is about 30 meters (98 feet).

A raster consists of a set of points or cells that cover the effects of the earth in a regular network and are addressed using row and column numbers. The smallest constituent of a raster is called a pixel or cell, the value of each of which represents the spectral or descriptive information of the terrestrial complication. Scanning data and satellite imagery have a raster structure. In addition, the unit of measured pixel data from the satellite and distributed models can be different. For instance, actual ET estimated based on RS data is watt/m^2 (energy balance equations) whereas it is a flux in hydrologic models with the unit of mm/day (Demirel et al. 2018; Guzinski et al. 2013). That's why we need bias-insensitive metrics like SPAEF and SPEM (Dembélé et al. 2020; Koch et al. 2018) to evaluate variables with different units.

5.2.7 Vector Data

In two-dimensional spaces, data are formed by combining geometric shapes such as dots, dashes, triangles, and other polygons, and in three-dimensional spaces, they are made by cylinders, spheres, cubes, and other multifaceted objects. In this model, the position of each point is precisely represented by a pair of coordinates in a given coordinate system. This means that its coordinates and one of the modes of representation [point (well, urban and rural points), lines (road, river and rail lines) and levels (lake and zoning) determine the position of each object or phenomenon.

5.2.8 Band

A range of electromagnetic spectra with a specified wavelength is called a band. Its value is different in each sensor and is a criterion for dividing and classifying meters.

5.2.9 Sensors

Any device that collects electromagnetic radiation reflected from various phenomena or other emitted energies (such as thermal infrared) and provides a suitable way to obtain information from the environment is called a sensor. Assessors are divided based on energy source (active and inactive) or information efficiency (visual and non-visual). Video information efficiency is divided into two groups: visual and numerical. It should be noted that each sensor is only sensitive to a range of electromagnetic spectra.

5.2.10 Multi Spectral Sensor

A sensor with a bandwidth of less than 50 bands is called a multi-spectrum sensor. The smaller the number of bands in an image, the greater the bandwidth of each band in the electromagnetic spectrum, which in turn reduces the spectral power.

5.2.11 Active Sensors

These sensors generate electromagnetic energy and already send energy to the desired phenomenon and collect and record their reflection.

5.2.12 Passive Sensors

These sensors do not generate electromagnetic energy and collect energy reflected from various phenomena on the Earth's surface to which the sun's electromagnetic radiation is emitted.

5.2.13 Pictorial Sensors

The information efficiency of these sensors can be converted to photos.

5.2.14 Non-imaging Sensors

The information efficiency of these meters is in the form of tables and diagrams and cannot be converted to photos.

5.2.15 Numerical Sensors

The information efficiency of these sensors is digital and they are converted to photos and used after certain steps.

5.2.16 Platform

Carriers of telemetry meters are called platforms and have different types such as satellite, airplane and balloon, etc. Radio-controlled aircraft and balloons are used for remote sensing at lower altitudes and satellites are used for remote sensing at higher altitudes. The height of the platform is a very important factor in choosing a platform because it depends on determining the resolution of the earth and the field of view of the moment being measured.

5.2.17 Projection

One of the most important basics of cartography and geodesy is how to transfer the spherical surface of the earth to a plane that called map projection. It is like peeling of the orange to a plane surface. Turning the geographical atlas shows that the orbits and meridians are not the same shape on all planes. On some pages, circuits are depicted as straight lines and in others as curves or even concentric circles. This deformation in the grid is the result of the transferring the spherical surface of the earth to the plane of the paper (map). One of the most common projection systems is Universal Transverse Mercator (UTM) coordinate system.

5.2.18 UTM

UTM is a coordinate system that assign coordinate to locations on the surface of the earth. UTM coordinate system process is like traditional methods of longitude and latitude. It means UTM represent locations of the map in a horizontal position. In other words, UTM ignore altitudes and treats the earth like a perfect ellipsoid. UTM is based on Gauss-Kruger projection system and in terms of longitude, the earth is divided into 60 zones of six degrees, numbered from west to east (from one to 60). Its scope does not include the North and South Poles: latitudes above 84° N and below 80° S have been removed.

5.2.19 World Geodetic System

The world Geodetic System (WGS) is a standard system used in navigation. WGS consists of, a reference for coordinates, a center of mass for the Earth, and a similar circuit used to move the earth in space. A pair of latitude and longitude is considered as the base or (zero) and the rest of the points are measured relative to that latitude and longitude. The location of this latitude and longitude for the reference could be considered anywhere on the Earth. Different versions of WGS presents up to now as are exist e.g. WGS 60, WGS 66, WGS 72 and WGS 84. WGS 84 system uses the same elliptical orbit around the Earth. The international elliptical dimensions of WGS 84 are determined by satellites and are very close to the Earth in the form of a globe. In WGS 84, third axis (z-axis) passes through the Earth's conventional pole (magnetic pole), and its X-axis is the interface between the Greenwich meridian plane and the equatorial plane. The Y axis is also adjusted so that the system is right-handed.

5.2.20 Spatial Resolution

The ability of each meter to identify the details of spatial phenomena at the ground level is defined as the power or spatial resolution. The smaller pixel's size, the greater number of networks (pixel density) and the better the detection of tolls, and consequently the larger the data volume. For example, if the pixels of each image are 15 × 15 meters, that is, each pixel covers 15 square meters of the earth's surface. Existing meters have a variety of spatial resolution, which are divided into three categories: low resolution (more than 60 meters), medium (between 10 and 30 meters) and high (between 30 cm and 5 meters).

5.2.21 Spectral Resolution

The ability of each meter to identify different spectra of electromagnetic waves is defined as the spectral power or resolution of a spectrum, in which the number of bands and the bandwidth of each image band are very important.

5.2.22 Radiometric Resolution

The number of bits that the meter assigns to receive energy is called the radiometric resolution, or in other words, the number of meters that the meter can assign to the energy received to create the image is called the radiometric resolution. The higher radiometric resolution, the more energy it can detect. For example, if the radiometric resolution is an 8-bit satellite image, it means that it can generate electromagnetic energy in a range of 0-255 nm (NM). Distribute the nanometer (0 for black and 255 for white) or if the image is 16 bits, it can divide the energy reaching the sensor in a range between 0-65535 NM. Therefore, it should be noted that if the radiometric resolution is higher, the image quality will increase, which is not detectable by eye, but these changes can be detected in pixel values.

5.2.23 Geosynchronous Orbit

In this circuit, the speed of the satellite is the same as the speed of the earth's rotation, and it takes one day to walk this orbit. The approximate height of this orbit is 36,000 to 3,000 km above the equator. This circuit is used for satellites that have meteorological, telecommunications and media applications.

5.2.24 Sun-Synchronous Orbit

In this orbit, the satellite is orbiting from north to south, and the orbit is variable over time. The orbit is about 500 to 1,000 km above the Earth's equator. This circuit is used in remote sensing satellites because the images taken by the satellite have the same time.

5.2.25 Field of View

The viewing angle of the whole sensor is called the field of view (the angle at which the meter sweeps the earth's surface), which can be calculated from Eq. (5.1).

$$FOV = 2 \times \arctan\left(\frac{W}{2H}\right) \tag{5.1}$$

where FOV = field of view; W = sensor width of the meter and H = height of the satellite.

5.2.26 Ground Field of View

The ground width of the field of view is actually the same as the imaged width, which can be calculated from Eq. (5.2).

$$GFOV = 2 \times H \times \tan(\frac{FOV}{2}) \tag{5.2}$$

where $GFOV$ = ground field of view.

5.2.27 Ground Resolution Element

The ground resolution element is the smallest angle that the sensor captures at any given moment because no satellite has a fully stable and stable orbit, and its altitude varies throughout the orbit. The field of view of the moment is a function of the orbital height of the satellite, the focal length of the optical system, and the size and dimensions of the sensor.

5.2.28 Classification

Classification can be thought of as a decision-making process in which visual data is transferred to a specific classroom space. Classification methods are traditionally divided into two categories: supervised and unsupervised classifications.

5.2.29 Monitored Classification

The monitored methods require basic information such as the number of classes, their characteristics, as well as the amount of known samples from each class.

5.2.30 Ncontrolled Classification

There are automated methods that do not require known examples, and pixels decide on their classification based on their values.

5.2.31 Visual Interpretation

If we analyze and separate data that is vector, it is called visual interpretation. Visual interpretation is considered as one of the traditional methods in remote sensing, but due to its high accuracy, its use is common and time and manpower are of great importance in this method.

5.2.32 Digital Interpretation

If raster data analyzed with optical tools (armed eyes) or with computer tools (GIS), it is called digital interpretation.

5.2.33 Fragmentation

The division of an image into continuous sections that are ideally aligned with the effects on the ground is called fragmentation. In image fragmentation, pixels that are similar to each other in terms of criteria (spectral characteristics) are considered as a fragment of the image, and therefore the location of a pixel in the fragmentation is important. Factors such as the integration (or elimination) of tolls, poor detection of toll boundaries, ambiguity in identified boundaries, and noise sensitivity have prevented fragmentation as a common method for extracting information. In addition, fragmentation methods usually do not work well for images with low spatial resolution and do not have the required accuracy.

5.2.34 Validation of Results

Evaluating results is one of the most important steps after information processing. Providing processing results without any parameters that express the quality or accuracy of these results reduces their value and in some cases makes them useless. Therefore, it is important to note that in addition to processing, the results must always be evaluated. There are several ways to evaluate the accuracy of results. The most common way to evaluate processing accuracy is to select a number of known sample pixels and compare their class with the processing results. These known data are called terrestrial realities or reference data. Accuracy assessment results are usually presented in the form of an error matrix, in which case the types of parameters and values that indicate the accuracy or type of error in the results (such as overall accuracy or Kapa coefficient) are extracted from this matrix.

Also we can use other similarity metrics to evaluate the model outputs and remote sensing data e.g. Structural Similarity Index (SSIM), Goodman and Kruskal's lambda (Goodman and Kruskal 1954), Theil's uncertainty, EOF, and Cramér's V (Cramér 1946; Koch et al. 2015; Rees 2008) and spatial efficiency metric (SPAEF) (Demirel et al. 2018; Koch et al. 2018).

5.3 Fundamentals and Basics

All materials are composed of atoms and molecules with a specific composition. Therefore, each material absorbs, reflects, or spreads electromagnetic radiation in a single form and under a specific wavelength that is related to its internal energy balance i.e. known as the unit material properties or spectral properties. In other words, the properties of energies created based on the conditions and type of materials on Earth are very different. Some objects have a reflective property in front of a particular wavelength. However, they have the property of absorbing and transmitting energy in another wavelength. The combination of such phenomena on different images creates special colors and allows the eye to distinguish between different shapes in the images. For example, sunlight is affected, reflected, or absorbed by molecules and suspended particles in the atmosphere as it passes through the atmosphere. This method of changing and analyzing the intensity of sunlight creates colors. For example, the blue color of the sky during the day is due to the emission of a blue spectrum in the atmosphere (all shorter wavelengths are released after a distance and only longer wavelengths reach the earth's surface) or because the leaves of some plants appear green. Chlorophyll absorbs the blue and red spectra and reflects the green spectrum. The reason for the green color of vegetation is the greatest reflection of the green spectrum.

Remote sensing systems operate in one or more parts of the visible, infrared, or microwave spectrum of the electromagnetic spectrum. In other words, each sensor system is sensitive to specific areas of the electromagnetic spectrum and records some of the spectral characteristics of objects. Of course, not all parts of the electromagnetic spectrum are used in remote sensing, that important causes of this problem include severe atmospheric absorption and distribution at certain wavelengths, the importance and usefulness of the data type collected, and technical considerations. These factors have caused some parts of the electromagnetic spectrum not to be used in general or to be used infrequently. Due to its high application, it is embedded in almost all multi-spectrum meters.

The amount of energy that reaches the sensor depends on how the energy and the body interact. If for each object the amount of energy reflected from the total energy reached to the object is measured at different wavelengths and plotted as a graph, the resulting curve is called the spectral behavior curve. The horizontal axis of this curve is the wavelength and the vertical axis of the graph represents the percentage of return energy. These energies can be measured in the laboratory or in real environments. After drawing the spectral behavior curve, a lot of information can be obtained about the object and how it appears in the image.

As mentioned, the sensors used in remote sensing collect information in different parts of the electromagnetic spectrum, and these sections are often not limited to the visible part of the spectrum. Since common displays use RGB color space to display colors, a combination of sensing bands should be assigned to three colors: red, blue, and green in the visible spectrum. Since conventional displays use RGB color space to display colors. Therefore, a combination of sensing bands should be assigned to three colors: red, blue, and green in the visible spectrum. One of the most common combinations used is the false color composite. Non-metallic bands are also used in this color composite. A non-false color composite is not seen in those natural colors for objects, but by knowing exactly the order of the assigned bands and how the tolls behave in these bands, it is possible to distinguish them.

Colored compositions are used to create visual mosaics, but their most important uses are in traditional visual interpretation and information extraction. By knowing the color combination as well as the type of sensor, an experienced interpreter can detect many side effects on the image. The best reference for creating an optimal color combination is the spectral behavior curve. By studying the spectral behavior curve of the desired effects, it is possible to identify the parts of the electromagnetic spectrum in which the mentioned effects appear more clearly, and then select the relevant bands from the bands in the sensor.

The RGB color system uses the three primary colors red, green, and blue to produce all colors. Each color is determined and can be produced by determining the value of these three components for it. Considering a three-dimensional coordinate system (in the form of a cube), it is at the origin of the black color coordinate system (0.0.0) and gradually other colors are produced by adding values in three axes. In the opposite corner of the origin, where the maximum possible numbers can be generated for the three primary colors, it will be white. The diameter of the cube that connects black to the white dot is called the gray line, and on it, the value of all three components of red, green, and blue is equal for each point, thus producing different levels of gray from black to white. The gray degree of the image is referred to as the brightness of the image. A histogram is a descriptive image of how the degrees of gray distribution are distributed. For analysis, the minimum and maximum brightness distances of each material in each band are determined and the abundance of matter is examined. The horizontal axis of the histogram indicates the intensity of the reflection from the surface in the desired band and the vertical axis also indicates the frequency of radiation reflecting materials.

Data from remote sensing systems, including aerial photographs and scanner images (satellite images), have a variety of errors and should be corrected before they can be interpreted and analyzed. These errors can be divided into geometric and radiometric categories. Geometric errors are related to the position of the pixels in the image relative to other phenomena and its absolute position, and the radiometric error is related to the amount of reflection recorded in the image.

Radiometric corrections are used to reduce or eliminate two major types of errors: atmospheric errors and device errors. Atmospheric errors are caused by the absorption and scattering of atmospheric particles, which blur the details of the image and thus reduce the power of sensor's spatial segregation. The greatest atmospheric effect is related to the scattering, which is highly dependent on the wavelength, so the effect of the atmosphere on different bands of a sensor is not the same. As the wavelength becoming longer, the atmospheric distribution impact is more effective. Atmospheric errors usually appear heterogeneous in images taken from a large viewing angle or with a large capture width. At the edges of the image, there are more atmospheric errors than in the middle of the image, which is due to the longer path that electromagnetic waves must travel through the atmosphere for side pixels. Atmospheric correction methods can be divided into two general categories, which are detailed correction and bulk correction methods. Device errors are those errors that occur due to the design or performance of the meter. These types of errors are varied and vary from one meter to another based on the type of system used. The two main device errors include missing line errors and tape errors.

On the other hand, satellite images have no coordinates and must be supplemented with additional geometric corrections with the help of GIS, and match a standard basis in terms of coordinates and then be analyzed. Satellite data can be geometrically corrected using one of the methods of using ground control points, using satellite orbital parameters and correlation, and matching a basis. The basis can be an image or a map. If the base is an image, the image-to-image correction, and if the base is a map, will be an image-to-map correction. If the ground of adaptation itself is the reference earth, that process is called absolute geometric adaptation and otherwise relative geometric adaptation. The most common geometric matching method is the use of ground control points. Their coordinates are known and used to validate data and find unknown parameters. After selecting the control points, the location of the points on the image and the map is determined by creating a spatial correspondence between the image and the map with the help of GIS. Then, the coordinates of these points are determined in the map coordinate system as well as the satellite image coordinate system. After that, the geometric error model of the coordinate system and its characteristics are selected. Due to the fact that there may be different errors depending on the type of sensor, the models used vary from one meter to another. In summary, the remote sensing function and GIS can be described as Fig. 5.2.

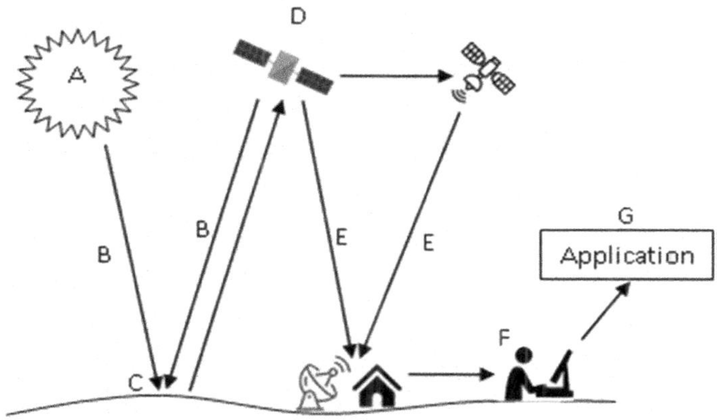

Fig. 5.2 The scheme of remote sensing

As you see in Fig. 5.2, RS components consist of seven different parts, which are located in part (A) of the electromagnetic energy source naturally (solar) or artificial (active sensors). In Section (B) of the energy sources, electromagnetic waves propagate toward the earth. In section (C), according to the material of the place where the waves hit the ground, they are divided into three states, which are absorption, reflection or diffusion. A number will be attributed to the amount of energy absorbed. Then (E) If the sensor is capable of sending information to Earth, the numerical information recorded by the sensor will be sent to Earth, otherwise the recorded digital information will be sent to a telecommunications satellite and transmitted to Earth by it. Section (F) examines the information received from the meter and applies the necessary corrections to it. And in section (G), the information evaluated by different users in the GIS is used to analyze the research and operations of their choice.

5.4 Importance and Necessity

Remote sensing as compared to other methods of information production, such as ground mapping, aerial photography, and local surveying have great advantages. Remote sensing rather than traditional methods has some advantages for instances, in remote sensing, scientists do not need to physical access to the study area, costly affordable and often are free and low time consuming. However, in traditional methods scientists must visit the study area physically and also these methods need lots of expertise for collecting and processing data and also time-consuming. This technology requires little (but specialized) manpower and very limited ground operations. However, it should be noted that if ground data are used to correct and validate data collected from remote sensing or GIS, it will increase the accuracy of the work.

Today, all data, processes, and outputs in remote sensing and GIS are digital, and this makes it possible to make the most of existing computer technology in performing analyzes and activities. Digital data is effective in creating a simple and easy connection between remote sensing and GIS, and speeds up processing and analysis in a variety of dimensions. This will examine all the various unknown aspects and phenomena whose effects have not yet been discovered, and identifying and being aware of them requires very costly and time-consuming research that may not be enough to last a lifetime.

One of the advantages of remote sensing is the availability of a variety of satellite imagery with different spatial and spectral characteristics, which allows experts to use this feature to obtain a more comprehensive and complete set of information in a shorter time than traditional methods. The presence of multiple sensors that perform a variety of imaging in different parts of the electromagnetic spectrum, in addition to the variety and increase of data, allows for multiple images for a specific location and more accurate, complete and multifaceted analysis can be performed simultaneously. In the remote sensing not only the analysis of the current and past situation can be examined, but also the future events can be predicted using different algorithms. The existence of these capabilities in remote sensing and GIS has led to getting out the monopoly of military and espionage systems and to serve practical and research purposes to control and be aware of what is happening.

5.5 Methods and Flowchart

In the studies, that uses remote sensing science and a GIS, an almost identical process is observed. The methods and process of remote sensing researches presented in Fig. 5.3.

In Fig. 5.3, the first step in remote sensing process is identifying appropriate input data. Then according to the conditions of case study, the optimal time interval (based on the required accuracy) and the optimal sensors (sensors that have images with desirable characteristics) are determined. As the next step, the required images are retrieved, the raw satellite images are corrected and the information is extracted and stored in the desired inputs of the GIS as digital layers. After verifying the results, the necessary maps are produced by the GIS and related software (such as ArcGIS, QGIS, ERDAS etc.).

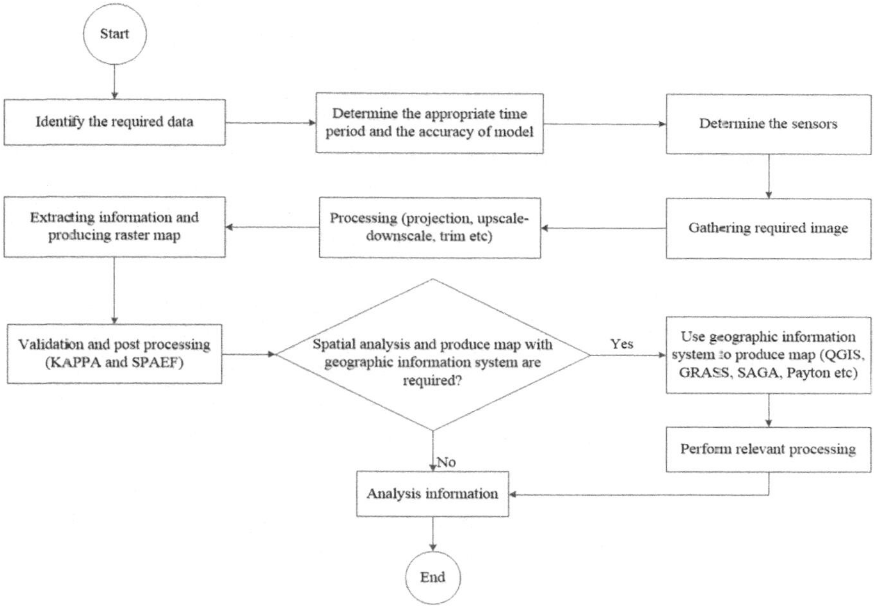

Fig. 5.3 The flowchart of RS and GIS process

5.6 Practical Examples

5.6.1 First Example

Demirel et al. (2019) collected groundwater point data and three different remotely sensed soil moisture products from the Advanced Microwave Scanning Radiometer on the Earth Observing System (EOS) Aqua satellite (AMSR-E), the European Space Agency Climate Change Initiative Soil Measure (ESA CCI SM v04.4), soil moisture active passive (SMAP), and remotely sensed total water storage anomalies from Gravity Recovery and Climate Experiment (GRACE) to constrain a lumped model (HBV) for the Moselle River Basin in Germany and France. Performance metrics ensured a good fit between observed and simulated streamflow as well as groundwater and soil moisture. Before model calibration, the most important parameters are identified using sensitivity analysis. Their comprehensive analysis showed substantial contribution of aggregated remote sensing products to the lumped model calibration. Additional new earth observations, such as groundwater levels from wells and satellite-based soil moisture data improved the model's physical behavior, while it kept a good water balance. Their results are in line with previous studies (Nijzink et al. 2018).

5.6.2 Second Example

Baghri (2016) investigated the identification of groundwater resources using satellite imagery. Groundwater resources, due to their cheap, easy and accessible extraction, are of good quality in case of sustainable use of the resource and are quite suitable for exploitation for sensitive uses such as drinking. In this regard, they are known as strategic water resources. The high volume of these resources, compared to surface water reservoirs and the low impact of climatic conditions, has made these resources a point of reference for water supply and development. In the current situation, a large percentage of the drinking water consumed by urban communities in the world is supplied by groundwater resources. Therefore, identifying and finding potential resources is essential for sustainable exploitation.

In this study, in addition to traditional methods of identifying groundwater resources such as geological surveys, geophysical and speculation, they used satellite imagery combined with new accreditation techniques. Researchers believe that soil characteristics, land use type, land slope and vegetation type can be used as a guide for locating groundwater resources, and the accuracy of such a guess has been investigated during this study. In this study, image of Landsat satellite and ASTER sensor, Terra satellite, ENVI and ArcGIS software have been used to process and access the required information. In order to identify the potential of groundwater resources, three methods have been used, which are artificial neural network, multivariate linear regression and analytical hierarchy process, and their accuracy has been evaluated. Talented areas for groundwater resources have been identified from data such as vegetation maps, land use, soil and slope.

Vegetation: Research shows that vegetation reduces rainfall and allows rain to slowly seep into the soil, and therefore the presence of dense vegetation in an area increases the likelihood of groundwater resources in that area. So the use of water resources, vegetation, meadows level increases and raises the water table. The vegetation map used in this study was extracted from the Landsat 8 satellite image and the vegetation density was determined by the vegetation index, which has the highest relationship with plant life volume among the vegetation characteristics.

The vegetation index is obtained from Eq. (5.3).

$$NDVI = \frac{NIR - RED}{NIR + RED} \tag{5.3}$$

where, *NIR* = near infrared; *RED* = red band.

In this study, first in ENVI software, radiometric correction was performed and in order to reduce the effect of opacity, atmospheric correction was applied and NDVI index was calculated based on two bands 4 and 5 Landsat image and then using classification tools in the software ArcGIS The data obtained were classified into different classes.

Slope: The slope map of the area is prepared from the ASTER image height digital model with a spatial resolution of 30 m. The slope map is obtained by applying the slope function in the ArcGIS environment. The slope plays an important role in the accumulation and infiltration of water into the lower layers. The higher the slope of the area, the faster the surface water flows and the less likely it is that water will seep into the lower layers. Accordingly, areas with lower slopes are predicted to increase the likelihood of groundwater.

Soil: Soil map is one of the most important parameters in identifying water resources. In this study, the soil of the study area has been divided and evaluated into several classes in terms of water absorption capacity.

Each of these four factors has a special weight in identifying water resources. In this study, they concluded that the neural network had the highest accuracy and the AHP method had the lowest accuracy, but the standard deviation values were acceptable in all three methods. Therefore, these four factors can be used to identify groundwater resources.

5.7 Summary

Human beings have always tried to know their surroundings better and more, and despite the efforts made, sometimes natural events (hazards etc.) have surprised them. The growing trend of human societies and population growth necessitates more and more sustainable use of natural resources and resources to meet human needs and greater coordination with the natural environment. At the same time, human beings are expanding their living space day by day and making changes in the areas and exploiting the natural resources that were previously known as pristine places and areas. Therefore, it is necessary to collect information from these phenomena. On the other hand, field harvesting from the globe is continuous, regular, up-to-date, and with adequate spatial coverage, it is time consuming, costly, and sometimes impossible. In this regard, remote sensing science and GIS are very effective and efficient in achieving these goals as quickly as possible, so that it has removed the barriers to understanding nature with full coverage from the Earth's surface and with short-term intervals. The high resolution and wide resolution of satellites have led to the evaluation of the information obtained as one of the most important measurement tools. Remote sensing output data is an irreplaceable source of information input to the GIS so that accurate and timely layers can be prepared by remote sensing science and examined by the GIS to achieve the intended goals.

New remote sensing techniques are high-performance new tools that, along with new computational methods and the development of various physical, statistical, and black box models that use remote sensing data as input, open up human curiosity to learn and see more. They are more visionary and add to his experience. In general, the applications of remote sensing techniques in water resources studies can be classified into the following groups:

- Studies of water balance components such as rainfall estimation, evaporation and transpiration. (This category includes land use determination and estimation

of drinking, industrial, agricultural, and environmental consumption, as well as hydrological studies of the body (catchment area) of catchments that lead to the modeling of resources and uses).
- Qualitative studies and water pollution.
- Groundwater identification and potential studies.
- Studies on river morphology and river engineering.
- Watershed management, erosion and sedimentation studies.

According to research in the field of water resources using remote sensing techniques and GIS, increasing the accuracy of methods and preparing guidelines and general principles for various studies seems necessary because the current process and results of studies are tested. Different methods and procedures indicate different goals, and recently there are signs of confidence in remote sensing techniques in the results that need to be accelerated. At the end, it should be noted that remote sensing science and GIS are very valuable due to increasing the speed of processing and collecting environmental information and data and need more investment and research in their use and application for full access and knowledge of the environment. That will increase the level of well-being, awareness of the danger, how changes will take place in the surrounding natural environment and the optimal use of available resources.

Acknowledgements The authors thank Iran's National Science Foundation (INSF) for the support of this research.

References

Ahmad I, Dar MA, Andualem TG, Teka AH (2019) Groundwater development using geographic information system. Appl Geomat:1–10

Ahmad MUD, Bastiaanssen WG, Feddes RA (2005) A new technique to estimate net groundwater use across large irrigated areas by combining remote sensing and water balance approaches, Rechna Doab, Pakistan. Hydrogeol J 13(5–6):653–664

Ahmadi B, Ahmadalipour A, Tootle G, Moradkhani H (2019) Remote sensing of water use efficiency and terrestrial drought recovery across the contiguous united states. Remote Sens 11(6):731

Alparslan E, Aydöner C, Tufekci V, Tüfekci H (2007) Water quality assessment at Ömerli Dam using remote sensing techniques. Environ Monit Assess 135(1–3):391

Ansari H, Marofi S (2017) Snow water equivalent estimation using AMSR-E and GLDAS model (case study: basins of northwestern Iran). J Water Soil 31(5):1497–1510

Baghri A (2016) Identification of groundwater resources using satellite data. M.Sc., Department of Civil Engineering, Ferdowsi University of Mashhad, Mashhad, Iran

Banerji S, Mitra D (2018) Geographical information system-based groundwater quality index assessment of northern part of Kolkata, India for drinking purpose. Geocarto Int 34(9):1–16

Burrough, PA, McDonnell RA (2000) Principles of geographical information systems, vol 1. Oxford University Press, England, United Kingdom

Chandra AM, Ghosh SK (2002) Remote sensing and geographical information system. Alpha Science, No, Oxford, United Kingdom

Clarke K (2001) Getting started with geographic information systems, vol 1. Prentice Hall, New Jersey, USA

Colwell RN, Estes JE, Photogrammetry ASO, Simonett DS, Fawwaz, TU (1983) Manual of remote sensing: interpretations and applications. American Society of Photogrammetry, No. 2, California, USA

Cramér H (1946) Mathematical methods of statistics. Princeton University Press, Princeton

Demirel MC, Mai J, Mendiguren G, Koch J, Samaniego L, Stisen S (2018) Combining satellite data and appropriate objective functions for improved spatial pattern performance of a distributed hydrologic model. Hydrol Earth Syst Sci 22(2):1299–1315

Dembélé M, Hrachowitz M, Savenije HH, Mariéthoz G, Schaefli B (2020) Improving the predictive skill of a distributed hydrological model by calibration on spatial patterns with multiple satellite data sets. Water Resour Res 56(1). p.e2019WR026085

Ganapuram S, Kumar GTVV, Krishna IVMM, Kahya E, Demirel MCC (2009) Mapping of groundwater potential zones in the Musi basin using remote sensing data and GIS. Adv Eng Softw 40:506–518

Ghazanfari moghaddam MS, Alizadeh A, Mousavi baygi SM, Faridhosseini AR, Bannayan awal M (2011) Comparison the persiann model with the interpolation method to estimate daily precipitation (a case study: North khorasan). J Water Soil 25(1):207–215

Goodman LA, Kruskal WH (1954) Measures of association for cross classifications. J Am Stat Assoc 49:732–764

Guzinski R, Anderson MC, Kustas WP, Nieto H, Sandholt I (2013) Using a thermal-based two source energy balance model with time-differencing to estimate surface energy fluxes with day–night MODIS observations. Hydrol Earth Syst Sci 2013:17–7

Kalhor K, Emaminejad N (2019) Sustainable development in cities: Studying the relationship between groundwater level and urbanization using remote sensing data. Groundwater Sustain Dev 9

Koch J, Jensen KH, Stisen S (2015) Toward a true spatial model evaluation in distributed hydrological modeling: Kappa statistics, Fuzzy theory, and EOF-analysis benchmarked by the human perception and evaluated against a modeling case study. Water Resour Res 51:1225–1246

Koch J, Demirel MC, Stisen S (2018) The SPAtial EFficiency metric (SPAEF): multiple-component evaluation of spatial patterns for optimization of hydrological model. Geogr Model Dev 11:1873–1886

Lyazidi R, Hessane MA, Moutei JF, Bahir M, Ouhamdouch S (2019) Management of water resource from semiarid area by elaborating database under GIS: case of Gareb-Bouareg aquifer (Rif, Morocco). Arab J Geosci 12(11):352

Ma Z, Yan N, Wu B, Stein A, Zhu W, Zeng H (2019) Variation in actual evapotranspiration following changes in climate and vegetation cover during an ecological restoration period (2000–2015) in the Loess Plateau, China. Sci Total Environ 689:534–545

McCabe MF, Rodell M, Alsdorf DE, Miralles DG, Uijlenhoet R, Wagner W, Lucieer A, Houborg R, Verhoest NE, Franz TE, Shi J (2017) The future of Earth observation in hydrology. Hydrol Earth Syst Sci 21(7):3879

Mohammadrezapour Tabari M, Montazeriun MH (2015) Delineation of the hydrological boundaries of the major dams' watersheds using satellite imagery and base maps. Water Eng 8(24):53–68

Nijzink RC, Almeida S, Pechlivanidis IG, Capell R, Gustafssons D, Arheimer B, Parajka J, Freer J, Han D, Wagener T, van Nooijen RRP, Savenije HHG, Hrachowitz M (2018) Constraining conceptual hydrological models with multiple information sources. Water Resour Res 54(10):8332–8362

Rees WG (2008) Comparing the spatial content of thematic maps. Int J Remote Sens 29:3833–3844

Sabins FF (1997) Remote sensing: principles and interpretation. WH Freeman Publications, No. 3, New York, USA

Smale A (2013) NASA's Imagine the Universe. www.imagine.gsfc.nasa.gov. Accessed 6 Aug 2019

Stisen S, Jensen KH, Sandholt I, Grimes DI (2008) A remote sensing driven distributed hydrological model of the Senegal River basin. J Hydrol 354(1–4):131–148

Stisen S, McCabe MF, Refsgaard JC, Lerer S, Butts MB (2011) Model parameter analysis using remotely sensed pattern information in a multi-constraint framework. J Hydrol 409(1–2):337–349
Usali N, Ismail MH (2010) Use of remote sensing and GIS in monitoring water quality. J Sustain Dev 3(3):228
Vinukollu RK, Wood EF, Ferguson CR, Fisher JB (2011) Global estimates of evapotranspiration for climate studies using multi-sensor remote sensing data: evaluation of three process-based approaches. Remote Sens Environ 115(3):801–823
Waqas MM, Awan UK, Cheema MJM, Ahmad I, Ahmad M, Ali S, Shah SHH, Bakhsh A, Iqbal M (2019) Estimation of canal water deficit using satellite remote sensing and gis: a case study in lower Chenab Canal System. J Indian Soc Remote Sens 47(7):1–10
Xin W, Can T, Wei W, Ji L (2019) Change detection of water resources via remote sensing: an LV-NSCT approach. Appl Sci 9(6):1223
Yilmaz MT, Hunt ER Jr, Jackson TJ (2008) Remote sensing of vegetation water content from equivalent water thickness using satellite imagery. Remote Sens Environ 112(5):2514–2522

Chapter 6
GIS Application in Water Resource Management

Arezoo Boroomandnia, Omid Bozorg-Haddad, Biswajeet Pradhan, and Amitava Datta

Abstract In this chapter, we consider the generic concepts of the Geographical Information System (GIS) and its role in water resources management issues. This chapter is divided into 7 sections. An overview of GIS emersion and its application in water science is provided in the first section. The second section remarks the common technical phrases and definitions in geospatial technology. The fundamental of GIS is expressed in the third section. The fourth section discusses the importance and necessity of employing GIS in water resource management. Finally, we covered the methodology of applying GIS in water management researches and discuss it through three case studies in detail in the last two sections. After studying this chapter, readers would gain an understanding of the problem solving in water resource management by the proper tools available on the GIS market.

Keywords Geographical Information System · GIS software · Surface water hydrologic modeling · Water supply and sewer systems modeling · Water quality and watershed protection modeling · Groundwater modeling

---

A. Boroomandnia · O. Bozorg-Haddad (✉)
Department of Irrigation and Reclamation Engineering, Faculty of Agricultural Engineering and Technology, College of Agriculture and Natural Resources, University of Tehran, Karaj, Tehran 31587-77871, Iran
e-mail: OBHaddad@ut.ac.ir

A. Boroomandnia
e-mail: aboroomandnia@ut.ac.ir

B. Pradhan
Center for Advanced Modelling and Geographical Information Systems (CAMGIS), School of Information Systems and Modelling, Faculty of Engineering and IT, University of Technology Sydney, CB11.06.106, Building 11, 81 Broadway, PO Box 123, Ultimo, NSW 2007, Australia
e-mail: Biswajeet.Pradhan@uts.edu.au

A. Datta
Faculty of Engineering and Mathematical Sciences Computer Science and Software Engineering, University of Western Australia, Perth, Australia
e-mail: amitava.datta@uwa.edu.au

O. Bozorg-Haddad (ed.), *Essential Tools for Water Resources Analysis, Planning, and Management*, Springer Water. https://doi.org/10.1007/978-981-33-4295-8_6

6.1 Introduction

Geospatial technologies particularly geographical information systems (GIS) focus on processing and interpretation of geographical data with the purpose of knowledge improvement in various fields. To become more familiar with the evolution of GIS technology over the recent decades and its role in water resource management, a summary of important historical development in GIS is explained in the following subsections.

6.1.1 GIS Development History

Application of spatially referenced data dates back before the "geospatial technologies" emergence. In the 1960s for the first time geographical information system was developed in Canada in response to the government's need for data handling (Shamsi 2005). For converting maps to numerical form, there were technical constraints such as limited computer storage capacity, slow processing speeds, and high costs of technology. Scientists overcame these obstacles in the seventies, which is known as the consolidation era rather than innovation in GIS knowledge. In this period, the U.S. Department of Home Affairs developed Map Overlay and Statistical System (MOSS), the first vector-oriented GIS software. In the 1980s, the demand for GIS experienced a continuous increase, particularly in natural resource management. In the early eighties, GIS became the key component in scientific researches to satisfy local, national, multinational, and global requirements. Indeed, the eighties could be considered a focal point in the development and applicability of the GIS (Tsihrintzis et al. 1996). In this period, companies like Environmental Systems Research Institute (ESRI), Computer Aided Resource Information System (CARIS), MapInfo Corporation and Earth Resource Data Analysis System (ERDAS) appeared as commercial vendors of GIS software. Also in 1982, U.S. Army developed Geographical Resources Analysis Support System (GRASS GIS), the first GIS software to utilize both vector and raster formats, and released it as open source from 1995. In the nineties, GIS industry had matured and took a significant place from commercial aspect (e.g. the number of companies) as well as academic aspect (e.g. researches and professional projects). For instance, the number of system installations experienced a twofold rise every two years; the GIS market reached 35 percent growth annually; MapInfo, the first desktop GIS product for Windows, was released; and the geographic information national research centers have been established (Tsihrintzis et al. 1996). By the end of the twentieth century, viewing GIS data over the Internet gradually became prevalent among users. More recently, the modern world witnesses the continuous growth in emerging of free, open-source GIS packages, which perform specific tasks and run on various operating systems. Besides, geospatial data and mapping applications have been accessible via the World Wide Web (Fu and Sun 2010). Some free and open source GIS software is listed in Table 6.1.

Table 6.1 Some of the free and open source GIS software (Jarar Oulidi 2019a)

Category	Software name
GIS Server	North 52°, MapGuide, MapServer, GeoServer, QGIS Server, deegree, PyWPS
Mapping framework	GeoTools, OpenLayers, MapBender, MapFish, GeoExt
Geospatial database	MySQL Spatial, PostGIS, Ingres Geospatial
Spatial analysis tools	R Project, PySAL, Open GeoDA
Mobile GIS	QGIS for Android, gvSIG Mobile, Geopaparazzi
Office GIS	QGIS, uDig, MapWindow, OpenJUMP, ILWIS, SAGA, GRASS
Remote Sensing	GDL, OSSIM, e-foto, Opticks

Advantages of free software are: declining budget expenditure, the source code openness, supervision and maintenance of geospatial data infrastructures, and the capability of the functionalities' development by professional implementation administrators (Jarar Oulidi 2019a).

6.1.2 GIS Application in Water Resource Management

Due to the spatial identity of data used in water-related environmental planning and management, data management is sophisticated. Therefore, engineers have benefited from the ability of GIS software for documentation, management, storage and visualization of spatial data. GIS in combination with simulation models and also without model interface has been employed in several areas, such as surface water and groundwater hydrologic modeling, water quality, water distribution and sewer network design, sustainable water resource management, flood hazard assessment, urban storm water, hydropower potential assessment, and water quality and quantity modeling in and agricultural areas. Some of these applications are discussed in detail in the following subsections. Also, the generic data used, analyzed, and produced by GIS tools in water-related programs are explained in Sect. 6.2.2.

6.1.2.1 Surface Water Hydrologic Modeling

In order to represent the effects of rainfall-runoff on surface water resources (e.g. lakes, rivers, and canals) various simulation software for surface water hydrologic modeling are released. Numerous parameters for defining local topography, soil type and characteristics, and land use are part of the required data as input for the hydrologic software. But analyzing torrent of data accurately, after sophisticated data gathering step, needs significant time and skill for preparation as input parameters. This process is simplified by the use of GIS in which analyzed data are automatically calculated and stored in the separate attribute tables, as defined by the modeler. For the first time in 1989, Terstriep represented the interface between GIS and hydrologic database. Afterwards application of GIS either as an independent tool or coupled with hydrologic models radically accelerated in twenty-first century. Some prevalent combinations of GIS and hydrologic software are summarized in Table 6.2 (Tsihrintzis et al. 1996; Martin et al. 2005).

6.1.2.2 Water Supply and Sewer Systems Modeling

Due to the spatial nature of the data in planning, design, analysis, operation and maintenance of water and sewer systems in cities, water and sewer specialists must employ technologies like GIS which aid in selecting minimum cost alternatives, and operating network in a smooth and efficient way. The key role of GIS in water supply management is its capability in applying the unique characteristics associated with consumers, which leads to accurate and up to date forecasting of water demands. Also, GIS network tools are powerful in allocation and routing difficulties analysis. The interface of GIS and water and sewer network models are summarized in Table 6.3 (Tsihrintzis et al. 1996; Martin et al. 2005).

6.1.2.3 Water Quality and Watershed Protection Modeling

Identification and prioritization of the polluted regions need the integration and visualization of various spatial information. GIS is a suitable tool which allows specialists to overlay coverages, evaluate and determine contaminant loading degrees efficiently. For instance, the runoff volume and consequent receiving stormwater quality is highly affected by the land use as a spatial variation. GIS makes simpler the process of simulating pollution distribution and declines the uncertainty stemmed of spatial averaging. Furthermore, erosion investigation, as the fundamental steps of watershed protection measures, is facilitated by GIS application. An overview of the combination of GIS and water pollution models are summarized in Table 6.4 (Tsihrintzis et al. 1996; Martin et al. 2005).

6.1.2.4 Groundwater Modeling

GIS technology presents a mechanism to collect, maintain, and make subsurface data available on demand for groundwater protection, resource investigations, and engineering design. GIS application in groundwater modeling is widespread, comprising groundwater quality and quantity. Some examples of coupled GIS and groundwater models are summarized in Table 6.5 (Tsihrintzis et al. 1996; Martin et al. 2005).

Table 6.2 History of GIS and hydrologic software interface

Time period	GIS tool/hydrologic model	Achievement
Twentieth century	GRASS	Creating landcover map to model stormwater runoff
	GIS MIKE-11	Producing flood inundation maps and of flood depths contours with preferred interwalls
	ArcInfo HEC-1	Deriving area-weighted hydrologic parameter
	RS ERDAS ArcInfo	Deriving the imperviousness percentage for diverse land cover classes
	ArcView SWMM	Estimating Subbasin curb length and modeling runoff
	GRASS SMoRMOD	Simulating rainfall-runoff
	ArcView HEC-1 HEC-2 Flood Hydrograph Package	Predicting flood hydrographs and perform drainage analysis
	GISHYDRO	Saving time and improving the accuracy of the hydraulic design process, thanks to the quickly assembling slope, landuse, and soil data
	ArcView HSPF MODFLOW	Providing the linkage mechanisms between models and calculating nonpoint sources loading
	ArcInfo HEC-1	Creating a database involving required hydrological characteristics for watershed rainfall-runoff model Calculating the basin area and average runoff curve numbers
	ArcInfo CMLS	Analyzing chemical movement through Layered Soils
Twenty-first century	ArcView/ArcInfo HEC-HMS HEC-RAS	Preparing geographic data Processing simulation results including profile data of water surface and velocity to create floodplain map, and compute flood damages, ecosystem restoration, flood warning response and preparedness
	ArcInfo HAZUS-MH	Estimating potential damages from hurricane winds, earthquakes, and floods throughout the United States
	ArcView PDTank	Simulating watershed management
	ArcVeiw MIKE 21	Modeling bays and estuaries

(continued)

Table 6.2 (continued)

Time period	GIS tool/hydrologic model	Achievement
	WMS TR-20	Modeling the rainfall-runoff process

6.2 Definition and Terms

As it has been discovered through various examples of GIS application in Sect. 6.2, the role of GIS, is to store, model, manipulate, combine, analyze and display vast volumes of geographical data efficiently. It also creates a link among geographical objects and semantic data. To gain a profound understanding of GIS, it is necessary to define the technical phrases in advance. These phrases are explained under five categories in this section.

6.2.1 Basic Definitions

Verifiable facts about the real world are defined as **data**. Organized data which expose patterns and ease search are considered as **information**. Information provides added value with regard to the data itself. In order to extract spatial information from spatial data, organizing data by spatial attributes is an essential primary step. Additionally, on account of the digital computers' nature, digital manipulation is feasible when data items are identified as **discrete spatial objects**. Hence, discrete representation of geographical space is obliged since spatially continuous fields cannot be stored digitally as continua (e.g. pressure and temperature fields). Points, lines, areas, volumes and surfaces are examples of discrete spatial objects which are used by spatial data models. Description of objects and their attributes in the digital form (both spatial and non-spatial) comprise spatial datasets. The collection of inter-related data, for a specific use, is called a **database**. Field digitizing and remote sensing image are examples of primary elements which creates a digital database. To store, edit and retrieve data in a database, a database management system (DMS) should be formed. **DMS** is a collection of software to achieve this purpose. Most geographical information systems separately organize spatial and non-spatial data. In terms of DMS, GIS are divided to two category for non-spatial data management: use an internal DMS or external DMS (Merriam 1994). These phrases are explained in details in following.

Table 6.3 History of GIS and water and sewer network models interface

Time period	GIS tool/hydrologic Model	Achievement
Twentieth century	ArcInfo WADSOP	Balancing looped water supply systems
	ArcCad KYPIPE	Providing water distribution system modeling
	ArcView EPANET	Simulating long-term changes of hydraulic behavior and water quality within pressurized pipe networks (known as AVNET)
	AutoCAD H_2ONET	Hydraulic modeling, optimizing network design, graphical editing, results presentation, database management, and enterprise-wide data sharing and exchange
	H_2OMAP	Combining mapping functions and spatial analysis tools with network modeling to organize strength management and business planning
	InfoWater	Integrating water network modeling and optimization functionality with ArcGIS
	ArcInfo/ArcFM, ArcView/MapInfo MIKE NET	Building a link to a GIS database structure, benefiting attribute mapping and geocoding, simulating steady flow, pressure distribution, and water quality of potable water distribution network over a long-term period
	ArcView Runoff Model	Simulating urban runoff
	ArcView SWMM	Simulating dry- and wet-weather flows on the basis of land use, demographic situations, meteorological data, hydrologic/hydraulic/treatment characteristics of the drainage and sewer network
	SWMMDUET	Providing interfaces for generating SWMM's input file and ploting SWMM's output file
Twenty-first century	ArcView MIKE BASIN	Analyzing water supply abilities while considering water rights for domestic, industrial, and agricultural sectors. Assessing multipurpose reservoir operations

(continued)

Table 6.3 (continued)

Time period	GIS tool/hydrologic Model	Achievement
	Sewer-CAD StormCAD	Evaluating current wastewater treatment system and collection network facilities, and future potential of extending facilities preparing a phased plan to implement and finance the essential improvement
	ArcView MIKE 11	Modeling the hydrodynamics and water quality of streams in detail
	H_2OVIEW Water	Enabling deployment and investigation of GIS row data and modeled results over the internal networks and Internet
	AGSWMM	Linking SWMM5 with ArcGIS while conducting the same functions as SWMM
	ArcGIS WaterGEMS	Using a geodatabase for integrated WaterCAD modeling within the ArcGIS platform
	SynerGEE Water	Integrating hydraulic simulation capability into a GIS
	ArcView WaterCAD	Direct building and maintaining the water and sewer networks inside a GIS
	MOUSE GIS	Simulating surface runoff, water quality, and sediment conveyance within the urban catchments and sewer networks

6.2.2 Geospatial Data

Spatial data in GIS are classified from different aspects. The two more common criteria in classification are the methods of data visualization and data storage, which are explained distinctively.

6.2.2.1 Visual-Based Classification

Spatial data can be shown as both **vector** and **raster** models in GIS. Visualizing the real world spatial objects via points, lines, and areas confined by lines is titled vector model. The simplest data format to manipulate which produces resulting files in smaller size is vector format. The most recent vector formats include (Jarar Oulidi 2019a):

Table 6.4 History of GIS and water quality and watershed protection models' interface

Time period	GIS tool/hydrologic Model	Achievement
Twentieth century	GIS QILLUDAS	Simulating urban runoff emphasizing runoff quality on urban watershed
	GRASS ANSWERS	Analyzing watershed erosion and deposition
	EDRAS QUAL2E	Calculating total maximum daily pollution load (TMDL)
	ArcInfo PSRM	Analyzing watershed runoff
	GIS AUTO-Q1	Determining nonpoint source pollution from an urban watershed
	GRASS WEPP	Reflecting the impact of surface soil conditions (effected by agricultural and forestry activities) on storm runoff and erosion
	ArcInfo ANSWERS	Estimating erosion, deposition and related hydrologic parameter
	GIS HEC-STORM	Characterizing pollutant concentration in stormwater
	GRASS SWAT	Generating topographic and model input parameters for sub-basin; editing input data's component, simulating and visualizing graphical results of watershed hydrology, water quality
	ArcInfo AGNPS	Controlling nonpoint source pollution
	ERDAS AGNPS	Supporting data for analysis and management of agricultural nonpoint source contamination
	GRASS AGNPS	Minimizing the sedimentation, nutrient and pesticide transport toward waterbodies
	ArcView HSPF,QUAL2E	Conducting water quality planning, flood mapping, and assessment of water erosion and sedimentation problems
	ArcInfo HSPF	Predicting, map producing, monitoring, and offering management strategies for agricultural pollutants
Twenty-first century	GIS	Detecting and monitoring the processes of waterlogging and salinization
	ArcView MIKE SHE	Simulating runoff quality and volume while considering interaction with surface water systems

(continued)

Table 6.4 (continued)

Time period	GIS tool/hydrologic Model	Achievement
	LF2000-WQX	Predicting the concentrations of the pharmaceuticals diclofenac and propranalol in catchments
	ArcView HSPF NPSM	Estimating land-use-specific NPS loadings for particular contaminants at a catchment
	ArcView HSPF TOXIROUTE	Calculating simple dilution and decay for a stream system (either mean-flow or low-flow conditions)
	MapInfo SCS runoff Model	Assessing pesticide load in runoff
	ArcView IDOR2D	Controlling water quality and pollutant transport
	IDRISI USLE	Evaluating soil erosion
	ArcView MIKE BASINS	Assessing loadings and receiving water impacts at different complexity degree on watershed scale

Table 6.5 History of GIS and groundwater models' interface

Time period	GIS tool/hydrologic Model	Achievement
Twentieth century	ArcInfo WHPA	Producing water-table and transmissivity contours
	GRASS DRASTIC	Identifying the correlation rate between the vulnerability of groundwater to pollution and the availability of nitrogen-based fertilizer
	ArcInfo GLEAMS	Evaluating effect of soil characteristics on pollution distribution
	ArcInfo MODFLOW	Evaluating changes in the hydrological processes of complex groundwater systems (e.g. water infiltration and hydraulic conductivity)
	IDRISI AgriFlux	Simulating groundwater nitrate transport
	FREEWAT LuKARS	Simulating the impact of land use changes by changing the area of dominant hydrotopes
Twenty-first century	ArcGIS DRASTIC	Finding the groundwater vulnerable zones
	ArcPRZM-3	Modeling of pesticide leaching potential from soil surface towards groundwater
	ArcGIS MODFLOW	Forecasting groundwater table depth from the soil surface and changes in groundwater systems

- Shapefile (*.shp): The ESRI company developed this format to store and exchange GIS data. Collection of files with the same base filename form a shapefile, for instance landuse.shp, landuse.shx (which covers the geometry and geometry index respectively) and landuse.dbf (presenting recipient data);
- coverage: another data format applied by ESRI to store the ArcInfo information type;
- simple geographic entities: An OpenGIS plan to store geographic data (e.g. points, lines, polygons) and related attributes.

Cells or pixels of the raster model represent spatial units. Cells are formed by row and column values, with the boundary of the grid recorded to benchmark spatial coordinates. A single grid cell remarks a point, a string of connected cells represent a line, and groups of adjacent cells show the areas. The raster data mode is mostly applied to store raster format images, including aerial photographs, digitized paper maps, and satellite images taken by Landsat, QuickBird, IKONOS and other satellites.

Each mode of data demonstration comes with benefits and drawbacks in terms of storage efficiency and the types of processing which can be completed. The nature of data in the real world and the user's experience and preferences highly determine the chose mode of data. Vector and raster modes are illustrated in Fig. 6.1. Additionally, some practical examples of vector and raster data commonly used in water resource management are listed in Table 6.6 (Korres and Schneider 2018).

- DEM data can be stored in all three forms of raster structure to meet different purposes, but grid-type is more common.

6.2.2.2 Store-Based Classification

Data are divided to **static** format and **dynamic** format from storing aspect. Vector and raster mode are categorized under static geospatial datasets. Static format crucially respond to the following features (Jarar Oulidi 2019a):

- geospatial objects: remark geographical and geometrical characteristics;
- semantic data: explain about geographical entities;
- metadata detail: provide information regarding the forecasting system used; data acquisition techniques, the quality of geographical data, the production and updating date; and
- style of data: visual characteristics of geographical data shown on a map (color or type of the line).

Dynamic format, in addition to aforementioned features, not only does make possible to store data and display them via a program during the implementation, but also offer to exchange geographical entities between two GIS applications. The well-known dynamic formats are summarized in Table 6.7 (Jarar Oulidi 2019a).

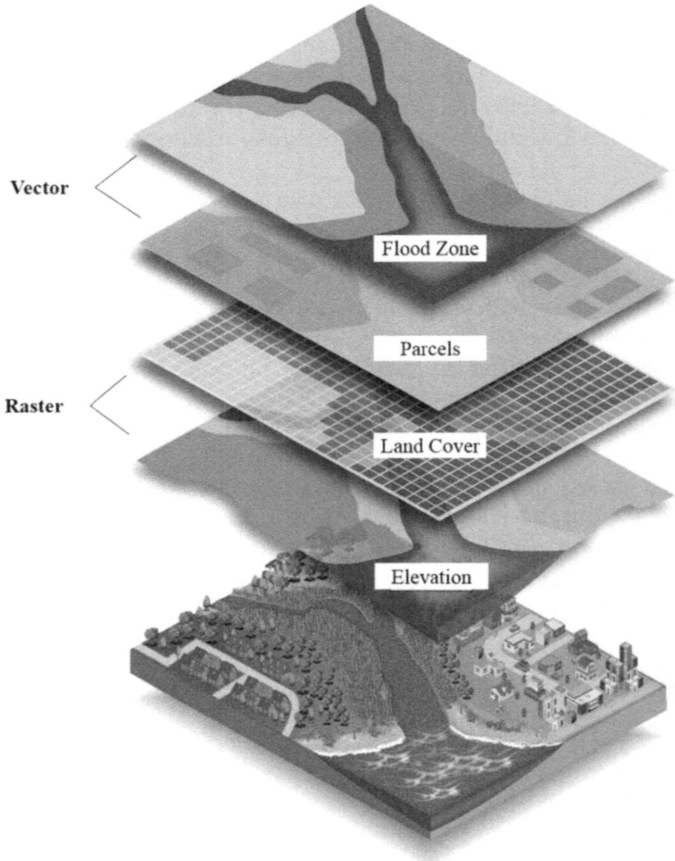

Fig. 6.1 Illustration of vector mode and raster mode visualization

6.2.3 Database Management System

The system of database management (DMS) is considered as a framework allowing users to archive and manipulate data in tables known as attribute table. DMS is also regarded as a system logically open to the clarity and flexibility of the general hierarchical, network, or relational data model. Data models store attributes as an inherent part of the graphics feature. Within the DMS, data is structured in the form of tables, containing records in rows and columns. Existence of only one common column between tables, makes it possible to couple rows of one table with those of another table (Jarar Oulidi 2019a).

Maybe the most challenging and expensive part of GIS applications' development is database designing. Providing appropriate information in a suitable and accessible

Table 6.6 Examples of various data structure types in water resource management

Data structure types		Example
Attribute data		• Explanations of locations, names, and properties like capacities of dams and reservoirs, pumps, valves, meters, hydrants, manholes or turbines • Time series data like river flow rates, streams level, reservoir releases
Vector	Point	Feature locations like measuring stations, point sources of pollution, sprits, wells
	Line	Network of stream, channel, pipe, road, sewer system
	Polygon	Boundaries of watersheds, aquifers, lakes, estuaries, drainage divides, dams
Raster	Grid	Continuous spatial parameters such as temperature, pressure, rainfall, elevation (DEM*), slope, water depth, groundwater level and surface, flow directions, closed depressions, land use, land cover, vegetation type and density, precipitation intensity and spatial patterns, impervious rate, population density
	Triangulated Irregular Network	Distribution of soil properties (organic Carbon, pH, water storage capacity, soil depth, hydraulic conductivity, cation exchange capacity, clay fraction, total exchangeable nutrients, lime and gypsum contents, sodium exchange percentage, salinity, textural class and granulometry) or the rasterization of soil group polygons, flood plain
	Network	Contour lines for majority of raster data

Table 6.7 Summary of well-known dynamic formats

Dynamic format	Description
GeoJSON	An open format enabling geographical data coding (the data exchange format JSON used as a basis)
Well-known text (WKT)	A format within a text mode, attending to demonstrate a geographical object (line, point or polygon)
Well-known binary (WKB)	Signify a single geographical object through a binary mode format
Geography markup language (GML)	An open regular XML to exchange the GIS data
Digital elevation model (DEM)	Format to store and show altitude data
digital raster graphic (DRG)	Format of storing digitized paper maps which are electronically scanned
Geotiff	Format to store raster data and couple georeferencing information with a tiff image

Table 6.8 Summary of spatial databases used by web applications in water resource management investigations

Free open source software for geographic information systems	Water resources web applications	Data format
SpatiaLite	BASHYT	Vector
MySQL	Automated Geospatial Watershed Assessment Flood Assessment Modeling Tool	Vector
PostGIS	Object-Oriented and OpenGIS Hydro Information System Cloud Framework for Hydro Information System Hydrogeological Information System Water Management Decision Support System Web-based Hydrologic Transport Model Web Application for Water Resources	Vector and Raster

method by the database could guarantee the success of GIS applications. Design of the database conduct through three stages: conceptual design, logical design, and physical design. Conceptual design is irrigated to both hardware and software, while logical and physical designs depends on software and hardware, respectively. Data dictionary documents the logical and physical structure of the layers and take part in designing a database. To carry out operations concerning tables and their data, the SQL language is made up. The SQL language comprise a series of operators in which function with generic relational data. H2/H2GIS, SQLite/Spatialite, SQL server, MySQL, and PostgreSQL are examples of the database management system. Some spatial database used in water-related researches are listed in Table 6.8 (Swain et al. 2015).

6.3 Fundamental Theory

The purpose of GIS creation had been providing efficient tools to support decision makers regarding spatial data. GIS is invaluable in relating collected geographical information to attribute data. Even though numerous GIS software offer different capabilities, majority comprise the following activities with spatial data to produce both standardized and customized cartographic products: **organization, visualization, manipulation, analysis and prediction** (Merriam 1994; Hall et al. 2000).

- **Organization**: organizing data, which is also known as data models, impose on examinations by spatial and non-spatial attributes. This step is done by a data storage and retrieval subsystem which allow users to quickly retrieve data

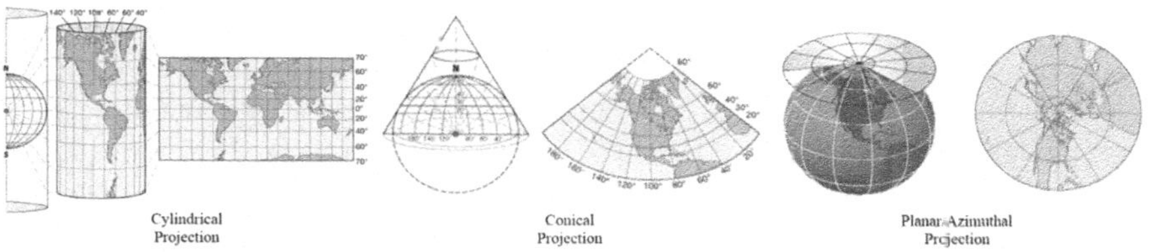

Fig. 6.2 Various types of projection

for subsequent analysis, and accurate updates. The efficiency and type of data organization has the fundamental importance since this step influences all the next steps.

- **Visualization**: GIS technology uses the graphical capabilities of computers to visualize geographical data. Thanks to georeferencing the real Earth's surface project into a two-dimensional map. By georeferencing firstly, every point on the Earth's surface will be delineated by a coordinate system identified as the geographic coordinate system (GCS). Secondly, the projected coordinate system will be employed to reflect points into two dimensions for creating a digital map. In a first step, latitude and longitude are used to represent specific locations on the Earth in the three-dimensional coordinate system. In the second step, mathematical transformation is applied to convert the real spherical surface of the Earth (3-D) onto a plain sheet of paper (2-D). Through the projection process one or more spatial properties (e.g. area, shape, distance, or direction) will be distorted (Selvam et al. 2019). Various methods of projection are shown in Fig. 6.2. The different countries have adopted different standard projections which leads to different map scale.
- **Manipulation**: Spatial query and data combination are complementary activity for realizing and interpretation of spatial phenomena in this step. By spatial query users can understand the special conditions of each case. Before manipulation, GIS answer two types of questions: "What are the features of the particular location?" and "Whereabouts do these characteristics occur?". The answers provided by efficient, organized data conduct users toward a proper combination of data to achieve the final solution. In combination process, spatial datasets from diverse sources would be merged. The results lead to a new understanding of spatial phenomena, while isolated spatial data types are unable to reveal that individually. For instance, overlaying a digitized geological map on a satellite image, can clarify the changes of a watershed over the particular time intervals. Data integration is also the other type of combining process in which a new map is created from a composite display of various spatial data. Indeed, integration models regarded as symbolic mathematical models which combine layers of data together by means of arithmetic and logical operations. A practical example is the combination of the maps of a lake and a radiometry pattern. The lakes map is result of digitizing a topographic base map containing drainage features. The radiometry map is generated from

processing an airborne image which contains 100 colors for exhibiting the intensity of the radio element measurement. While only two colors indicate the present and absent of the lake. Thanks to "map algebra" which is written in a programming language, a combination of these two maps would illustrate the new map in the color classification of the radiometry map where lakes are present. Therefore, the result can be stored as a new map, relating the spatial data of water bodies to radiometry patterns. This example presents the ability of one map algebra statement in producing a new practical map. However, GIS is equipped to the powerful features of linking several map algebra statements together to form more complex algorithms. Combination of numerous maps and attribute tables is feasible via one processing step, which is named "map or cartographic modelling".

- **Analysis**: The spatial analysis functions of GIS distinguish it from the other databases and information systems. Measuring distances, statistical calculations and graphs, fitting models to data values and other processes are analyzing tools in GIS. Sometimes for specific analyses other computer programs are required to be joined to GIS. Analysis is applied on exported data prepared as maps or tables.
- **Prediction**: GIS is a proper tool to determine the result of making a specific set of assumptions in a research with the goal of assessing a model's performance. By the use of map algebra, symbolic models can be created which combines data layers under defined specific rules. Indeed, in geographic analysis the rules are imposed on map with vector or raster data formats and the related attribute data to find answers to specific problems. For example, selection of the most suitable sites for artificial groundwater recharge is feasible as a function of changes in water table and soil hydraulic conductivity can be applied in map algebra (Paramasivam 2019). Also interpolation techniques can be applied to the exploration, and prediction of groundwater level or even to estimate hydrological parameters such as precipitation in location with lack of data. Another example of GIS application in prediction is forecasting hazardous locations for flood. The modelling tools of GIS improves its application further than display of geospatial information, and introduces it as a functional application in problem solving (Merriam 1994).

6.4 GIS Importance and Necessity

Before GIS technology invention, the process of modeling in water resource management depended on paper-format maps. Owing to the complexity of the natural environment, processing spatial data was tedious, required considerable workload and had a significant margin for error. GIS, a platform providing geospatial data in digital formats, significantly decline the workforce demand in data pre-processing and save time and money. In a generic manner, quick evolution of electronic spatial data in the geosciences, make the GIS an essential tool in water resources science because it accomplishes the following functions (Merriam 1994):

- providing a tool for homogenizing data extracted from different sources (e.g. DWG, DAT, SHP);

- advancing the data related progress (e.g. management, visualization, processing, analysis);
- enabling the visualization of 3D data;
- enabling the extraction of photo-aerial images;
- utilizing the huge volumes of hydrological remote sensing data gathered daily and prone to remain useless without database management system;
- managing and updating statistical, geostatistical and digital modeling;
- improving data accessibility by use of either GIS office automation or Internet applications;
- increasing the perception of water and the natural environment interactions;
- improving the rationalization of data capture and the facilitating the interpretation of the simulation results;
- Providing the Power of Integration through manipulation and analysis of layers and modelling the interrelationships between layers
- Offering a Decision Support Framework

Totally, data collected from various sources including geophysical devices, aerial tools, satellite, ship-borne, ground-based, and underground produce vast volumes of numbers which are almost worthless before proper organizing and displaying. Besides, analyzing data of waterbodies, sediments, soils, and plants produces huge amounts of new spatial data which are in urgent demand for efficient spatial data systems to be assessed. Therefore, GIS is a vital digital system which is essential for processing, displaying, planning and managing the invaluable spatial data in water resource management (Selvam et al. 2019).

6.5 Methodology

In order to apply GIS in the decision-making process in water engineering, following the steps shown in the flowchart of Fig. 6.3 is recommended. Generally, water resource investigations which benefit GIS application include six main steps: defining the study objectives, gathering geographical database, designing or selecting a database management system, selecting GIS software and required model, and finally presenting results in the form of maps. More explanation for each stage is expressed in the following sub-sections.

6.5.1 Defining the Objective

It is important to make a clear definition of objectives or the problem which should be solved. The objectives as the cornerstone of decision-making process, profoundly impact choosing the required spatial and attribute databases. The objective could

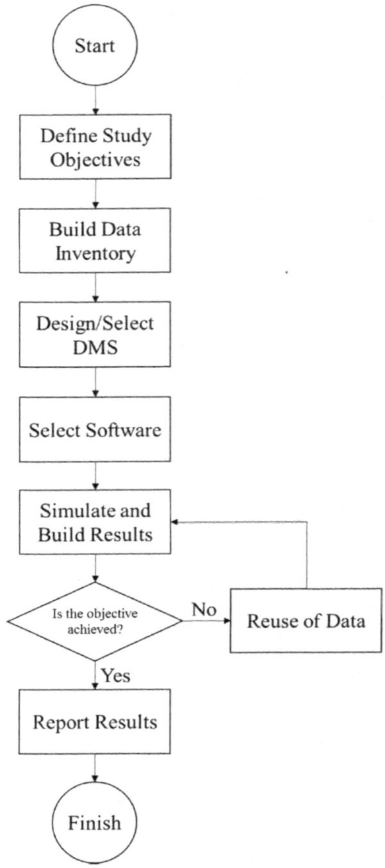

Fig. 6.3 Steps of utilizing GIS in decision-making process

be the initial designing, problem solving, or predicting future condition of water-related systems. After precise defining of the problem, the relevant data inventory like attribute data and map layers are recognized.

6.5.2 Building Data Inventory

The GIS data can be provided by frequent sources including governmental, private, and academic. Also, data are gathered from several sources including Global Positioning System (GPS), aerial photographs, digitized hard copy maps, satellite images, laboratory analysis, field measurements and investigations, and etc. For instance, applying various statistical methods (like Fisher's linear discriminant analysis) on satellite images driven from IKONOS, resulted in generating soil impermeability

Table 6.9 Examples of parameters used in water resource management

Water data class	Prevalent parameters
Climatological	Temperature, evapotranspiration, precipitation, wind speed and direction, solar irradiation
Water quality	Nutrients (Nitrate, Phosphorus, Sulphur.), urban pollutants, pH, pesticides, turbidity
Hydrological	Surface stream flow and height, flow speed, flood plain, physico-chemical quality
Hydrogeological	Piezometric level, aquifer thickness, stratigraphic logs, permeability, hydraulic conductivity, transmissivity
Complex data	Output of hydrological models, water storage evaluations, complex physicochemical calculations
Administrative data	The charges, action limit of a hydraulic basin agency

map. The six main category of water data parameters which commonly used in GIS analysis are listed in Table 6.9 (Jarar Oulidi 2019b).

Among data supplier platforms some are developed on the web. When all the required data are gathered, the database should be designed. In the database designing process the following factors should be identified: study area borders, data layers, coordinate system, features of each layer and its associated attributes, attributes' coding and organization. During the data-entry process attribute data are entered in the form of database files with one common field with the spatial database. Spatial data with the vector format creates topology via the general digitization process, while making topology from raster data requires converting to vector at first. Scanning and digitization are examples of data conversion methods. Data conversion techniques, resolution, source, and scale of the map layers, are considered as the foundation of the system and take the share of 75% of typical GIS costs.

6.5.3 Database Management

Database management involves three main tasks: coupling the digitized map into real world coordinates, discovering coverages for analysis, and preserving the database. This step is completely explained in Sect. 6.2.3. The design of the database management system is compelled by application needs.

6.5.4 Software Selection

Water resource management models benefit GIS contribution in water-related issues through interchange, interface, or integration techniques. In interchange method, GIS is applied to extract model input files, and then they are copied into the model

Table 6.10 Software Selection Matrix Example

Application	AV	SA	NA	3DA	AI	SDE	AO
Planning and engineering	*			*	*	*	*
Operations and maintenance	*		*		*	*	*
Construction	*		*		*	*	*
Infrastructure management	*		*		*	*	*
Water resources and hydrology	*	*			*	*	
Finance and administration	*				*	*	

Notes AV: ArcView; SA: Spatial Analyst; NA: Network Analyst; 3DA: 3D Analyst; AI: ArcInfo; SDE: Spatial Database Engine; and AO: ArcObjects

manually. The hydraulic model output files are copied into the GIS software manually too, in order to display model output data. The "open" architecture of GIS software provides the sharing of geographic data. This characteristic enables all GIS software to be used in the interchange method. Advantages of an open GIS are:

- the ability to run on various operating systems, database management systems, and platforms
- supporting expansion application requirements, ranging from an office engineer using GIS on a desktop, to a portable field technician using a handheld device

In the interface method, the data interchange process is basically automated. In this method, the added model-specific menus or new buttons in the GIS software automate the data transfer stage. Creating the model input file automatically is the main improvement of the interface methods in comparison with the interchange methods. Whereas, both mentioned methods do not allow to edit data and launch the model from within the GIS software, integration offers both GIS and modeling functions through a combination of model and GIS. This method signifies the closest relationship between a GIS and a model.

Selecting an appropriate GIS software depends on the several factors, including user's ability and interest, software price, method of model and GIS combination, project scale and etc. For instance, in terms of the users' needs, a list of available software using an ESRI GIS platform are compared in Table 6.10 (Shamsi 2005). According to this table, ArcView, ArcInfo, and Spatial Database Engine carry out all the GIS applications of institutes.

6.5.5 Presenting Results

The final results are often represented in form of map. Depends on the method of GIS and model combination, resulted maps might be the final export of GIS or part of the pre-processed data considered as model input parameters. The recent development in GIS technology leads to a presentation of some results in the form of video. Yet, most of the depicted data are map in raster format. The open architecture of GIS software allows users to simply share digital maps with others. To take a profound understanding, some examples of presented results are expressed in Sect. 6.6.

6.6 Practical Examples

As it has been remarked in Sect. 6.1.1, GIS is an inseparable part of water engineering issue. In this section the GIS application in three different water-related problems, including groundwater recharge assessment, water availability analysis, and urban stormewater management, is explained in detail.

6.6.1 Assessment of Groundwater Recharge Potential

Groundwater characteristics assessment such as groundwater recharge is possible through direct and indirect methods. Geological and geophysical investigations, drilling tests, gravimetric and magnetic methods are considered as a direct method. Indirect methods, however, is dominated by computer modeling and GIS software combination with field studies. The required accuracy level, the project implementation, and the accessible resources determine to apply direct or indirect method. The title of case study chosen for GIS application in groundwater investigation is "Groundwater Recharge Potential for Sustainable Water Use in Urban Areas of the Jequitiba River Basin, Brazil" (Costa et al. 2019).

Estimating aquifer recharge potential within the Jequitiba River basin for the purpose of identifying privileged areas for restoration, recovery, and preservation was the main goal of this study. To achieve this goal: firstly, geospatial data (temperature, precipitation, geologic, hydrometric, relief, land use, and soil characteristic) are compiled; secondly, aquifer replenishment potential at catchment scale is estimated via a physical-based spatially distributed model; thirdly, the aquifer recharge capability map is produced as the model output; and finally, stream flow recession analysis is applied for model validating. The materials used in this study, in the form of map and attribute table, are shown inthered from different websites (Costa et al. 2019).

Table 6.11 These materials are gathered from different websites (Costa et al. 2019).

The climatology information was interpolated by the IDW technique which decreases the effect of the values recorded in the farthest stations from the JRB, with higher weights for the values of the adjacent stations. When required data are analyzed surface runoff factor (RF) and water percolation factor (PF) are calculated. Runoff factor is calculated by assessment of hillside lengths and slopes (LS factor) and runoff coefficients (C) derived from the DEM. The calculation of LS_{fuzzy} carried out by tools embedded in the GIS platform (QGIS). Soil characteristics including

Table 6.11 Materials used in the recharge potential evaluation model (Costa et al. 2019)

Data Type	Use in the replenishment estimation
Digital elevation model	Calculation of slope length and steepness factor
Land use and cover map	Calculation of runo_ parameter
Soil map containing porosity and hydraulic conductivity data	Calculation of percolation factor
Rainfall and evapotranspiration data	Calculation of recharge potential
Stream flow data	Validation of recharge potential using na independent method
Geologic map	Information for discussion
Administrative data Population data	Additional information

total porosity (n) and hydraulic conductivity (Ks) are utilized in calculation of water percolation factor. Groundwater replenishment is calculated by GIS for both individual points of the catchment and an average of the entire catchment. Ultimately, results are validated by making comparison between previously calculated average recharge and a counter value projected by the hydrograph recession method based on stream flow analysis (Costa et al. 2019).

The final recharge potential map is generated by a combination of maps (precipitation's spatial distributions, evapotranspiration, RF factor and PF factor) employing the appropriate tools of QGIS. The results showed that the aquifer recharge potential varies from 0 to 4626.4 m^3/ha.year (Costa et al. 2019).

6.6.2 *Assessment of Water Availability*

Spatial information plays a very significant role in management of sustainable water resources. GIS provide assessment and monitoring of the spatial data over time within the watershed to evaluate the available water resource and water demand. In this section, GIS application for water availability assessment is clarified via a case study in Indonesia titled "Application of GIS for Assessment of Water Availability in the Cianten Watershed, West Java" (Mirrah and Kusratmoko 2017).

The goal of this study is analyzing the water shortage problem which is common in the Cianten watershed, particularly in the dry season. Therefore, main purpose of this study is revealing the spatial pattern of the water availability index in the Cianten watershed, particularly over the dry period. The next step is data gathering. For water availability calculation in both annual and dry season, climate factors and watershed characteristics are utilized. Whereas water demand calculation is done

Table 6.12 Collected data explanation (Mirrah and Kusratmoko 2017)

Data	Explanation
DEM	Resulted from processing the SRTM data, which are achieved from USGS Glovis
Land use	Extracted from a digital Land use map (scale 1: 100,000) issued by the Ministry of Land and Spatial Planning
Soil type	Derived from the soil type digital map (scale of 1:100.000) issued by Soil Research Institute, Ministry of Agriculture of the Republic of Indonesia
Climate (rainfall and temperature)	Provided by the Meteorology, Climatology and Geophysics Agency for the period of 2007-2016. 8 stations collected the average monthly rainfall data, and one station recorded the average monthly temperature data
Drainage network Data	Stemmed from the processing of the digital topography map (scale of 1:25.000) issued by Geospatial Information Agency
Total Population	Secondary data achieved from the Central Bureau of Statistics, Bogor Regency, in 2016

using the population and land use data. The spatial data used in this research are described in Table 6.12. Also, the workflow for computing the water availability by GIS technology is depicted in Fig. 6.4.

In this study, the GIS database system is created with layers of rainfall and temperature, sub-catchment boundary, slope, stream network, and land use Delineating the Cianten watershed to the sub-catchment units is achieved by interfacing ArcSWAT software with ArcGIS. Then, the hydrological unit for each sub-catchment is driven by overlaying the slope, land use and soil type layers. Thiessen Polygon method is applied to calculate the spatial distribution of the rainfall in the study area among the eight rainfall stations (for the annual and seasonal time scale). The GIS modeling process finally created two water availability maps (annual and the dry season) which are illustrated in Fig. 6.5. The water availability varies from 9266 m^3/ha to 15,991 m^3/ha for annual assessment, and from 2285 m^3/ha to 4147 m^3/ha for dry season. Comparison between water demands and water availability indicate the various levels of water deficit during the dry season in the most sub-catchments.

6.6.3 *Urban Stormwater Management*

Choosing among various urban stormwater measures depended on the precise collection of spatial data and modeling. The considerable share of impervious area in urban areas leading to routing huge runoff volume on one side, and noticeable financial damage of a probable flood in cities on the other side, raise the importance of stormwater modeling. SWMM, as one of the oldest and most popular software, is

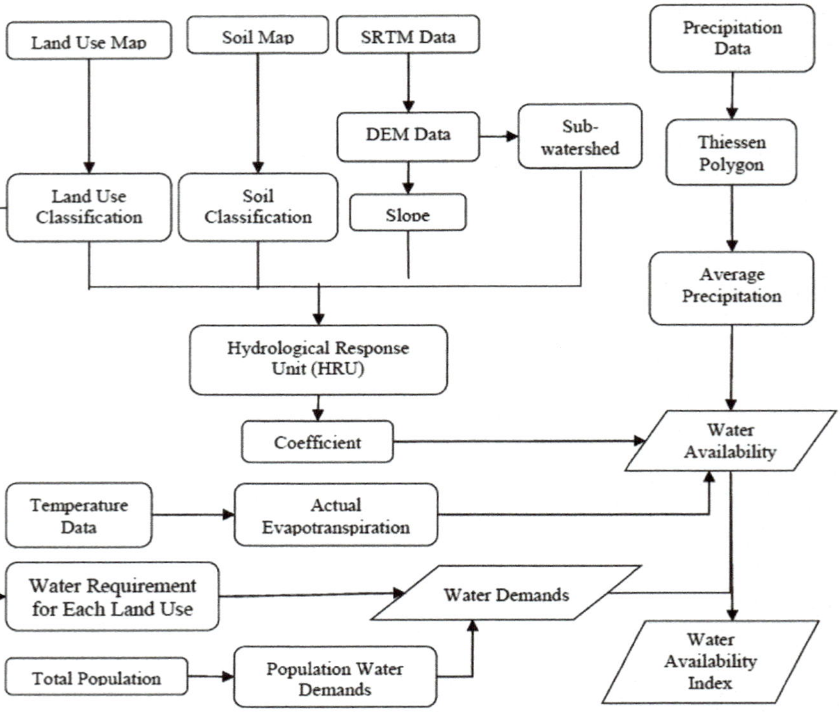

Fig. 6.4 Workflow to calculate water availability via GIS tools (Mirrah and Kusratmoko 2017)

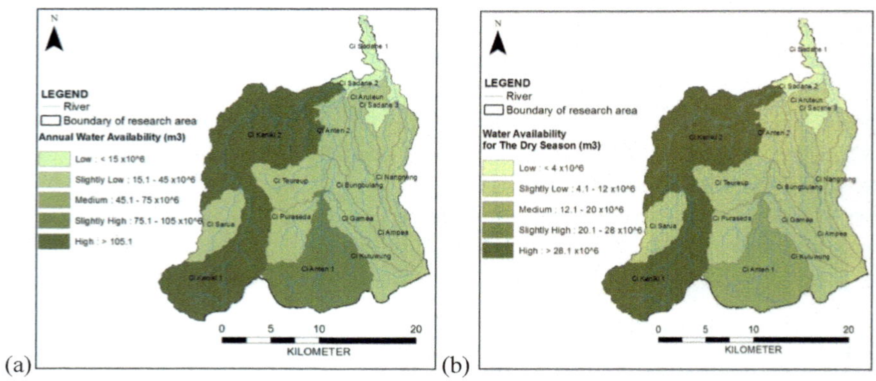

Fig. 6.5 Water availability in Cianten watershed, **a** annual, **b** dry season (Mirrah and Kusratmoko 2017)

Table 6.13 Collected data explanation (Kong et al. 2017)

Data	Explanation
Topographic	In CAD format, date: 2011 DEM 5 m * 5 m resolution
Aerial photograph	Scale 0.1 m * 0.1 m, date: 2012
Regulatory planning	Land use map, road planning map, runoff collection network map in CAD format, date: 2013–2030
Rainfall	daily and hourly distribution data for a 10-year return-period rainfall event, date: 2015

broadly utilized for simulation, analysis, and design of runoff control infrastructure in urban areas like catchment planning, drainage piping network, and runoff control via low impact development (LID) structures. In comparison with other hydrological models, however, SWMM application in large scale regarded as a drawback. For the purpose of applying SWMM on large urban catchments, GIS utilization is assessed by some researchers. In the selected case study, the interchange technique is applied between ArcGIS and SWMM5.0 model to combine urban planning data, geospatial and hydrological information for simulating stormwater runoff at the municipal scale. The title of this investigation is "Modeling stormwater management at the city district level in response to changes in land use and low impact development" (Kong et al. 2017).

The study objective is to find out hydrological responses to changes in land use and potential hydrological effects of LID implementation. The spatial data used in this research are described in Table 6.13.

Data processing step starts with converting the CAD data to a GIS shapefile dataset (the Universal Transverse Mercator (UTM)-projected Xi'an 80 is chosen as the projected coordinate system). Next, the aerial photograph data are rectified and georeferenced to the UTM coordinate system via the reference topographic map in ArcGIS software (Version 10.2, ESRI). The eCognition (Trimble Inc.) software (version 8.7) is also applied to create the land use map from photograph data by manual delineation and interpretation of landscape polygons. Lastly, all the regulatory planning data were transformed to GIS shapefile datasets and utilized in generating the land use, road, urban drainage network maps for the planning scenario analysis.

To delineate sub-catchments, which are the basis unit of any hydrological model, DEM is used by the ArcHydro extension in ArcMap (9.3 ESRI). This step includes producing depression-less DEM, determining the flow direction, calculating the flow accumulation, and then creating the outlet of the stream networks. The result map is shown in Fig. 6.6.

The GIS output files should be transformed to proper format to be readable by SWMM as an input file. Therefore, the boundary map of sub-catchments and rainwater conduit map were generated by GIS in vector format and converted to the .inp format to be readable by SWMM. Firstly, the sub-catchment polygon in vector

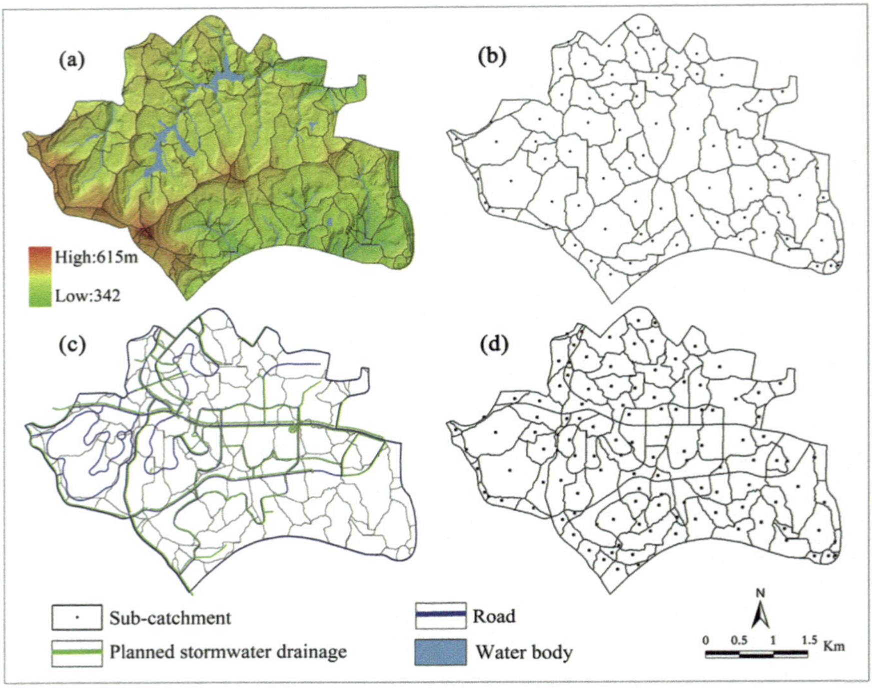

Fig. 6.6 Discretized sub-catchments in the study area: **a** digital elevation analysis; **b** current sub-catchment layout; **c** planned road and drainage networks; **d** discretized sub-catchments under the future planned land use. (Kong et al. 2017)

format were changed to point datasets, where all the vertices of the original polygons were conserved. Next, other related data layer needed as the input data in the model was exported as a .txt file. Finally, the file extension of the TXT file (.txt) was altered to .inp, which meant the SWMM inputs criteria for modeling. So the runoff routing modules in the SWMM are prepared for employing on a large scale.

Land use type and the imperviousness of each sub-catchment determine the type and area of the proposed LID controls. To intersect the impervious land with the sub-catchment boundaries, firstly, whole the merged impervious land areas were overlaid with the layer of sub-catchment, and their attributes were allocated based on the attributes of each sub-catchment in the ArcGIS environment. Then, the Location Selection tool in ArcGIS is used to intersect the impervious area in sub-catchments with the drainage network system. Thus, the resulting impervious area chosed by the drainage networks denoted directly connected impervious area (DCIA) with the attributes of individual sub-catchment, declaring the impacts of LID controls on DCIA reduction and consequent hydrological processes.

6.7 Summary

GIS is a computer-based platform utilized for storing, analyzing, and visualizing the geographical information. In GIS technology, the database of spatial data and their attributes are held separately and linked to create favored maps in diverse layers, while each layer comprising information about one feature. Snice majority of the basic required data in water resource management has spatial nature and their management is sophisticated, scientist and water engineers enormously applied GIS technology in problem-solving, designing, forecasting and decision-making. For each of the mentioned purposes, GIS is employed autonomously or in combination with simulation models through interchange, interface, or integration techniques. Generally, GIS software (both open source and commercial) facilitate and simplify the analyzing process of data with different types and sources (climatological, hydrological, hydrogeological, water quality, administrative) in several fields including: surface water hydrologic, groundwater, water supply and sewer network, sustainable water resource management, flood hazard assessment, urban stormwater, hydropower potential assessment, and water quality and quantity.

Acknowledgements The authors thank Iran's National Science Foundation (INSF) for the support of this research.

References

Costa AMD, Salis HHCD, Viana JHM, Pacheco FAL (2019) Groundwater recharge potential for sustainable water use in urban areas of the Jequitiba River Basin, Brazil. Sustainability 11

Fu P, Sun J (2010) Web GIS: principles and applications. ESRI Press, Redlands

Hall CAS, Hall M, Taylor M (2000) Chapter 7 - Geographical modeling: the synthesis of a GIS and simulation modeling. In: Hall CAS, Perez CL, Leclerc G (eds) Quantifying sustainable development. Academic Press, San Diego

Jarar Oulidi H (2019a) 1 - Theoretical framework. In: Jarar Oulidi H (ed) Spatial data on water. Elsevier

Jarar Oulidi H (2019b) 2 - Technical framework: spatial data infrastructure for water. In: Jarar Oulidi H (ed) Spatial data on water. Elsevier

Kong F, Ban Y, Yin H, James P, Dronova I (2017) Modeling stormwater management at the city district level in response to changes in land use and low impact development. Environ Model Softw 95:132–142

Korres, W, Schneider K (2018) 2.04 - GIS for hydrology. In: Huang B (ed) Comprehensive geographic information systems. Elsevier, Oxford

Martin PH, Leboeuf EJ, Dobbins JP, Daniel EB, Abkowitz MD (2005) Interfacing GIS with water resource models: a state-of-the-art review. J Am Water Resour Assoc

Merriam DF (1994) Series editor's foreword. In: Bonham-Carter GF (ed) Geographic information systems for geoscientists. Pergamon

Mirrah AA, Kusratmoko E (2017) Application of GIS for assessment of water availability in the Cianten Watershed, West Java. In: Geoinformation science symposium

Paramasivam CR (2019) Chapter 2 - Merits and demerits of GIS and geostatistical techniques. In: Venkatramanan S, Prasanna MV, Chung SY (eds) GIS and Geostatistical techniques for groundwater science. Elsevier

Selvam S, Manisha A, Vidhya J, Venkatramanan S (2019) Chapter 1 - fundamentals of GIS. In: Venkatramanan S, Prasanna MV, Chung SY (eds) GIS and geostatistical techniques for groundwater science. Elsevier

Shamsi UM (2005) GIS applications for water, wastewater, and stormwater systems. Taylor & Francis

Swain NR, Latu K, Christensen SD, Jones NL, Nelson EJ, Ames DP, Williams GP (2015) A review of open source software solutions for developing water resources web applications. Environ Model Softw 67:108–117

Tsihrintzis VA, Hamid R, Fuentes HR (1996) Use of geographic information systems (GIS) in water resources: a review. Water Resour Manage 10:251–277

Chapter 7
System Dynamics Approach for Water Resources Systems Analysis

Arya Yaghoubzadeh Bavandpour, Hamed Nozari, and Sajjad Ahmad

Abstract The world is currently facing extremely high water scarcity, due to the limitation of available water resources, droughts, and increasing water demand follows population growth and changing consumption patterns. In such circumstances, it is necessary to determine the optimal and sustainable operational policies of this vital source. In addition, considering the role of simulation and optimization approaches as one of the management tools, the existence of a comprehensive, integrated, and dynamic notify system is necessary regarding the type and combination of the costs. This tool helps experts and users to compare different scenarios in specific time periods, and to be able to adopt measures to manage water consumption. In this regard, system dynamics approach is one of powerful management and simulation tool in solving, supporting, and decision-making complex issues. This approach refers to the computer simulation method for simulating a dynamic and complex system with feedback process inclusion and make system users a better understanding of the dynamic behavior of systems during time. This method has been successfully applied in a wide variety of business and socio-economic fields. Furthermore, in recent years, this method has been used in water resources research, such as flood management, operation of reservoirs, management of catchment basin, planning management, and analyzing the decision-making policies of water resources management, and the results have been very significant. Considering the importance of this issue and the benefits of using System Dynamics analysis approaches in solving problems, in addition to introducing this method, examples of different studies conducted in the simulation of water systems using system dynamics analysis technique are presented, in this chapter.

A. Y. Bavandpour (✉) · H. Nozari
Department of Water Science Engineering, Faculty of Agriculture, Bu-Ali Sina University, Hamedan, Iran
e-mail: a.yaghoubzadeh@agr.basu.ac.ir

H. Nozari
e-mail: h.nozari@basu.ac.ir

S. Ahmad
Department of Civil and Enviromental Engineering and Construction, University of Nevada, Las Vegas, USA
e-mail: Sajjad.ahmad@unlv.edu

O. Bozorg-Haddad (ed.), *Essential Tools for Water Resources Analysis, Planning, and Management*, Springer Water, https://doi.org/10.1007/978-981-33-4295-8_7

Keywords System dynamic · Water resources · Flood management · Operation of reservoirs

7.1 Introduction

Population growth, urban development, agricultural development, and the rapid growth of various industries, and on the other hand, increasing water demand and water resources limitations have made serious problems for supplying water. These crises in alignment with its subsequent problems can only be decreased through correct management and planning. However, the existence of multiple decision-makers and consumers with different preferences and priorities in water resources management and planning issues raise considerable disagreements and concerns about the allocation of water resources, which will be of utmost concern for managers and planners in this segment. Therefore, water resources management is interdisciplinary and multi-component management and requires comprehensive decision making, which is now one of the most important challenges. Nowadays, with the rapid development of computational tools and the appearance of new optimization and simulation models, the application of systematic approaches in planning and management of water resources systems has been expanded, making it possible to make complex systems more accurate by identifying components and analyzing the relationships between them. One of these approaches is the System Dynamic (SD) approach. This method is one of the most effective methods available for a comprehensive evaluation of system performance. System dynamics was first devised by Forrster (1961) to better understand strategy issues in complex dynamic systems. Models written in this way, with insights into the feedback processes, help system users to better understand the dynamic behavior of systems over time. The application of this method is widespread, which is more emphasized in socio-economic issues.

The application of this method in water research, such as the planning of river basin, management, and planning of water resources systems, management of reservoirs, urban water systems, flood, irrigation, and drainage have been developed and had excellent results. Over the years, several System Dynamics models have been developed for various water (Chen et al. 2017; Ahmad 2016) and environmental management (Amoueyan et al. 2017; Rusuli et al. 2015) applications. A comprehensive review of system dynamics applications has been provided by Winz et al. 2009; Mirchi et al. 2012). System Dynamic has been used for the management and planning of river basins and water resources in several studies such as integrated analyses of water resources in Canada (Simonovic and Rajasekaram 2004), SD analysis for Zayandeh-rud river basin management (Madani and Mariño 2009), water management in complex systems (Hassanzadeh et al. 2014), SD simulation model for sustainable water resources management and agricultural development (Kotir et al. 2016), SD application in integrated water resources modeling (Zomorodian et al. 2018), hybrid SD and optimization approach for sustainable water resources planning (Li et al. 2018), analysis of water management scenarios using coupled hydrological model and SD (Qin et al. 2019), dynamic management of a water resources-socioeconomic-environmental system (Dong et al. 2019).

There are other applications of SD in hydropower studies, which were mentioned in hydropower generation assessment using SD (Sharifi et al. 2013) and power generation simulation of a hydropower reservoir (Jahandideh-Tehrani et al. 2014). SD has also been used to investigate the impact of water demand priorities on downstream by Qashqai et al. (2014). In the last three studies, SD has been used in research involving a reservoir or a set of reservoirs.

SD is also used in studies of irrigation and drainage networks and studies of irrigation water management. Vaez Tehrani et al. (2013) used SD to model irrigation network rehabilitation. Nozari and Liaghat (2014) used SD to simulate drainage water quantity and quality and showed that the SD model can be used to manage and control drainage salinity to prevent environmental damage. Nozari et al. (2014) simulate irrigation network and crop pattern by using SD, and the results showed that SD is suitable for simulation. Matinzadeh et al. (2017a, b) used SD to simulate nitrogen dynamics in agricultural fields with drainage systems. Fluchinotta et al. (2018) used SD to manage irrigation water in southern Italy.

SD also has been used for flood management (Ahmad and Simonovic 2000, 2001, 2004, 2006), water security studies (Chen and Wei 2014) and water security assessment in Rafsanjan, Iran (Bagheri and Babaeian 2020), Urban water system management (Karimlou et al. 2019), groundwater modeling (Bates et al. 2019), surface water quality management (Elshorbagy and Ormsbee 2006), integrated system dynamics model (Liu et al. 2015), the impact of global climate change on water quantity and quality (Duran-Encalada et al. 2017), water quality modeling (Amoueyan et al. 2019 a and b; Venkatesan et al. 2011a, b; Nazari-Sharabian et al. 2019), water allocations (Wu et al. 2015; Qaiser et al. 2011, 2013; Kandissounon et al. 2018), climate change impact on water resources (Dawadi and Ahmad 2012, 2013; Zhang et al. 2016), carbon footprint of water projects (Shrestha et al. 2011a, b, 2012; Bukhary et al. 2018), water conservation (Ahmad and Prashar 2010; Dow et al. 2019), rainwater harvesting (Tamaddun et al. 2018), and energy planning (Moumouni et al. 2014). Models have also been developed (Stave 2003; Nussbaum et al. 2015) for Lake Mead and the Las Vegas water supply system to educate public about water conservation.

7.2 Basics and Logic of the Subject

Issues in a system have both dynamic characters and feedback structure. Based on the dynamic characteristics, the quantity and quality dimensions of the system change over time, and according to the system's feedback structure, different elements of the system are influenced by each other at any time. The system dynamics method is based on feedback control theory and nonlinear dynamic theory which enables the construction of a real-world model for a better understanding of phenomena (Sterman 2000). This method can be used when the human mind is incapable of analyzing the structure, relationships, and behavior of a phenomenon (Bala et al. 2017). In fact, system dynamics is a method based on systematic thinking which can provide the possibility to describe complex systems based on reality, and also user participation in model development in addition to the description.

7.3 Definitions and Terms

In this part, common expressions in creating and using SD is going to be described.

7.3.1 *System*

System is a set of elements and components interacting with each other for a specific purpose. Systems may be classified as (a) open systems and (b) feedback systems. In open systems, the output is defined by input, and the output has no impact on the input. However, feedback systems are closed-loop systems in which the inputs are influenced and changed by outputs. Figure 7.1a shows an open system and Fig. 7.1b shows a feedback system.

A feedback system is classified as a positive feedback system or a negative feedback system. Positive feedback systems have brought growth and increase, while negative feedback systems will bring a reduction (Bala et al. 2017; Sterman 2000). Connector signs describe the structure of the system and are unable to determine variables' behavior which defines how changes in the cause will lead to changes in effect (Table 7.1).

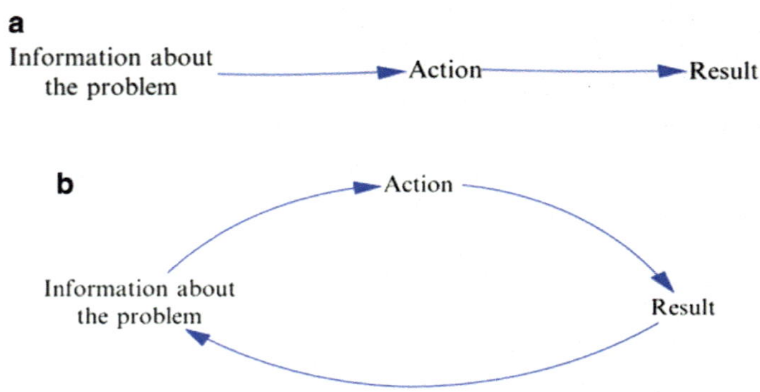

Fig. 7.1 **a** Open system (Bala et al. 2017). **b** Feedback system or closed-loop system (Bala et al. 2017)

Table 7.1 Positive and negative feedbacks definition

Figure	Describe	Computational equations
	If x increase (decrease), than y will increase (decrease)	$\frac{\partial y}{\partial x} > 0$
	If x increase (decrease), than y will decrease (increase)	$\frac{\partial y}{\partial x} < 0$

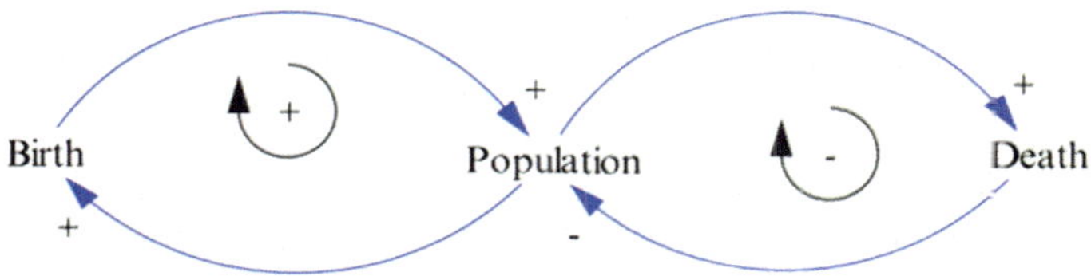

Fig. 7.2 Schematics of positive and negative feedback loops (Bala et al. 2017)

7.3.2 Causal Loop Diagram

All dynamic systems are created by a positive and negative loop. In other words, positive loops help the system to improve and grow, and negative loops play neutralizing and balancing roles in the system (Bala et al. 2017). In general, the feedback process consists of negative and positive loops. These loops illustrate the causal relations of a system, which, in fact, are the main structure of a dynamic system. The primary purpose of these diagrams is to represent causal hypotheses during modeling to express the entire structure interconnectedly. These diagrams help the creator to communicate quickly with feedback structure and basic defaults. Figure 7.2 shows an example of positive and negative feedback loops.

As it represents, with population growth, the birth rate increases, which leads to population growth again creating a positive feedback loop. On the other hand, population growth leads to an increase in the death rate, which causes population reduction, creating a negative feedback loop. Thus, birth in positive feedback leads to an increase in population, and death, in negative feedback, leads to a decrease.

7.3.3 Stock and Flow Diagram

Stocks are systems accumulations and represent system statuses, and system decisions and activities are governed by them. Flows indicate changing rates which means they demonstrate the processes which increase or decrease stocks. It can be said that in a system, decisions are made based on the stock variable, and through flow variables, those changes are implemented.

To draw stock and flow diagrams, four essential elements of stocks, flows, connections, and converters are used (Table 7.2). Stock is displayed as a block in the model to indicates the status or conditions of the model at any given time. Flows are divided into inflows and outflows; inflows are shown by arrows in towards stock variable, and outflows are also shown by arrows in the opposite direction going out of stock variable. Connections are used to show the relationships among model variables. In fact, connections carry information from one part to another part of the model which is shown by arrows. In the end, converters connect the input to output which can be manifested as mathematical equations or diagrams and tables. shows elements to be used in stock and flow diagram.

The mathematical structure of stock and flow diagram with one stock variable with an input and an output can be seen in Fig. 7.3.

Table 7.2 Elements for creating stock and flow diagrams

	Stocks
	Flows
	Connections

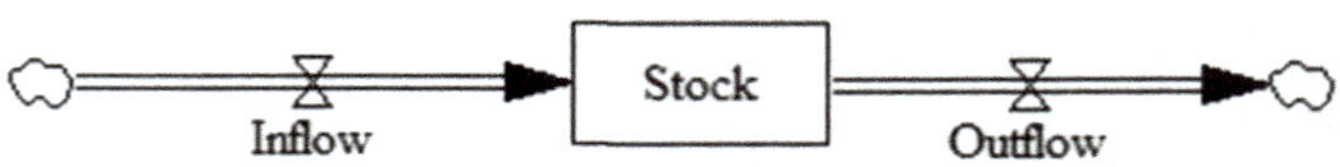

$$Stock(t) = \int_{t_0}^{t} (Inflow(s) - Outflow(s))ds + Stock(t_0)$$

$$ds/dt = Inflow(t) - Outflow(t)$$

Fig. 7.3 Mathematical structure of stock and flow diagram

7.4 Importance and Necessity

The ever-increasing demand for water every day is the result of agricultural development, overpopulation, and the rapid growth of industries. At the same time, the limitation of controllable water, as well as continuous demand growth, requires accurate planning for better water management and operation regarding this limited source. The existence of comprehensive understanding and the ability to predict circumstances help planners use these water resources more appropriately and achieve an optimal sustainable operation based on seasonal and climate conditions. However, multiple decision-makers and consumers with different priorities and interests in water resource planning and management issues create considerable disagreement and tensions over water resources allocation, which will be of much concern for managers and planners of this section. Today with the rapid development of computing tools along with the emergence of new optimization and simulation models, the application of systemic views has expanded in water resource planning and management and it has been made possible for complex systems to be analyzed better by recognizing and analyzing components and their connections. One of these methods is the System Dynamic (SD) approach. This method has been developed for the simplification of interaction between managers' models and analysts' official models. The main feature of this tool is to recognize and to show the feedback processes of consumers, flow structures, and time delays and considering nonlinear relationships to show system dynamics.

7.5 Materials and Methods

In the process of using system dynamics, there are five steps for each problem: 1. Statement of the problem; 2. development of a dynamic hypothesis; 3. mathematical expression of the simulation model; 4. testing the simulation model; and 5. policy or strategy of designer, experimentation, and analysis (Sterman 2000). This section describes these five steps.

7.5.1 *Statement of the Problem*

In order to have successful modeling first, problems must be identified and desired objectives must be specified. That is, the following should be specified in this stage.

- What is the problem? What is to be investigated, and why is it important?
- What are the essential variables and main concepts of the system?
- What is the time specifications for the implementation of the current study?
- How the system and essential variables behaved in the past and how is it predicted to behave in the future?

It can be seen that to recognize the problem; a complete description must be provided based on available reports, previous and current studies, expert opinions, historical and statistical records, and past behavior of the system. A thorough description of the problem provides a proper understanding by showing important and effective components for the researcher.

7.5.2 Development of a Dynamic Hypothesis

After identifying the problem, the next step is to develop a dynamic hypothesis based on the basic and initial behavior of the problem over time. The dynamic hypothesis is a conceptual model consisting of causal loop diagrams and state-flow diagrams or their combination. Therefore, at this step, policy structure diagrams, causal loop diagrams, and stock-flow diagrams should be defined.

7.5.3 The Mathematical Expression of the Simulation Model

At this step, the initial structure of the model is created by the above-mentioned tools. In other words, at this step, the relationship between these important and effective components is recognized and related parameters are formulated. Also, in this part, the initial value of variables and their possible estimation are determined, and the simulation model is ready to be implemented.

7.5.4 Testing the Simulation Model

After the simulation, it is necessary to check whether the model structure complies with the rules and decision-making processes of the system or not. Are the dimensions of equations used in the model compatible with each other? With changing in parameters, boundaries and, time intervals, will there be significant changes in numerical values, behavior, and policies?

One of the important tests is the comparison with historical data. In this test, the accuracy of the model in the simulation of a historical event can be determined. System dynamic modelers have prepared a variety of specific tests that help users to achieve a better understanding of the model. In following these tests, and their tool will be introduced.

7.5.4.1 Boundary Adequacy

This test examines model behavior change by stabilizing boundary assumptions and policy changing duo to model boundaries extension. This test also determines whether the basic and essential concepts can address the endogenous problems of the model or not. Using model charts, diagrams, model equations, use reports, expert opinions, archived documents, direct inspection or participation in system processes along with modifying the model to add appropriate additional structures and change constants and exogenous variables endogenous will help users in this test.

7.5.4.2 Structure Assessment

The purposes of this assessment are to examine the compatibility of model structure with descriptive principles and concepts of the system, level of aggregation, conformity of model to the physical laws, and the ability of decision rules to capture the behavior of system components. For this assessment, tools such as model charts (policy structure diagrams, causal diagrams, stock, and flow diagrams), model equations, reports, expert opinions, archived documents, and direct inspection or participation in system processes can be used. By conducting partial model tests and laboratory experiments, the rationality of decision rules can be evaluated, and the mental models and decision rules of system participants can be elicited. The development of disaggregate models for comparing with aggregate formulations and disaggregation of suspicious structures are the remaining tools and procedures of this assessment.

7.5.4.3 Dimensional Consistency

The purpose of this test is to evaluate the dimensional consistency of model equations without the using parameters which have no real-world meaning. For this test, dimensional analysis software and model equations investigation for finding suspect parameters can be used.

7.5.4.4 Parameter Assessment

The purposes of this assessment are to examine the compatibility of parameter values with descriptive and numerical concepts and principles of the system and inspect real-world counterparts of all parameters. For this assessment, statistical methods for parameters estimation, partial model tests for calibrations, judgemental methods, and development of disaggregate submodels are the primary tools.

7.5.4.5 Extreme Conditions

In this test, model equations will be tested by extreme values as input in order to evaluate their response to extreme conditions like extreme policies, shocks, and parameters. In this test, each equation must be investigated, and their response to extreme values of inputs, alone and in combination, must be analyzed. By subjecting model to large shocks and extreme conditions and implementing tests for examining model conformance to basic principles, model and system response to extreme conditions will be evaluated.

7.5.4.6 Integration Error

This test examines results sensitivity to time step choices or numerical integration methods. By changing the time steps and using different numerical integration methods, model behavior will be tested.

7.5.4.7 Behavior Reproduction

In this test, the model ability to reproduce favorite behavior of the system in terms of quantity and quality and generating different modes of observed behavior in the real system will be evaluated. Another purpose of this is to figure out the accordance between variables and data. Computing statistical measures between model outputs and observed data such as descriptive statistics, time-domain methods, frequency-domain methods, comparing model output and data qualitatively, and examing model response to test inputs, shocks and noises are the essential tools for this test.

7.5.4.8 Behavior Anomaly

In this test, the possibility of the model for resulting in abnormal behavior due to changing or deleting assumptions will be examined. By loop knockout analysis and replacing equilibrium assumptions with disequilibrium structures, this test can be implemented.

7.5.4.9 Family Member

In this test, the model ability to generate observed behavior in other instances of the same system will be evaluated by calibrating the model to a wide range of related systems.

7.5.4.10 Surprise Behavior

In this test, the ability of the model for generating previous unobserved or recognized behavior, and success anticipate of the system response to novel conditions will be tested. For this test, the user must keep accurate, complete, and comprehensive simulations records and also use the model to simulate the future behavior of the system. For better results, all disagreements between model behavior and user understanding of the real system must be resolved, also documenting components mental model prior to starting modeling.

7.5.4.11 Sensitivity Analysis

Numerical, behavioral, and policy sensitivity analysis are the main ones for dynamic models. Univariate and multivariate sensitivity analysis, analytic methods, model boundary and aggregation tests, and optimizations approaches are practical methods for performing these analyses.

7.5.4.12 System Improvement

The purpose of this test is to evaluate the modeling process effect on system improvement and designing, developing and creating instruments and appropriate controlling tests for modeling process evaluation, monitoring, and treatment are the essential tools in this test.

The mentioned tests were adopted and extended from Sterman (2000).

7.5.5 Policy or Strategy Designer, Experimentation, and Analysis

After validating the model, different scenarios are defined and implemented by the model for solving the problem. The purpose of these scenarios is to analyze different situations to achieve the best solution for the problem. In this point, the sensitivity of each parameter and the impact of policies on important variables are evaluated by investigating the connections between policies and different decisions.

7.6 Practical Examples

7.6.1 Simulation of Drainage Water Quantity and Quality

Nozari and Liaghat (2014) used SD to simulate water movement in soil and solute transport in saturated and unsaturated conditions in the presence of a drainage system. The predictive variables were water table fluctuations, drain discharge, drain water salinity, and groundwater salinity. In this study, the conceptual model which was defined for saturated and unsaturated zones and boundary conditions consists of infiltration and evapotranspiration from the soil surface, lateral seepage and, deep seepage. The results from this model can be used to manage and control drain water salinity and also to prevent environmental damage. Figure 7.4 represents the flowchart of the above-mentioned system.

The causal loop diagrams in this study are shown in Figs. 7.5, 7.6 and 7.7.

In the stock and flow structure of this model, important variables are layer, water table, and groundwater.

Figure 7.8 shows the stock and flow structure of the related problem.

In order to validate the model results, in addition to boundery condition testing and sensitivity analysis, a case study was conducted on ARC1-18, 25 ha research site,

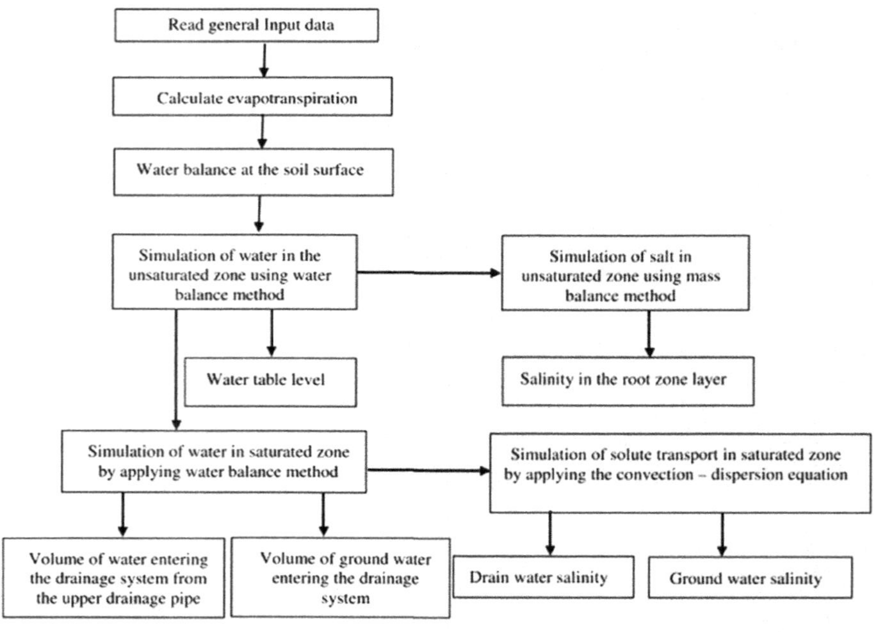

Fig. 7.4 Flowchart for the water table depth, drained volume, salinity of drain water and salinity of groundwater calculations

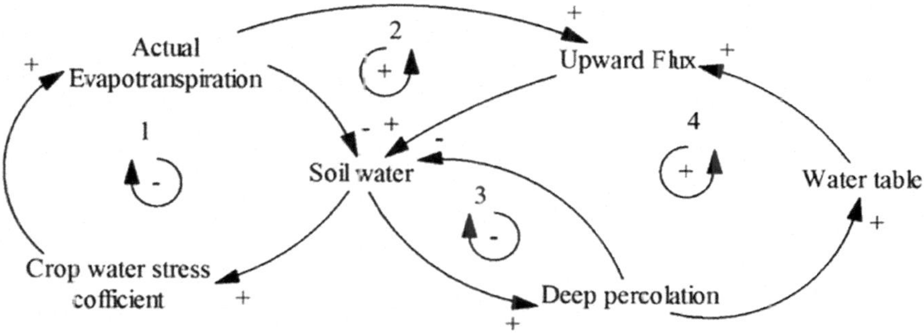

Fig. 7.5 Unsaturated zone causal loop diagram

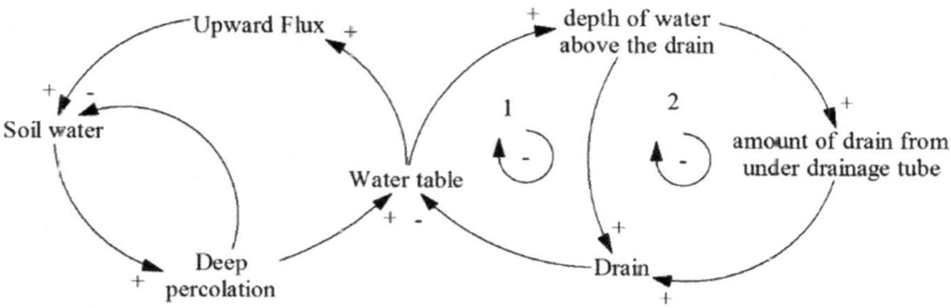

Fig. 7.6 Drainage performance causal loop diagram

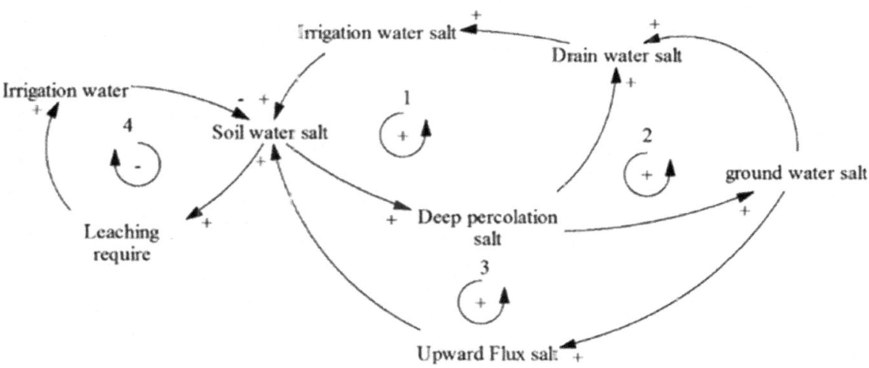

Fig. 7.7 Causal loop diagram for dynamic salinization model

located on Amirkabir Sugarcane Research Center, which is one of the seven units of the sugarcane development plan in Khuzestan, Iran. To evaluate water table fluctuations and groundwater salinity, three rows of piezometers were installed at distances of 100, 250, and 375 m from the collector. In every row, there were three piezometers

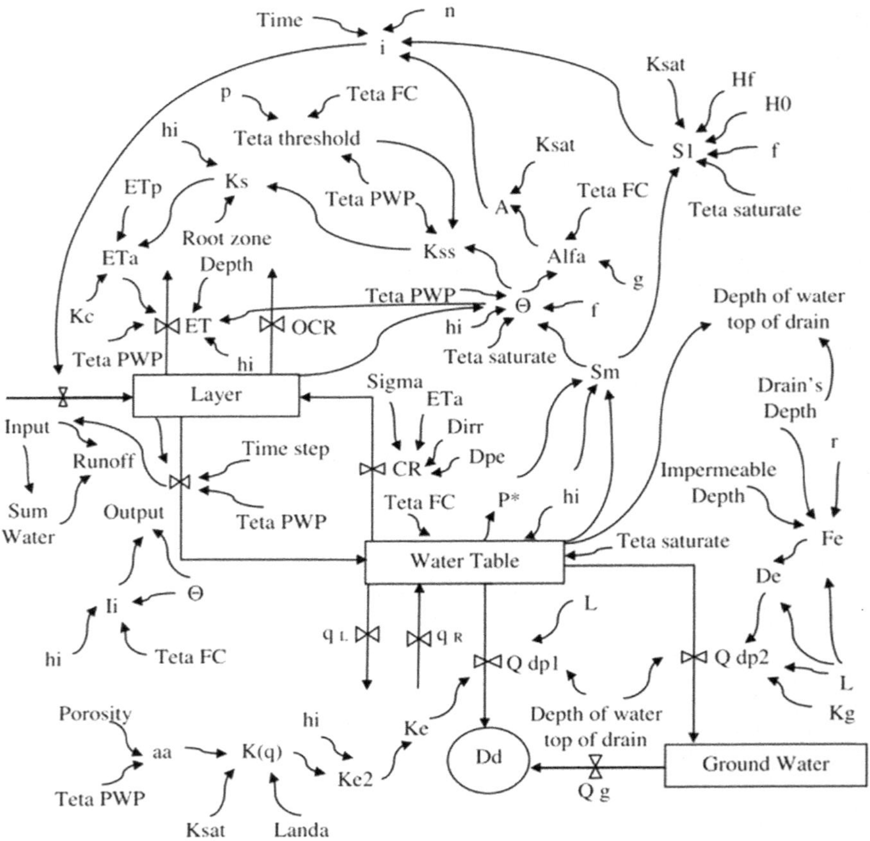

Fig. 7.8 Stock and flow structure

at depths of 0.2, 0.6 and, 1 m below the drain tube for collecting samples. During the irrigation period (April to September), parameters such as daily water table fluctuations, drain discharge, irrigation water salinity, piezometers water salinity, and drain water salinity have been recorded. In Table 7.3, the statistical indices show the comparison of the model results with the observed data. The results show an effective

Table 7.3 Standard error and relative standard error for different variables

Variable	Water table depth (cm)	Drainage flux (L/s)	Drain water salinity (dS/m)	Ground water salinity (dS/rn)		
				220 cm	260 cm	300 cm
SE	14.4	0.43	2.8	0.49	0.29	0.36
RSE	8	20	19	12.9	7.5	8.2
R^2	0.7	0.76	0.2	0.63	0.7	0.64

and acceptable performance of the model in simulating water table level, drainage discharge, and groundwater salinity.

7.6.2 Power Generation Simulation of a Hydropower Reservoir System: Case Study of Karoon Reservoir System

Jahandideh-Tehrani et al. (2014) used SD for calculating energy production in a hydropower reservoir system in several operational scenarios. Complex and nonlinear relationships characterize hydropower reservoirs, and variables such as storage volume, release, power production, and reservoir water level depend on each other. Therefore, the use of trial-and-error simulation or other methods based on a repeating loop is needed. This research has studied a three-reservoir system consisting of Khersan 1, Karoon 3, and Karoon 4 in Karoon river basin in Khuzestan, Iran. This system consists of a storage reservoir under study called Khersan 1 and two storage reservoirs under operation called Karoon 3 and Karoon 4. Figure 7.9. shows the location of these reservoirs.

The water released from the upstream reservoirs goes directly into the downstream reservoir and affects its energy production. Therefore, this study focused on hydropower reservoir systems and the effects of upstream reservoirs on the power generation of downstream reservoirs. Eight operational scenarios that involve different combinations of reservoirs are defined to calculate average values of power generation over a 44-year operational period. From eight defined scenarios, six scenarios are considered for an individual reservoir, and upstream reservoirs impact on downstream reservoirs and the other two scenarios are considered for multi-reservoir systems. These scenarios are as follows:

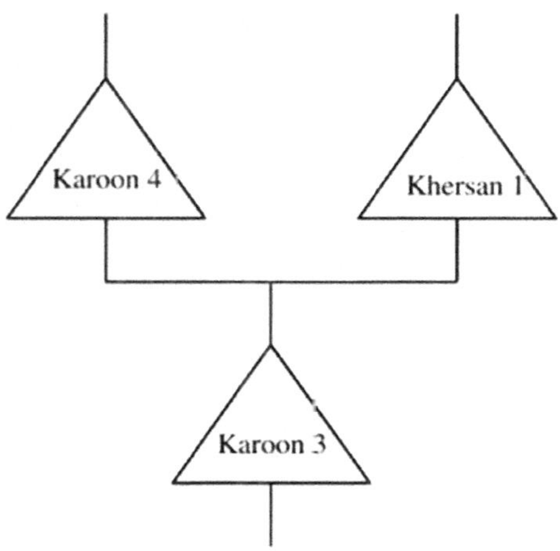

Fig. 7.9 Location of studied reservoirs Results

- Operational scenario 1 considers Khersan 1 individually with the condition that there are no other upstream or downstream reservoirs,
- Operational scenario 2 considers Karoon 4 individually with the condition that there are no other upstream or downstream reservoirs,
- Operational scenario 3 considers Karoon 3 individually with the condition that there are no other upstream reservoirs,
- Operational scenario 4 considers Karoon 3 individually with the condition that both Khersan 1 and Karoon 4 are located upstream,
- Operational scenario 5 considers Karoon 3 individually with the condition that only Khersan 1 is located upstream,
- Operational scenario 6 considers Karoon 3 individually with the condition that only Karoon 4 is located upstream,
- Operational scenario 7 considers power generation of the two reservoirs Karoon 4 and Karoon 3.
- Operational scenario 8 considers power generation of the three reservoirs Khersan 1, Karoon 4, and Karoon 3.

In this study, performance criteria such as reliability, vulnerability, and resiliency for each scenario are defined at three performance thresholds (100, 90, and 70%).

Figure 7.10 represents the stock-flow model diagram of a reservoir and the effect of variables on each other, and Fig. 7.11. represents the stock-flow model diagram of a reservoir at downstream.

Table 7.4 shows the results of performance criteria and average annual energy production of each operational scenario.

The results illustrate that scenario 8 has more average energy production in comparison with other scenarios. Although Khersan 1 is under study, comparison of

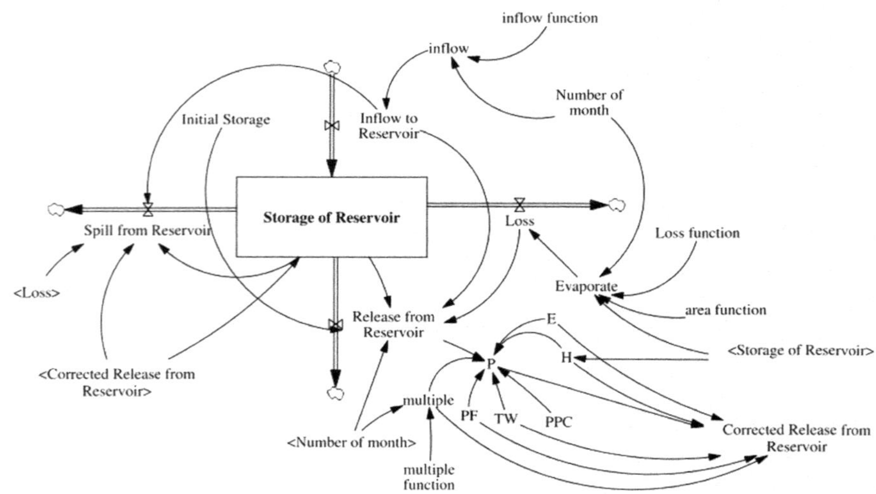

Fig. 7.10 Model of each single reservoir

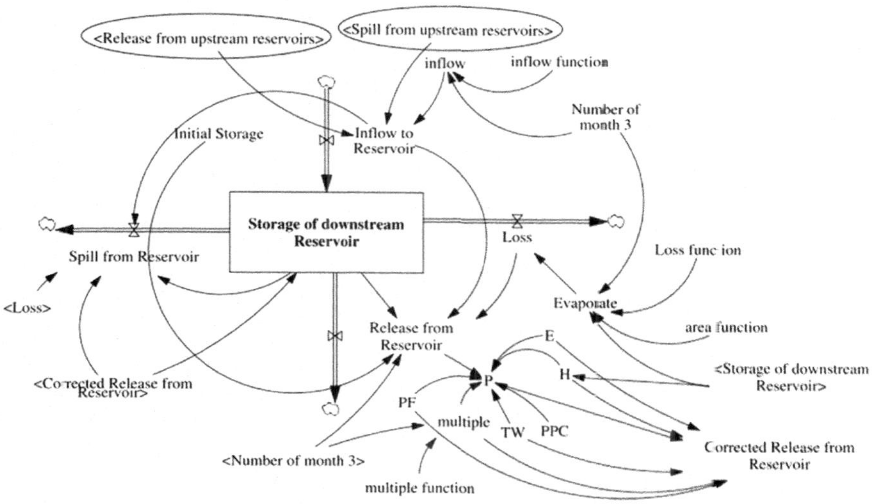

Fig. 7.11 Model of downstream reservoir in reservoir system model

Table 7.4 Performance criteria and average annual energy production of each operational scenario

Operational scenario number								
Performance criteria (%) and average annual energy production (109 W • h)	1	2	3	4	5	6	7	8
Time-based reliability 100%	41	87	36	32	36	31	31	19
Time-based reliability 90%	44	88	44	39	43	38	46	45
Time-based reliability 70%	53	90	56	51	56	51	61	63
Volumetric reliability 100%	72	94	70	70	70	69	78	77
Volumetric reliability 90%	75	95	73	73	74	73	82	82
Volumetric reliability 70%	83	96	79	81	79	81	91	90
Vulnerability 100%	28	6	30	30	30	31	22	23
Vulnerability 90%	25	5	27	26	26	27	18	18
Vulnerability 70%	18	4	21	19	21	19	9	10
Resiliency 100%	12	28	9	7	9	7	7	6
Resiliency 90%	15	28	14	10	13	11	12	12
Resiliency 70%	19	31	18	14	17	14	20	20
Average annual energy production	921	1,651	3,069	3,062	3,082	3,052	4,703	5,635

scenarios 7 and 8 show the importance of Khersan 1 construction. In other words, the addition of Khersan 1 to Karoon 4 and Karoon 3 reservoir system leads to higher average energy production and relative stability of performance criteria. Results of Table 7.3 shows that the average energy production in scenario 4 is less than that in scenario 3 because of water regulation and evaporation from upstream reservoirs, which reduce the average energy production in scenario 4 compared to that of scenario 3. Due to the presence of Khersan 1 in upstream of Karoon 3 in scenario 5, the average energy production of this scenario has increased compared to scenario 3. The presence of Khersan 1 in upstream of scenario 5, which stores water in the wet season and releases it during the dry season, has caused an increase in power generation and water storage in Karoon 3. In the wet season, due to the limitation of reservoir volume, scenario 3 loses more water by spillage than scenario 5.

The results of this study indicate that system dynamic has the ability for using in hydropower systems. It can also be highlighted that construction and addition of the Khersan 1 to Karoon 3 and Karoon 4 reservoir system increases energy generation by 20% during the 44-year simulation period, and regardless of construction costs, Khersan 1 can help to meet more energy needs during peak consumption periods.

7.6.3 Modeling of Reservoir Operations for Flood Management

(Ahmad and Simonovic 2000) prepared a general framework for modeling reservoir operations using the SD approach. Prepared model, developed for a single multipurpose reservoir with a focus on flood management role of the reservoir in order to develop a reservoir policy for high-flow years for minimizing flooding. Also, this model can be used as a tool for studying the impacts of changing reservoir storage allocation and temporal distribution of reservoir levels and outflows. This approach has been used for modeling reservoir operations of Shellmouth reservoir, located on the Assiniboine River, close to the Manitoba/Saskatchewan border in Canada. The flooding in the Assiniboine River, mainly caused by heavy spring runoff, has resulted in extensive damage to residential, agricultural, and industrial property. The Shellmouth reservoir was developed primarily to protect the cities of Brandon and Winnipeg from floods on the Assiniboine River, and supplementary benefits of the project include flood control to agricultural land. Release from the reservoir, which exceed the channel capacity, causes flooding at several locations along the river downstream of the reservoir. Due to the lack of existence of control structure on spillway for regulating reservoir outflow, the objectives of the simulation modeling study were defined as:

- Developing a reservoir operational policy for high flow years to minimize flooding.
- Exploring the impacts on the reservoir flood management capacity by installing gates on an existing unregulated spillway.

- Developing a tool for evaluating alternative operating rules by changing the reservoir storage allocation, the reservoir levels at the start of the flood season, and the reservoir outflow.

Figure 7.12 represent the schematic location of the Shellmouth reservoir, Assiniboine River and, cities at downstream of the reservoir.

Two causal loop diagrams were implemented in the structure of the reservoir dynamic model Fig. 7.13. Figure 7.13a illustrates the positive influence of inflow on the reservoir and an increase in reservoir storage causes an increase in the reservoir level, which causes flooding at the reservoir upstream. Figure 7.13b represent the storage reallocation, and by increasing flood storage zone in the reservoir, flooding will be reduced; however, it will also reduce the supply of water for other uses.

The implemented dynamic model for flood management can be divided into three main sectors: the reservoir, the upstream area, and the downstream area. A schematic diagram of the reservoir and its three sectors is shown in Fig. 7.14.

The results proved that system dynamics is a successful, user friendly, and effective approach for reservoir modeling. Alos researchers acknowledged that by revising

Fig. 7.12 Study Area

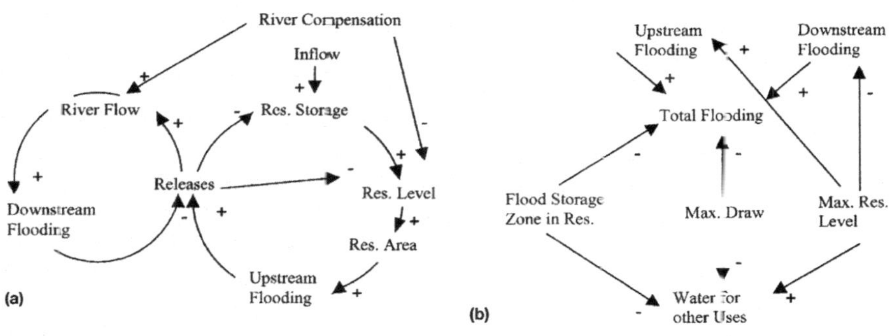

Fig. 7.13 Causal loop diagrams

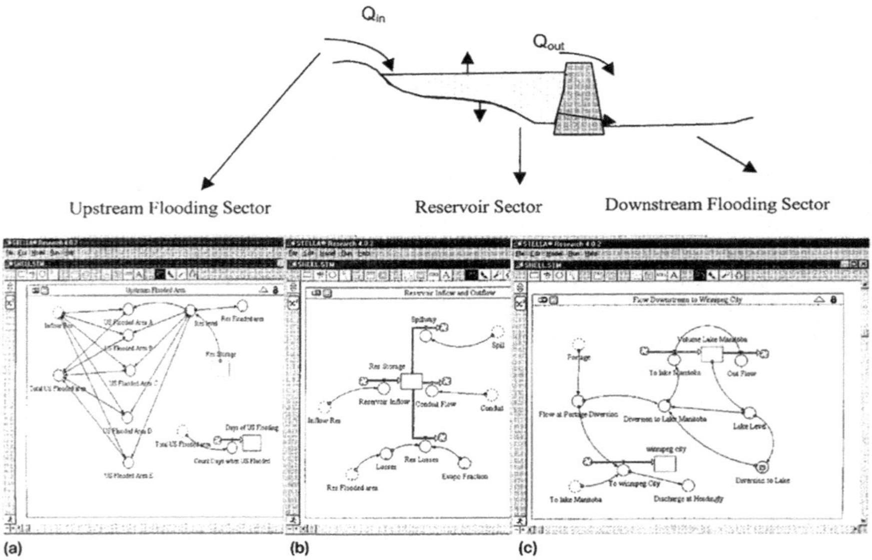

Fig. 7.14 Schematic diagram of reservoir with different sectors

operating rules, the capability of the Shellmouth reservoir for flood management can be improved. Due to the revision of the operating rules, the contribution of the Assiniboine River towards the flooding of Winnipeg City is negligible. At the end of this study and by considering the simulation of the shellmouth reservoir operation, it was recommended that the installation of gates on the spillway will improve reservoir flood management capacity, especially for large floods.

7.7 Summary

Water is one of the biggest challenges of this century, which can be sources of many positive and negative changes in the world. Due to the limitation of water resources as well as the development of industrial and agricultural projects along with population growth, which has made this vital resource unsustainable in many parts of the world. Thus, the optimal use and management of these resources have become particularly important. However, occasionally the adopted management solutions are not effective in improving the situation, and even our performance in solving the problem may lead to new problems after the implementation of the adopted policy. Because all possible feedbacks ranges are not considered in one system. In order to prevent adverse reactions to adopted policies and achieve the dynamic interrelationship between existing management systems, it is necessary to expand model boundaries by a comprehensive approach. For solving such problems, an analytical system with a decision-making system is required to model all involving processes in a complex system systematically. The characteristic of this approach is a system that can be divided into multiple subsystems to work together to achieve a specific goal. Also, by assuming the surrounding environment as a variable, the combination of solutions for solving the problem can be considered as variable, and with a systematic insight can solve such problems.

Acknowledgements The authors thank Iran's National Science Foundation (INSF) for the support of this research.

References

Ahmad S (2016) Managing water demands for a rapidly growing city in semi-arid environment: study of Las Vegas, Nevada. Int J Water Resour Arid Environ 5(1):35–42

Ahmad S, Prashar D (2010) Evaluating municipal water conservation policies using a dynamic simulation model. Water Resour Manag 24:3371–3395. https://doi.org/10.1007/s11269-010-9611-2

Ahmad S, Simonovic SP (2000) System dynamics modeling of reservoir operations for flood management. J Comput Civ Eng 14:190–198. https://doi.org/10.1061/(ASCE)0887-3801(2000)14:3(190)

Ahmad S, Simonovic SP (2001) Modeling dynamic processes in space and time – a spatial system dynamics approach. World Water and Environmental Resources Congress. Orlando, FL. May 20–24, 2001. pp 1–20. https://doi.org/10.1061/40569(2001)88

Ahmad S, Simonovic SP (2004) Spatial system dynamics: new approach for simulation of water resources systems. J Comput Civ Eng 18:331–340. https://doi.org/10.1061/(ASCE)0887-3801(2004)18:4(331)

Ahmad S, Simonovic SP (2006) An intelligent decision support system for management of floods. Water Resour Manag 20:391–410. https://doi.org/10.1007/s11269-006-0326-3

Amoueyan E, Ahmad S, Eisenberg JNS, Pecson B, Gerrity D (2017) Quantifying pathogen risks associated with potable reuse: a risk assessment case study for Cryptosporidium. Water Res 119:252–266. https://doi.org/10.1016/j.watres.2017.04.048

Amoueyan E, Ahmad S, Eisenberg JNS, Gerrity D (2019) Equivalency of indirect and direct potable reuse paradigms based on a quantitative microbial risk assessment framework. Microb Risk Anal 12:60–75. https://doi.org/10.1016/j.mran.2019.06.003

Amoueyan E, Ahmad S, Eisenberg JNS, Gerrity D (2020) A dynamic quantitative microbial risk assessment for norovirus in potable reuse systems. Microb Risk Anal 14:100088. https://doi.org/10.1016/j.mran.2019.100088

Bagheri A, Babaeian F (2020) Assessing water security of Rafsanjan Plain, Iran – Adopting the SEEA framework of water accounting. Ecol Indic 111:105959. https://doi.org/10.1016/j.ecolind.2019.105959

Bala BK, Arshad FM, Noh KM (2017) System dynamics. modelling and simulation

Bates G, Beruvides M, Fedler CB (2019) System dynamics approach to groundwater storage modeling for basin-scale planning. Water (Switzerland) 11. doi:https://doi.org/10.3390/w11091907

Bukhary S, Ahmad S, Batista J (2018) Analyzing land and water requirements for solar deployment in the Southwestern United States. Renew Sustain Energy Rev 82:3288–3305. https://doi.org/10.1016/j.rser.2017.10.016

Chen C, Ahmad S, Kalra A, Xu ZX (2017) A dynamic model for exploring water-resource management scenarios in an inland arid area: Shanshan County, Northwestern China. J Mat Sci 14:1039–1057. https://doi.org/10.1007/s11629-016-4210-1
Chen Z, Wei S (2014) Application of system dynamics to water security research. Water Resour Manag 28:287–300. https://doi.org/10.1007/s11269-013-0496-8
Dawadi S, Ahmad S (2012) Changing climatic conditions in the Colorado River Basin: Implications for water resources management. J Hydrol 430–431:127–141. https://doi.org/10.1016/j.jhydrol.2012.02.010
Dawadi S, Ahmad S (2013) Evaluating the impact of demand-side management on water resources under changing climatic conditions and increasing population. J Environ Manag 114:261–275. https://doi.org/10.1016/j.jenvman.2012.10.015
Dong Q, Zhang X, Chen Y, Fang D (2019) Dynamic management of a water resources-socioeconomic-environmental system based on feedbacks using system dynamics. Water Resour Manag 33:2093–2108. https://doi.org/10.1007/s11269-019-02233-8
Dow C, Ahmad S, Stave K, Gerrity D (2019) Evaluating the sustainability of indirect potable reuse and direct potable reuse: a southern Nevada case study. AWWA Water Sci 1:1–16. https://doi.org/10.1002/aws2.1153
Duran-Encalada JA, Paucar-Caceres A, Bandala ER, Wright GH (2017) The impact of global climate change on water quantity and quality: a system dynamics approach to the US–Mexican transborder region. Eur J Oper Res 256:567–581. https://doi.org/10.1016/j.ejor.2016.06.016
Elshorbagy A, Ormsbee L (2006) Object-oriented modeling approach to surface water quality management. Environ Model Softw 21:689–698. https://doi.org/10.1016/j.envsoft.2005.02.001
Forreser JW (1968) Principle of systems. MIT Press, Cambridge, MA
Ghashghaie M, Marofi S, Marofi H (2014) Using system dynamics method to determine the effect of water demand priorities on downstream flow. Water Resour Manag 28:5055–5072. https://doi.org/10.1007/s11269-014-0791-z
Hassanzadeh E, Elshorbagy A, Wheater H, Gober P (2014) Managing water in complex systems: an integrated water resources model for Saskatchewan, Canada. Environ Model Softw 58:12–26. https://doi.org/10.1016/j.envsoft.2014.03.015
Jahandideh-Tehrani M, Bozorg-Haddad O, Mariño MA (2014) Power generation simulation of a hydropower reservoir system using system dynamics: case study of karoon reservoir system. J Energy Eng 140:1–12. https://doi.org/10.1061/(ASCE)EY.1943-7897.0000179
Kandissounon GA, Kalra A, Ahmad S (2018) Integrating system dynamics and remote sensing to estimate future water usage and average surface runoff in Lagos, Nigeria. Civil Eng J 4(2):378–393
Karimlou K, Hassani N, Rashidi Mehrabadi A, Nazari MR (2019) Developing a model for decision-makers in dynamic modeling of urban water system management. Water Resour Manag. https://doi.org/10.1007/s11269-019-02428-z
Kotir JH, Smith C, Brown G, Marshall N, Johnstone R (2016) A system dynamics simulation model for sustainable water resources management and agricultural development in the Volta River Basin, Ghana. Sci Total Environ 573:444–457. https://doi.org/10.1016/j.scitotenv.2016.08.081
Li Z, Li C, Wang X, Peng C, Cai Y, Huang W (2018) A hybrid system dynamics and optimization approach for supporting sustainable water resources planning in Zhengzhou City, China. J Hydrol 556:50–60. https://doi.org/10.1016/j.jhydrol.2017.11.007
Liu H, Benoit G, Liu T, Liu Y, Guo H (2015) An integrated system dynamics model developed for managing lake water quality at the watershed scale. J Environ Manag 155:11–23. https://doi.org/10.1016/j.jenvman.2015.02.046
Madani K, Mariño MA (2009) System dynamics analysis for managing Iran's Zayandeh-rud river basin. Water Resour Manag 23:2163–2187. https://doi.org/10.1007/s11269-008-9376-z
Matinzadeh MM, Abedi Koupai J, Sadeghi-Lari A, Nozari H, shayannejad M, M (2017a) Development of an innovative integrated model for the simulation of nitrogen dynamics in farmlands with drainage systems using the system dynamics approach. Ecol Modell 347:11–28. https://doi.org/10.1016/j.ecolmodel.2016.12.014

Matinzadeh MM, Abedi Koupai J, Sadeghi-Lari A, Nozari H, shayannejad M (2017b) System dynamic modeling to assess the effect of subsurface drain spacing and depth on minimizing the environmental impacts. Int J Environ Sci Technol 14(3):563–576

Mirchi A, Madani K, Watkins D, Ahmad S (2012) Synthesis of system dynamics tools for holistic conceptualization of water resources problems. Water Resour Manag 26:2421–2442. https://doi.org/10.1007/s11269-012-0024-2

Moumouni Y, Ahmad S, Baker JR (2014) A system dynamics model for energy planning in Niger. Int J Energy Power Eng 3(6):308. https://doi.org/10.11648/j.ijepe.20140306.14

Nazari-Sharabian M, Taheriyoun M, Ahmad S, Karakouzian M, Ahmadi A (2019) Water quality modeling of Mahabad Dam watershed-reservoir system under climate change conditions, using SWAT and system dynamics. Water (Switzerland) 11:1–16. https://doi.org/10 3390/w11020394

Nozari H, Heydari M, Azadi S (2014) Simulation of a right abshar irrigation network and its cropping pattern using a system dynamics approach. J Irrig Drain Eng 140:1–7. https://doi.org/10.1061/(ASCE)IR.1943-4774.0000777

Nussbaum EM, Owens MC, Sinatra GM, Rehmat AP, Cordova JR, Ahmad S, Dascalu SM (2015) Losing the lake: simulations to promote gains in student knowledge and interest about climate change. Int J Environ Sci Educ 10(6):789–811

Pluchinotta I, Pagano A, Giordano R, Tsoukiàs A (2018) A system dynamics model for supporting decision-makers in irrigation water management. J Environ Manag 223:815–824. https://doi.org/10.1016/j.jenvman.2018.06.083

Qaiser K, Ahmad S, Johnson W, Batista JR (2011) Evaluating the impact of water conservation on fate of outdoor water use: A study in an arid region. J Environ Manag 92:2061–2068. https://doi.org/10.1016/j.jenvman.2011.03.031

Qaiser K, Ahmad S, Johnson W, Batista JR (2013) Evaluating water conservation and reuse policies using a dynamic water balance model. Environ Manag 51:449–458. https://doi.org/10.1007/s00267-012-9965-8

Qin H, Zheng C, He X, Refsgaard JC (2019) Analysis of water management scenarios using coupled hydrological and system dynamics modeling. Water Resour Manag:4849–4863. https://doi.org/10.1007/s11269-019-02410-9

Rusuli Y, Li L, Ahmad S, Zhao X (2015) Dynamics model to simulate water and salt balance of Bosten Lake in Xinjiang, China. Environ Earth Sci 74:2499–2510. https://doi.org/10.1007/s12665-015-4257-2

Sharifi A, Kalin L, Tajrishy M (2013) System dynamics approach for hydropower generation assessment in developing watersheds: case study of Karkheh River basin. Iran. J Hydrol Eng 18:1007–1017. https://doi.org/10.1061/(ASCE)HE.1943-5584.0000711

Shrestha E, Ahmad S, Johnson W, Shreshta P, Batista JR (2011a) Carbon footprint of water conveyance versus desalination as alternatives to expand water supply. Desalination 280:33–43. https://doi.org/10.1016/j.desal.2011.06.062

Shrestha E, Ahmad S, Johnson W, Batista JR (2012) The carbon footprint of water management policy options. Energy Policy 42:201–212. https://doi.org/10.1016/j.enpol.2011.11.074

Shrestha E, Ahmad S, Johnson W, Batista JR (2011b) The carbon footprint associated with water management policy options in the Las Vegas Valley, Nevada. J Nevada Water Resour Assoc 6(1):2–9

Simonovic SP, Rajasekaram V (2004) Integrated analyses of Canada's water resources: a system dynamics approach. Can Water Resour J 29:223–250. https://doi.org/10.4296/cwrj223

Stave KA (2003) A system dynamics model to facilitate public understanding of water management options in Las Vegas, Nevada. J Environ Manag 67:303–313. https://doi.org/10.1016/S0301-4797(02)00205-0

Sterman JD (2000) Business dynamics systems thinking and modeling for a complex world. McGraw-Hill Higher Education

Tamaddun K, Kalra A, Ahmad S (2018) Potential of rooftop rainwater harvesting to meet outdoor water demand in arid regions. J Arid Land 10:68–83. https://doi.org/10.1007/s40333-017-0110-7

Vaez Tehrani M, Monem MJ, Bagheri A (2013) A system dynamics approach to model rehabilitation of irrigation networks case study: Qazvin irrigation network. Iran. Irrig Drain 62:193–207. https://doi.org/10.1002/ird.1729

Venkatesan AK, Ahmad S, Johnson W, Batista JR (2011a) Salinity reduction and energy conservation in direct and indirect potable water reuse. Desalination 272:120–127. https://doi.org/10.1016/j.desal.2011.01.007

Venkatesan AK, Ahmad S, Johnson W, Batista JR (2011b) Systems dynamic model to forecast salinity load to the Colorado River due to urbanization within the Las Vegas Valley. Sci Total Environ 409:2616–2625. https://doi.org/10.1016/j.scitotenv.2011.03.018

Winz I, Brierley G, Trowsdale S (2009) The use of system dynamics simulation in water resources management. Water Resour Manag 23:1301–1323. https://doi.org/10.1007/s11269-008-9328-7

Wu G, Li L, Ahmad S, Chen X, Pan X (2013) A dynamic model for vulnerability assessment of regional water resources in arid areas: a case study of Bayingolin, China. Water Resour Manag 27:3085–3101. https://doi.org/10.1007/s11269-013-0334-z

Zhang F, Ahmad S, Zhang H, Zhao X, Feng X, Li L (2016) Simulating low and high streamflow driven by snowmelt in an insufficiently gauged alpine basin. Stoch Environ Res Risk Assess 30:59–75. https://doi.org/10.1007/s00477-015-1028-2

Zomorodian M, Lai SH, Homayounfar M, Ibrahim Sh, Fatemi SE, El-Shafie A (2018) The state-of-the-art system dynamics application in integrated water resources modeling. J Environ Manag 227:294–304. https://doi.org/10.1016/j.jenvman.2018.08.097

Chapter 8
Application of Agent Base Modeling in Water Resources Management and Planning

Mohsen Agha-Hoseinali-Shirazi, Omid Bozorg-Haddad, Melinda Laituri, and Don DeAngelis

Abstract The complexity of water demand management issues due to the existence of interconnected natural, political, social and economic areas makes it important to know the behavior of the agents. This complexity influences modeling and simulation that predict the behavior of agents as consumers. National and international disputes contribute to water management problems, especially in countries suffering from a water crisis. Researchers use a variety of frameworks to solve complex problems. One of these is the Complex Adaptive System (CAS) that has been the focus of attention in the last two decades. One approach used in this framework is the use of agent-Based Modeling (ABM). Agent-based modeling has been used to address problems in water demand management. This section discusses the importance, structure, limitations and advantages of agent-based modeling. Finally, examples of how to use agent-based modeling in addressing water demand management problems are illustrated. In first example, the ABM used to simulate and understand the effects of agent interactions and behaviors in urban water supply and demand on water scarcity condition. For second problem the ABM simplifies the complexity of all conflicting views and interactions of competitors in social network and provides a powerful tool for test new managerial scenarios to understand the consequences of decisions in a simple way.

M. Agha-Hoseinali-Shirazi · O. Bozorg-Haddad (✉)
Department of Irrigation and Reclamation Engineering, Faculty of Agricultural Engineering and Technology, College of Agriculture and Natural Resources, University of Tehran, Karaj, Tehran, Iran
e-mail: OBHaddad@ut.ac.ir

M. Agha-Hoseinali-Shirazi
e-mail: mshirazi8030@yahoo.com

M. Laituri
Department of Ecosystem Science and Sustainability, Colorado State University, Fort Collins, CO, USA
e-mail: Melinda.laituri@colostate.edu

D. DeAngelis
Wetland and Aquatic Research Center, Department of Biology, University of Miami, Coral Gables, FL, USA
e-mail: don_deangelis@usgs.gov

O. Bozorg-Haddad (ed.), *Essential Tools for Water Resources Analysis, Planning, and Management*, Springer Water, https://doi.org/10.1007/978-981-33-4295-8_8

Keywords Agent-based modeling · Water demand management · Modeling and simulation · Agent · Complexity

8.1 Introduction

Water is one of sustainable development's factors, which is of great importance for the management of water resources for any country due to its vulnerability. Human activities such as agricultural development, population growth and urban development, industrial development, and natural factors such as climate change have increased the consumption of renewable and non-renewable water resources. Increasing water use in various sectors (agricultural, industrial and urban) have led to water resources crisis. Oftentime, conflicts arise between users who use water for different purposes. Therefore, consideration of the different goals of users in water resources management requires technical and human analysis. However real-world testing may not be feasible due to the high cost and lack of access to the real environment. For these reasons, the use of modeling and simulation is common to contuct an analysis on a sample environment.

Modeling and the derived outputs provide a means to examine complexities of the real world to test hypotheses and model different scenarios.

The purpose of this paper is to illustrate the principles of modeling, simulation and analysis in the field of water resources management and planning. A review of modeling and simulation is presented. The role of social networks in the context of water resource management is discussed. Agent-based modeling is described in relation to water resources management and planning. Finally, examples of how agent-based modeling is applied in water resources management and planning is described.

8.2 Modeling and Simulation Overview

Models represent and simulate real or fictitious scenarios of a target or application for examination to better understand real systems. Model apply rules or algorithms that govern the system in a mathematical form. By constructing a model, the analyst can gain a better insight into the problem and test the validity of his/her hypothesis.

Models are classified into three main categories:

1. Scale models that reduce the detail and complexity of the model. For example, analysis and management of the dam reservoir using the standard operation policy model (SOP).
2. Ideal models magnify some of the target features. Water resource management modeling to change the area under cultivation and the type of agriculture for more valuable economic products.

3. Analogical models are applied to systems that are not well understood, however have a proper representation of the real world and estimate how the real world or system might work. Conflict management between different farmers in the catchment with common-pool water resources and water rights for some farmers that have existed from the past.

Modeling is used to explore real world problems and find solutions. However, models are always less complex than the real world but models can be run using different parameters to emulate the real world.

Simulation is a method that describes the behavior of a system using mathematical concepts and models to imitate the performance of the real-world process (system) over a certain time interval. Simulation attempts to provide, an accurate description of changes to the status of the system. That is, changes in the current state to that of the future can examined in accordance with the rules in the simulation. Thereby; simulation allows the analyst to obtain a real visualization of the research in question. It is very important to know what the model consists of and what is determined in the simulation. Thus, simulation can be defined as experimentation or model implementation.

Because of the complexity of today's world, modeling and simulation modeling are an interdisciplinary field that facilitates and integrates databases across disciplines. Modeling is used in different areas, when it is necessary to create new knowledge in specific fields. Models therefore act as mediators in the decision-making process of politicians, and in such cases, the models can be the interface between scientists and politicians and make the findings of scientific research available to decision makers (Kolkman and van der Veen 2005). In this case, decision makers with less expense but at a faster and more secure rate can predict real-world results and have a deep insight into the good and bad consequences of their decisions.

Several modeling and simulation examples are described to understand how these methods can be used in different contexts. The first simulators used were in martial arts of ancient Rome. They were among the most prepared armies in the world at the time. With the help of the simulations they were able to fight in unknown areas with contrasting tactics (Sokolowski and Banks 2009).

Leonardo da Vinci was a prominent artist-scientist who used modeling to test many of his inventions and projects. By recognizing the different components of the machine, he knew how the machine worked, and by modifying its components, he was able to improve the performance of existing machines or build a new machine. The same type of modeling has been used to understand the human body and that of animals, as it can explain the shape and composition of muscles to medical scientists.

Modeling and simulation is also used in computer games. Chess is one of the games in which humans can play with their computer. The humans can upgrade their playing level because computers are able to provide good simulations of the how their opponents might play.

In the military, flight simulations of warplanes are another modeling and simulation application. Before the war in Iraq, soldiers trained in various conditions of war with the supposed Iraqi troops and thus completed their operations in a short period.

This type of training and war was the basis for the later methods used in the Afghan wars, terrorist groups, and so on (Sokolowski and Banks 2009.

Nowadays there are various types of software for modeling in water resources engineering calculations. In order to achieve reliable results, the engineer needs to master the computational assumptions and processes in the software and, in addition, have practical experience in applying them. Many models such as HEC, MODFLOW, GIS and … series use hydrological and hydraulic data in surface water and groundwater modeling (Reeves et al. 2010). Due to the complexity of water resource issues, the importance of stakeholder roles and social networks, a number of software packages have been developed to explore the relationship between stakeholders and the environment. The outputs of these models are used in water resources management. The advantage of these software packages is that they are linked to hydrological and hydraulic software.

The use of modeling and simulation is very widespread and has been incorporated in almost all fields.

8.3 Agent Based Modeling (ABM)

Due to the presence of humans in an environment called a community and the interactions between humans and humans and their surroundings, the analysis of these two parts separately makes the results slightly unrealistic. Therefore, these two sections should be considered in a common context. For example, common-pool water resources have this property. Surface water and groundwater users in one area will affected by the amount of water consumed by users in another, and thus the amount of water use will affect the volume of water and will eventually have a positive or negative impact on system stability (Ostrom 2007). Considering water users, whether urban, industrial or agricultural consumers (individually or in tandem), the benefits, approaches, interactions between them and their environment can produce models that lead to results that are close to reality. Successful implementation of policies adopted by decision makers are largely guaranteed by considering the users, who play a key role in carrying out these policies. That is why water resources systems are considered as complex systems.

Agent-based modeling has been a powerful tool in simulation that has been used in many scientific fields and simulation studies over the past two decades. This modeling simulates different groups and their interactions based on local (behavioral) rules. Many researchers have been fascinated by the technical capabilities that agent-based modeling can offer and how they can achieve their goals using this type of modeling (Chalabi and Lorenc 2013; Israel and Wolf-Branigin 2011; Morell et al. 2010). In agent-based modeling, through its considering of the relationships and interactions between humans and humans with the environment, the model attempts to simulate a virtual environment close to the real environment by approximating the model conditions as far as possible. Another advantage of this type of modeling, in addition to the social dimension, is that agent-based modeling can be linked to and used simultaneously with hydrological and hydraulic models. Therefore, there is no restriction on the definition and drawing of interactions between individuals and the linkage between water resources models and agent-based modeling. The task then is to use the agent-based models, data, and field data collected by the researcher all together, which can pose a major challenge for the modeling exercise.

8.3.1 Agent-Based Modeling in Water Resources Management and Planning (Where)

Due to the increasing management bottom-up view in recent years, this type of modeling has had a growing trend in the water systems sector due to its ability to communicate between users and water resources systems and provide a realistic insight into the complex issues involved.

ABM can be applied to simulate the complex behavior of consumers of water. Consumer behavior influenced by interaction with other factors, water suppliers, water price, meteorological information, water resources conservation training campaigns, inventions and expand them throughout the community and their belief in interacting with authorities for had better manage resources, etc.

In some countries, the export of agricultural products plays an important role in their economy. Surface irrigation is of great importance in the agricultural sector of these countries due to its direct relationship with the production of agricultural products. If the country's water laws give water rights to trade water, there is the possibility of hoarding in order to increase the price of water and benefit more profit from the water owners, which causes problems in managing water resources. drought years and climate change is also making it difficult to access water in arid and semi-arid regions. The result of the lack of water supply in the agricultural sector is the reduction of income and profit of the farmer. Policies such as the use of reuse of surplus water and the return flow to provide irrigation water are proposed. This policy must be economically viable in order to convince the farmer (as a factor) to follow this policy. Here the agent-based model should be used. Arnold et al. (2015) presented a simulation model based on the hydro-economic model for the complex irrigation system of south-central Chile. To this end, the integrated agent-base model was parameterized from a wide range of quantitative data in the irrigation system. The proposed model is based on income from manufactured products, ancillary costs, fixed costs of machinery and irrigation infrastructure, wages paid or received, and the level of profit paid or received, and the objective function based on the maximum benefit by the farmers was made. In this study, the reuse of surplus water and the return flow were examined economically and it was shown that it had a significant impact on the added value of agriculture. This finding, however, has not been evaluated by the use of hydrological-like models at the time of analysis, and its important role in crop production and increasing farmer's income has been overlooked. In the simulation, it was shown that farmers' extensive use of reuse of surplus water or returned water reduces the effects of water scarcity. The findings show that the use of excess or return water should not be overlooked when changing policies or adapting to climate change related to irrigation water management.

Access to irrigation water is limited in arid and semi-arid regions, and the agricultural sector is constantly facing problems in providing the water it needs. Therefore, negotiations between water suppliers and farmers are very important. The agent-based model is used to evaluate and predict the possible outcome of negotiations between the two groups. This model can provide insights into how to optimize water allocation for water managers. Belaqziz et al. (2016) proposed the agent-based model based on negotiations to use the flood distribution system at the beginning of the planting season. The flood distribution method has limitations such as environmental and agricultural conditions, the actual need of each product for water, possible limitations of the irrigation system, etc., which makes it necessary to prepare a predictive model of factor behavior in order to optimize water distribution. The proposed model simulates revenue improvement and water resource optimization to assess farmers' decisions about whether or not to choose the flood method.

Changing farmers' behavior in order to change landuse practices is problem that managers need to consider in order to determine different landuse policies for the benefit of the environment. It can be said that one of the reasons for the failure of many environmental policies is the lack of knowledge about how farmers behave towards the policies adopted by managers. Identifying classes of landowners, land use patterns, and assessing how to respond to policies is very helpful for decision makers because they can evaluted at different options. Sethuram et al. (2008) presented the agent-based model for watershed planning in order to produce ecosystem services and agricultural products according to price and policy scenarios based on three agents with different goals. This model helps policy and decision makers to identify different factors in the watershed area and evaluate different political options based on the information available and the interaction between the agents and their environment. This study showed that the agent-based model can predict a 75% map of land use in real conditions.

Agricultural economists see decentralization, innovation, and market change as three important factors of development. Therefore, the development of economic analysis to market-oriented innovations, structural evolution and the necessary changes in the use of resources are important to change the approach. Berger (2001) provided a spatial multi-agent approach to modeling heterogeneous economic behavior and political responses from the perspective of the farm and households in Chile. Traditional farmers were studied here. The proposed model shows a higher income by considering innovation and increasing the intensity of work on the farm. As a result, decision makers can hope for a change in the structure and a desire of farmers to change the structure. Also, unwillingness to make trade agreements will not improve their income situation. In this study, it was shown that GIS-based integrated multi-agent models are a powerful tool for policy analysis and natural resource management.

One of the problems faced by water managers is the consumption of urban water. There are several things to consider when managing the city. Darbandsari et al.

(2017) introduced the agent-based model for urban water management to simulate the behavioral characteristics of water consumers and social communication. climatic conditions, water price, and advertising policies are use to build the framework for the agent-based model. The results of the model showen that water prices and the type of advertising are strategies that affect water consumption.

As another example, concerning domestic water resources management, Schwarz and Ernst (2009) studied effective scenarios in three water saving methods (for example showerheads, toilet flushes and rain harvesting) in southern German households used ABM. They said that these technologies are widely used even without advertising and promotion among people. They created scenarios to explore lifestyle changes and the use of inventions and innovations in this area.

Complex systems can be divided into several types of management and be evaluated by type. The first ones are aquatic systems with constant equilibrium conditions such that behaviors and interests of stakeholders are managed ignoring any unexpected dynamics. Riegels (2007) designed a framework for integrated model between hydrological, economic, and ecological phenomena with a constant optimization approach. This framework used water pricing policy to achieve ecological and groundwater sustainability goals. The model simulated the behavior of farmers to know what the response to the change in water price was, and the model was used to determine the effect of pricing strategies based on system conditions for one-time interval. The second type are aquatic systems that are managed based on an understanding of the dynamics of multiple systems and populations. For example, water resource management naturally affects ecological systems, and resources and populations may be at risk when interventions and resource and population reductions affect the performance of these strategies. In this case, dynamic simulation is used for aquatic systems with stable equilibrium conditions to understand that water quality cannot be restored or water resources cannot recover. As mentioned, multiple equilibrium states are very important. The transition for some systems may be fast or it may be sensitive to new input data and positive feedback loops. Systems may have difficulty keeping stable equilibrium conditions, and can transition to an alternative equilibrium as the environment changes; this situaion is called a critical transition point or threshold. For water resource systems, critical thresholds are the states of complex systems. In response, agents interact with each other to collect and exchange information or resources and make individual and independent decisions. These agents can change their behaviors over time and adapt to the changing conditions and, based on their last experiences, make new profit-making decisions. These complex adaptive systems are display emergence and form how individual behaviors lead to behavior.

A series of studies address water supply and demand management, where water demand is influenced by interactions, social factors, and supply. Frameworks have been developed to link ABM with water system simulation models to investigate decentralized decisions and interactions in water supply dynamics. For example, Zellner (2007), Reeves and Zellner (2010) used ABM to simulate changes in water consumption and land use in the region and study the impact of zoning on groundwater storage by linking the groundwater model with ABM.

Communities and decision makers interact with water infrastructure systems that are central to the design and management of water resources. In some communities, families or neighborhoods interact with the choice of centralized and decentralized service providers, and their decisions about adopting new and alternative technologies can affect infrastructure performance as well as water consumption. Studies can provide a feedback loop between technology expansion, consumer behaviors, infrastructure performance, and management coordination strategies. Here are some examples of these strategies. Klots and Hiessl (2005), for example, created an ABM for households that use highly water-efficient technologies. When households entered less sewage to the network, operators considered lower costs to improve the efficiency of the sewage system (system maintenance costs), which creates a feedback loop to increase adoption of efficient water technologies and thus reduce flow into the sewage network. Here, the ABM does not connect to a hydraulic model of the sewage network but reports the volume of sewage. However, by connecting to a hydraulic model of the sewage network, not only the volume but also the velocity can be predicted. The research results emphasize the interactions between consumers and infrastructure in different parts of urban water systems. An ABM was created to study the adoption of green infrastructure to control flood and runoff (Montalto et al. 2013). Households were selected as agents and two rules compared. The first rule was based solely on economic interests, and the second rule applied a model of trust, confidence and cooperation based on theoretical and interview data. The model results predicted the adoption and time needed to achieve community green infrastructure goals.

The consumer's reaction changes quickly with changes in water distribution systems regarding the quality of water supplied. An ABM framework was designed to simulate this reaction. The agents were introduced as individual consumers, which defines water use and/or water reduction rates based on their knowledge. The ABM is linked with a hydraulic simulation model to evaluate the social and hydraulic health consequences. The results showed that the release of a polluting column could dramatically change the course of some events, thus endangering an unknown number of consumers.

Managers and engineers, as individual actors, interact dynamically because of design to increase the capacity of infrastructure systems. The ABM framework was created to examine the infrastructure development based on demand and supply plans. The results showed that the cost efficiency of an adaptive planning approach is more preferable to a long-term forecast-based approach of water demand. The model was used to represent households, politicians and financial managers as actors to simulate the dynamics of infrastructure expansion. The model results compared with historical data on pipe life, infrastructure investment, and water prices. The above examples were cases of reactive agents.

ABM is used to simulate water consumers, including farmers, environmental systems, urban and industrial uses, etc., who share water resources (shared pool resources). These models are harvested at each stage of time to select the volume of water harvested and use the maximum profit. That is, factors affect other factors, and the maximum benefit for one factor may be influenced by another factor (e.g., water scarcity or scarcity). Another application of ABM in water resources management is the simulation of a group of farmers who simulate a decision on water consumption and the type of crop to be planted according to the common water basin. Ng et al. (2011) simulated total profits based on climate forecasting and optimal crop performance using a dynamic planning approach, where ABM is associated with a hydrological model that simulates the results of a farmer's historical behavior. he does. This showed that real farmers decide to be cautious in dangerous situations. Other issues include dynamism and interaction in the water trade market, which considers water licensing as a factor. Here, representatives can be seen as companies that use trade, negotiation, and law enforcement to trade with other companies. ABM relates to a water quality model for evaluating authorized river trading systems and uses the results to demonstrate the efficiency of the regional water trade market.

8.3.2 Agent-Based Modeling in Water Resources Management and Planning (Why, When)

Sustainable development needs to consider when to use agent-based modeling. In short, the definition of sustainable development is "to meet the needs of the resources available now without neglecting the needs of the future". In sustainable development, the political system must support the decisions made, an economic system must supply new resources with attention to sustainable development, the social system must respond to underdevelopment, the consumer system must protect the ecology by relying on regulations and rules, and a management system must be capable of automatic correction. Sustainable development is a set of actions, principles, and ways of thinking about human activities that explore different global mechanisms using different approaches and current knowledge. Sustainable access to safe water, for example, is one of the essential foundations of the next generation's well-being.

The world's water cycle and the factors affecting surface and ground water create the natural capacity of water resources. In recent years, the concept of sustainable development has entered into water resources management. Water resources management attempts to find ways to supply and use water resources, but some of the actions have caused significant damage to water resources. Therefore, it seems important to develop criteria and indicators for water resources management. That is to say, due to the limited resources of water and the need for future generations based on a sustainable development perspective to better manage these resources, social issues, economic values and stakeholder relationships must be taken into account. The importance of this issue is that the extent to which these relationships are not considered makes suitable management impossible. Therefore, models should be used to consider these issues, as they are able to reveal cause and effect relationships. Agent-based Modeling (ABM), by presenting a cause and effect paradigm, can provide accurate insights into resource management, thereby bringing managers, decision makers and planners closer to solving problems. In addition, agent-based modeling, by taking into account socio-psychological approaches, can provide a good insight into stakeholder collaboration with management strategies.

The question that arises here is; when it is useful to use agent-based modeling. In short, the following points can be expressed:

- When interactions between agents are complex, nonlinear, discontinuous, or discrete. This means that the agent's behavior can be significantly changed by factors, such as neighbors interacting with each other.
- The location of the agents is not constant and the space between the agents is very important, such as in traffic, firefighting, shopping at the supermarket, etc.
- The population is heterogeneous and each individual agent is different. Each may act differently in evacuating an area in times of crisis and natural disasters, in the stock markets, regarding online stores, etc.
- Agents exhibit complex behavior, such as learning and adaptation
- Interaction situations are heterogeneous and complex. When interactions are homogenous (all individuals behave in the same way), there is no need to use agent based modeling, but this is rarely the case in social networks. One of the characteristics of social networks is that there is strong deviation of individuals from the norm.

"For which systems is it useful to create an ABM?" is another question that arises, which asks what are the advantages and disadvantages of ABM. Hare and Deadman (2004) have presented some requirements. Systems that should be modeled with ABM have the following characteristics:

- Their focus is not on stable equilibrium but on the phenomena and behaviors that lead to it. Therefore, unstable dynamics must be analyzed.
- They would require many necessary assumptions for equilibrium, such as homogeneity of space, decision uniformity, rationality and so on, which in fact cannot be fulfilled.
- They include intelligent human behaviors such as techno-social systems. Here the flexible tasks of the group or team must be modeled.
- They include educational or evolutionary processes at the individual and demographic levels.
- They need to need to include heterogeneity of the states and the rules of conduct.
- They are multilevel systems that need to observed at multiple levels. This is a situation where there can be emerging phenomena at the higher levels.
- Decisions are made at different levels of society. Low-level decision-making relates to the behavior of individuals, while supervisory bodies or authorities carry out high-level decision-making. Feedback loops occur between the different demographic levels.
- The systems derive their dynamics from spatial and flexible interactions. The size, structures, and interactions of variable people can be difficult to describe with other simulation models.

8.3.3 *Software Used in ABM*

This section introduces a number of tools used in agent-based modeling. The focus here is on the overall goals of the software. JASSS1 also publishes occasional reviews of current and new tools.

8.3.3.1 REPAST

This software based on the Java program. It provides grammar packages for the most common tasks related to implementing agent-based modeling. Repast was originally developed for use in the social sciences. Therefore, it is very useful for network analysis. Now with the introduction of Repast Symphony this software has become a visual software with the ability to draw charts.

8.3.3.2 Swarm

It is one of the primary tools for implementing agent-based modeling and complex systems. In fact, like Repast, it creates the central core of the model and tells the modeler where to build the model, as well as collecting, analyzing data and displaying and controlling the model. Programming is done with Java. Here, people with low programming skills have to spend some time learning coding.

8.3.3.3 MASON

The commands are based on the Java program, which facilitates large-scale simulation programming. The great feature of this software is in its high performance, but it requires programming skills. The software supports multi-partitioning and supports not only two dimensions but also 3D images from the simulator core.

8.3.3.4 Netlogo

It is designed with consideration for the end user, and has three basic aspects. The first aspect is editing of the model, which is similar to the Starlogo programming language. The second aspect is the visualization of the environment and its parameters, which allows the user to change the parameters in the model with the help of controllers. The third aspect is that the software consists of commands. Netlogo is increasingly popular due to its extensive instructions, good training, and a large library of pre-existing models.

8.3.3.5 SeSAM

This software provides a completely intuitive space for creating ABM. Unlike Swarm, which requires users to know the programming language, the software does not require knowledge of the programming language. Codes (even for displaying data, mapping and any analysis) are assembled using graphical representation. The software kernel is a simulator of its own system model, built using active shapes and tables. Initial predefined commands can be redefined for actions, perceptions, and process functions.

8.3.3.6 Anylogic

Anylogic Simulation Software, a product of XJ Company, is the first and only dynamic simulation software to enable users to simultaneously model system dynamics, discrete simulation, and Agent-based modeling and be able to simultaneously develop their model in the same software environment using the above methods. Programming in this software is so flexible that it can simulate the most complex business, economic, and social systems processes, taking into account details at every level of interest. Using pre-designed components, many systems such as manufacturing and logistics, business processes, human resources, and customer behavior can be simulated in a way that is most consistent with the behavior of real systems. This software has been chosen by thousands of users worldwide, covering all business units, government, universities, and is currently used as standard simulation software in many companies around the world.

Many other tools are used for agent based modeling, most of which are made for specific purposes and are therefore not widely used. These include CORMAS and MadKit.

These software products can differentiate in terms of programming language, fields and aspects of application, predefined programs, number and type of agents, and the main factors involved in the simulation, including the agents, the environment and the interactions between the agents are compared.

The primary programming language used is Java, except for Netlogo and SeSAm. For all the factors involved in the simulation (agents, environment, and inter-agent interaction), in all programs, the modeling is first performed and then evaluated from the bottom-up using the ABM. In Swarm, for example, the influencing factors in the pre-environment model are identified, while in SeSAm, the focus is on the division of environments and agents (defining characteristics, goals, activities, etc.). All programs use predefined instructions for ABMs but apply ABM, depending on the program, in different domains. The graphical dimension helps the user build the model. When the graphic dimension does not exist, most of the modeling load falls on programming, which limits non-expert use. Such a graphical dimension is very evident in the SeSAm model, but in NetLogo, this graphical dimension is used as the editor where the code should be written. The agent's behavior demonstrates all their reactions, but does not directly support graphic design and presentation. Although the modeling is done with specific tools, not general frameworks, it is very desirable for social simulation of anthropological systems.

8.3.4 The Challenges of Using ABM

It is important to understand that the degree of design freedom can pose great challenges for model users (especially those with less experience). The main challenge is knowing what elements should include in the model or which modules can be removed, added, or partially incorporated into the model. The main challenge is knowing what elements should be included in the model or which modules can be removed or added, or partially incorporated into the model. A model with few details has fewer assumptions for adaptability. There are also fewer calibration parameters in such models, which makes it easier to run the model. However, although these models are very simple, the degree of reliability of the results is relatively low. Some researchers focus specifically on agent interactions and begin modeling from the detailed analysis of interactions between agents.

Other challenges of ABM include calibration, Verification, and validation. Difficulties in calibrating and adjusting the model make the results of the model fragile. The change in the value of a parameter affects all the factors and thus results in the change in the overall results. These parameters described by Polhill and Izquierdo (2006) as "knife edge parameters". This is especially true when ABM is associated with an ecological model. Ecological processes such as water resources systems, surface and ground water, precipitation, runoff, etc. must calibrated. In the calibration, the relevant variables and coefficients must be modified so that the results match the variables measured in the real environment (observations). Due to the complexity of the systems for which ABM is performed, the number of processes and parameters is higher than usual, which increases the complexity and workload, taking into account the relationship between the two social and ecological sectors. After calibration, one must first verify the concept of model adaptation with the conceptual model and then validate the concept of ABM with the actual system (Feuillette et al. 2003). Most of the work has been done solely on the validation section. Validation can be done in both qualitative and quantitative ways. The quantitative method is to compare the model results with the observations or documentation available in the study area, to perform sensitivity analysis, and to model behavior with maximal data. The qualitative method of validation, due to the lack of quantitative data, is through interviews with local users, specialists and experts. However, it should be noted that any model that is tested in a variety of ways is more reliable. Nevertheless, the nature of ABM limits this. On the other hand, it should be noted that the nature of the models is that in order to achieve the main objective it is necessary to evaluate it based on details of the whole phenomenon. so it is very difficult due to the complexity of the social behavior of the subject. This same complexity ultimately leads to serious scientific and practical problems in calibration, which makes it difficult to interpret the results.

Another challenge is the link between the social and ecological sectors. Although this section is one of the strengths of water management modeling and planning, but these strengths is faced with challenges when updating the model. In the issue of groundwater utilization, for example, assuming no impact of the water losses on the water level in the adjacent wells (Mulligan et al. 2014), the groundwater model can be implemented and updated with the social sector at the same time. However, on a much shorter time step, the impact of the water drop on the adjacent well should be considered (Madani and Dinar 2012). As a result, it is necessary to calculate, at least at the beginning of each social step, the physical parameters such as rainfall, river flow and biophysical parameters such as water balance, groundwater level, volume of water in dam reservoir, plant growth and so on. Then this information should be entered to social section. Therefore, depending on the precision, the required time steps could be annual, planting season, or even on 10-days intervals should be used for implementation and updating.

8.4 The Framework for Creating the Agent-Based Model

Three important components must have taken into account when constructing an agent-based model.

- The first component comprises individuals known as the agents. However, what is the Agent? It has characteristics such as self-sufficiency, autonomy, social dynamics (change of state and condition over time), can modify through time and is goal-oriented. In fact, the agents, based on their rules and behaviors, interact very strongly with other agents as well as with the environment. Agents have two very important aspects as characteristics and behaviors or actions.
- The second component of the agent-based model is the interaction between the agents and environment. Interaction refers to the behavioral rules of agents and their environments. These interactions are not necessarily visible, but they must be taken into account when running the model.
- The third component of the agent-based model (ABM) is the environment that is simulated. This environment should include all the elements involved. In this environment, interactions occur between agents introduced into a communication network known as the social network.

Many water management issues have the characteristics of complex systems. Many studies have conducted on the discovery and use of ABM in water resources management and planning. ABMs vary in scope, case study in the field of water, management questions, and extent of social and physical interaction. Modelers first ask themselves questions, then formulate hypotheses, select variables and parameters, build a model, and finally analyze the model. They can also bring up new questions and hypotheses and analyze them.

Because of the structure of the ABM method, it is difficult for researchers to relate this method to analytical models because they do not have a specific structure, relationship, or formula, and are qualitative-descriptive and have rules governing them. For this reason, a standard protocol was proposed by Grimm et al (2006) to organize the writing of research ABM. This protocol is called Overview, Design and Details (ODD), which tested by 28 modelers. The protocol consists of three main blocks under the themes of overview, design and details. Overview includes general information about the purpose, structure, environmental and individual processes of the model. The Design section describes a set of general concepts associated with designing the principles of complex systems. The model results represent the output and the observations the modeler describes how the output is evaluated. In addition, agent behaviors are described, including adaptation, learning and interaction, and the types of decisions that agents use, based on predictions, feelings, and functions. The compilation should include a description of the cumulative level in the model with respect to the source and the stochastic effect. In the Details section more technical implementation details including initial values, inputs, and subsets explains that are include in the Overview section.

Based on the description given, these three general sections can be subdivided into seven more minor elements, namely: Purpose, State variables and Scale, Process Overview and Scheduling, Design concepts, Initialization, Input, and Sub-models. The following are briefly discussed below:

1. Purpose: The purpose of any investigation should be clear. Without a purpose, research cannot follow properly. Thus, by defining the purpose of the research correctly, it becomes clear why some aspects are considered and others are not.
2. State variables and Scale: This section describes the structure of the model system. The institutions involved in the model, the hierarchical levels that exist, the temporal and spatial accuracy of the system, and so forth. State variables are variables that specify entities such as individuals and habitats. For individuals, characteristics such as age, gender, habitat, etc. and habitat are defined as location, soil type, risk, etc. At this stage, information collected from entities known as covariates, and finally, time step, time horizon, and size of the spatial cells are determined.
3. Process Oreview and Scheduling: To better understanding the ABMs, it is important to know what processes built into the model. Examples are processes such as food production, nutrition, growth, turbulent events, and management. This section describes the design and planning of the model process. Here, the time step, the processes that are performed concurrently, the random variables, and so on can also explained.
4. Design concepts: Here is a general framework for designing and building relationships within the model. Which system phenomena in particular emerge from individual characteristics and which ones are imposed? What adaptation features respond directly or indirectly to environmental changes or change on their own? What types of functions are used? How the consequences predicted? What kind of interactions existence between agents? Is there stochasticity the model?
5. Initialization: This section includes questions such as: How are environments and individuals or groups created at the start? How the initialization of the state

variable done? Is the initial value constant or changing over time? Is the initial value arbitrarily selected or based on available data?

6. Input: The dynamics of many models are created by the conditions that change with time and place. Precipitation is an example that varies with time and place. All environmental conditions are considering as inputs. Model output is model response to input. In fact, the dynamics of each model is due to the dynamics of the state variables.
7. Sub-models: In this element, the details of the model components are defined. In order to better explain the details of the models below, it is advisable to operate in two ways. First is the mathematical skeleton of the model, which means the equations or rules governing the model. The explanations of the equations and the rules in this section should be minimal and the parameters should have the necessary explanations. The second is a complete description of the model. That is, the equations and rules expressed in the mathematical framework are clearly explained and answered the questions such as "What are the equations and rules hypotheses?", "How are parameter values determined?", "How is the calibrated model?"

With attention to the ODD framework and the use of this framework in created the ABM, the challenges described above still arise.

According to the review of resources, there are generally seven steps in building ABM:

1. Define the purpose of a model and what kind of output it will expected.
2. Identify the agents or institutions that are modeled.
3. Identify the relationship between agents and the agents with the environment.
4. Identify the environment (such as the structure of the social network) in which agents exist and interact with each other.
5. Determine the variables needed to indicate the behavior of each agent, including the objectives of each agent and its relative priorities.
6. Determine and construct the structure of agent's logic.
7. Verify the agent behavioral model.

Two examples described to better understanding the ABM.

8.5 Practical Examples

Examples grouped into active and reactive agent groups as representatives of human actors and human organizations, and these described in the discovery of research questions. Reacting agents considered to simulate the dynamics of a large population of simple actors. Studies examine the impact of reactive behavior and its feedback on water resources and infrastructure.

Examples that can exemplified for this type of agent are: The urban water demand system, which is in the small dimension of updating agent demand for water and in its large dimension, the level of community demand. Urban water demand system and its supply source are the small dimension of updating agent demand for water and in its large dimension the level of community demand and storage behind dams and aquifers. A sanitary sewer pipe network in small dimension updated the volume of sewage inputs into the network and in large dimension is sewer network' flow. The problem is low-impact development to cope with the flood, the small dimension of the acceptance of green infrastructure by agents and the large dimension of installing these infrastructures at the community level. The issue is the pollution of the water distribution system, which in the small dimension updated agent demand for water and in large dimension is the number of potential hydraulic network users. Finally, the urban water demand system, which is the small dimension of updating the agents' decisions to expand the infrastructure and, in its larger dimension, the cost of infrastructure development.

Active agents used to simulate system performance when actors strive to achieve goals. One of the cases that can mentioned is the river basin, the small dimension of which is updating the agent's decisions regarding water harvesting and landuse and in large dimension the satisfaction of the water users. The quality of the river water, the small dimension of which is water-trading permits and the larger dimension is the qualitative effects of water.

8.5.1 Reactive Agents and Urban Water Supply

Here, the ABM described and created to simulate urban water supply and demand. The proposed model used to understand the effects of individual interactions and behaviors on the dynamics of hydrological flows and water scarcity reduction. Imagine a city that supplies the volume of water it needs from a dam. Urban water demand comprises both residential and non-residential sectors (industrial and commercial). The dam's task is also to collect existence water in the basin and preserve it for the downstream ecosystem and flood control. As the population increases and the demand for water, the volume of the dam decreases. To tackle water shortage, the utility director approves the "Drought Stage Plan" to limit outdoor water use. Urban consumers are also by awareness of water shortage problem try to use water-efficient appliances that reduce water demand and thus they can confront to water shortages.

ABM created to simulate the interaction between the urban consumer, the program manager and the water resources in the urban water supply system. Consumers and program managers seen as reactive agents that interact with each other and the water system includes a dam and catchment. The hypothetical community is also a city in southeastern America. The urban household population assumed at 100,000 that increases with growth rate and simulated the supply dynamics of a 50-year horizon.

8.5.1.1 ABM Framework

Here, the ODD (Overview, Design and Details) protocol used as the formal method in building ABMs to describe the case study.

1. Purpose: The purpose of the model is to evaluate the impacts of population growth and their adaptation. This is the stage of identifying drought event and accepting the use of water-efficient appliances on the volume of water storage in the dam reservoir and the ability to supply water to meet demand.
2. State variables and Scale: Households are the reactive agents. Each agent represents a household and determined by the number of inhabitants, the monthly household demand, and non-residential demand volume. Households represent a community that consumes water storage. To simulate population growth, the number of household agents increases each year. The water manager is the reactive agent who implements the drought process based on the reservoir water storage. Household and water consumption agents linked to basin simulation models and reservoir systems. The state variables representing the changes in the characteristics of agent's systems and water system summarized in Table 8.1. The 50-year planning horizon simulated on a monthly basis. The model parameters based on water consumption and supply data for Raleigh City, North Carolina (Mashhad Ali 2014) adjusted and explained in a simple formula (Fig. 8.1).

Table 8.1 State variables for the ABM framework of urban water supply

State variable	Definition	Potential values
Household agents		
$D_{\text{indoor,init},t}$	Demand indoor uses at time ***t***	$\{D_{\text{indoor,init},t}, 0.5 \times D_{\text{indoor,init},t}\}$
$D_{\text{oudoor,init},t}$	Demand for outdoor uses at time ***1***	$\{D_{\text{indoor,init},t}, 0.67 \times D_{\text{outdoor,init},t} \times D_{\text{outdoor,init},t}, 0.0\}$
contacted	Agent has been contacted by three consumer agents	{true, false}
adopter	Agent has adopted water-efficient appliances	{true, false}
P_{adoption}	Probability of adoption	{0.0, 0.25, 0.50)
Utility manager agent		
DS_t	Drought stage at each time step	{0, 1, 2, 3}
Water system		
R_t	Runoff at time ***l***	Eq. (8.7)
L_t	Release from reservoir at time ***t***	Eq. (8.8)
SP_t	Spill from reservoir at time ***I***	Eq. (8.9)
RD_t	Residential demands at time ***1***	Sum of consumer-agent demands
V_t	Volume of water in reservoir at time ***t***	Eq. (8.10)

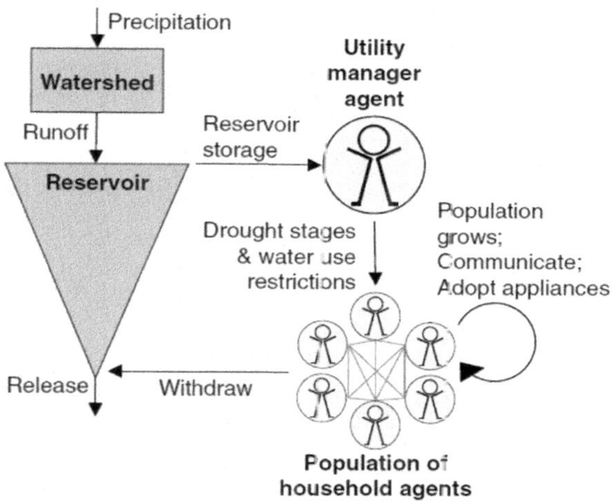

Fig. 8.1 ABM Framework for Simulating Urban Water Supply, the arcs represent the flow of information between different parts of the framework

3. Process Oreview and Scheduling: The order of application of ABM given for each period in below. In the first step, the water utility agent for supplying drought stages considers the water storage in the reservoir. The water utility agent adapts with the limitation of water consumption in response to reservoir water storage. Reservoirs are for purposes such as sediment trapping, downstream release, water supply and flood management. At each time step, the Drought Stage Operation Management (DS_t) factor is determined at three levels 1, 2 and 3. Drought Stage 1 when the volume is 109.8 billion liters (29,000 million gallons or 60% of the reservoir reserves remain). Second-stage Drought when the volume is 102.2 billion liters (27,000 million gallons or 47% of the reservoir reserves remain) and Drought Stage 3 when the volume is 94.6 billion liters (25,000 million gallons or 33% of the reservoir reserves remain).

The second step is consumer agents that regulate non-household demand based on the drought stage. The Response of the household agent to the drought stage under a set of rules is stated below:

$$\text{IF } DS_t = 1 \text{ THEN } D_{outdoor.t} \leftarrow 0.67 \times D_{outdoor.init.t} \quad (8.1)$$

$$\text{IF } DS_t = 2\text{THEN } D_{outdoor.t} \leftarrow 0.33 \times D_{outdoor.init.t} \quad (8.2)$$

$$\text{IF } DS_t = 3\text{THEN } D_{outdoor.t} \leftarrow 0 \quad (8.3)$$

where $D_{outdoor}$ consumer demand for non-residential purposes at time t, $D_{outdoor,\,init,\,t}$ every consumer's primary demand for non- residential purposes (For each month or time step assigned t-shown in Table 8.2). If a drought stage is lifted, consumers will use water source volumes (Doutdoor, init, t) for non-residential purposes.

Table 8.2 Monthly input data

Month	Rainfall [cm (in.)]	Indoor demand $D_{\text{indoor,init},t}$ [L/household (gallons/household)]	Outdoor demand$D_{\text{outdoor,init},t}$ [L/household (gallons/household)	Watershed coefficient, Cj [millionL/cm (million gallons/in.)j	Watershed coefficient, S_t [cm (in.)]
January	7.6 (3.0)	23,939 (6,324)	0	11,326 (7,600)	5.6 (2.2)
February	7.6 (3.0)	23,939 (6,324)	0	18,145 (12,175)	5.8 (2.3)
March	8.9 (3.5)	23,939 (6,324)	0	18,964(12,725)	7.1 (2.8)
April	7.6 (3.0)	23,939 (6,324)	0	16,170(10,850)	6.6 (2.6)
May	7.6 (3.0)	23,939 (6,324)	5,758 (1,521)	15,797 (10,600)	16.0 (6.3)
June	7.0 (2.75)	23,939 (6,324)	5,758 (1,521)	8,681 (5,825)	17.5 (6.9)
July	10.2 (4.0)	23,939 (6,324)	5,758 (1,521)	7,750 (5,200)	19.1 (7.5)
August	10.2 (4.0)	23,939 (6,324)	5,758 (1,521)	7,973 (5,350)	18.3 (7.2)
September	10.8 (4.25)	23,939 (6,324)	0	18,927 (12,700)	30.0(11.8)
October	7.6 (3.0)	23,939 (6,324)	0	8,495 (5,700)	14.2 (5.6)
November	5.9 (2.33)	23,939 (6,324)	0	12,817 (8,600)	1 1.7 (4.6)
December	5.1 (2.0)	23,939 (6,324)	0	11,624(7,800)	6.4 (2.5)

In the third stage, water-use agents adopt the use of water-efficient appliances based on the drought stages and in relation to other agents. Households are also responding to droughts by using water-efficient appliances. The $P_{adaption}$ parameter is the probability of adaption for each agent. Its value is set at zero at each time step and increases based on the stages of drought and related factors. Each factor that receives messages from three or more households in the previous steps will act in accordance with the message received:

$$\text{IF } DS_t \geq 2\text{THEN } P_{adaption} \leftarrow P_{adaption} + 0.25 \tag{8.4}$$

$$\text{IF contacted} = \text{true THEN } P_{adaption} \leftarrow P_{adaption} + 0.25 \tag{8.5}$$

At each time step, each agent selects his or her household appliances according to family members. If the agent chooses a device, he designated as the daptor and acts to reduce according to Formula 6 for domestic demand:

$$\text{IF adaptor} = \text{true THEN } D_{indoor.t} \leftarrow 0.5 \times D_{indoor.init.t} \tag{8.6}$$

where $D_{indoor,\,t}$ demand exerted to each consumer for household goals in step t, and $D_{indoor,\,init,\,t}$ the initial demand of each consumer with household goals per month or step t.

Any agent that considered as adapter remains for the adapter simulation period.

In step 4, consumer agents communicate with other consumers about adopting water-efficient appliances. Each agent randomly sends messages to the other two agents about these appliances.

Step 5, the basin runoff is calculated. For calculate runoff from rainfall and monthly reservoir storage used mathematical models. Basin runoff calculated by the following formula:

$$R_t = C_t \frac{(P - 0.2 \times S_t)^2}{P_t + 0.8 \times S_t} \tag{8.7}$$

where R_t is runoff (million liters/month or million gallons/month), P_t is monthly precipitation (centimeter or inch), S_t (centimeter or inch), And C_t is the monthly runoff coefficient. Formula 7 based on the command curve method and contains the C_t variable to calculate the volume of runoff produced in the basin. The runoff coefficient is change monthly to better represent the seasonal variations in the rainfall-runoff relationship.

Stage 6, Reservoir volume calculated based on demand, runoff, spill and release. The equilibrium volume for the tank (formulas 8 to 10) used to calculate V_t at each step. Release is the amount of water that flows out of the dam each month for downstream demand. Formula 8 shows the relationship between storage and water outflow:

$$L_t = 6 \times 10^{-10} \times V_{t-1}^{2.8478} \quad (8.8)$$

where V_t tank volume at time step t, Sp_t, the amount of water removed from the tank to ensure that the tank volume does not increase.

The spill calculated as the volume of water that the reservoir does not overtop:

$$S_{P_t} = \max(V_{t-1} + R_t - NRD_t - RD_t - L_t - MV.0) \quad (8.9)$$

where Maximum volume, MV = 284 billion L, RD_t is sum of household demands, NRD_t is sum of non-residential demands. The volume of the tank, V_t, calculated at each time step from the following formula:

$$V_t = \max(V_{t-1} + R_t - NRD_t - RD_t - SP_t - L_t.0) \quad (8.10)$$

where the monthly time step increases one by one and the steps are repeated from the beginning of the first step.

4. Design concepts: Here, the model is designed to implement and communicate with the principles of complex systems. Therefor, the following concepts should consider:

Emergance: The sum of the total system demand and the volume of water stored in the tank are system characteristics that determined by the interactions between agents and subsystems. In addition, when the system is in a drought stage, it is influenced by the interaction between household agents, the interaction of household agents with water managers and their decision on the amount and type of water consumption.

Interactions and Adaptations: Household agents associated with the water management agent by obtaining information from the drought stages. they interact with other agents by sending a message that increases the likelihood of accepting water-efficient appliances. Instead of using a social network such as the World Wide Web, to simulate connection, each agent can communicate with other agents in the population. A message shows that someone has bought water-efficient appliances. The agent receives the message and so the number of messages increases and as a result the message receiving agents increase. The water supply manager does not receive messages directly from household agents, but they monitor the water resources system to determine the amount of water stored in the dam. Household agents adjust their consumption based on the received signals, and the Water Management Agent also adjusts the drought stage schedule based on reservoir storage.

Sensing: Agents are simulated to sense for system parameters. The water manager directly measures the water level of the reservoir, and household agents receive information from other agents' messages, although not calculating environmental conditions.

Stochasticity: stochasticity present in agents' behavior because they adopt to water-efficient appliances. So, they interact and communicate about adaptation. Other mechanisms and behaviors simulate as deterministic processes.

Table 8.3 Modeling scenarios

Model	Scenario	Agent	Agent behavior
Urban water supply	Scenario 1	Household agent	Use water for indoor and outdoor purposes
	Scenario 2	Household agent	Reduce outdoor water use based on drought stages
	Scenario 3	Household agent	Reduce outdoor water use based on drought stages; adopt water –efficient appliances based on drought dages and communication with household agents
Water quality management	Scenario 1	Polluter agent	Select random polluter agent to trade
	Scenario 2	Polluter agent	Select polluter agent within netwoifc
	Scenario 3	Polluter agent	Select polluter agent within netwoik: update preferences for polluter agentsbased on successful traces

Observation: The model reports the total system's demand, the volume of water stored, and the level of the drought stage at each time step.

5. Initialization: The initial population of household agents is estimated at 100,000, representing 300,000 persons, with a 4% increase each year. The reservoir initially was 145.7 billion liters, the highest level during the drought. The volume of non-residential demand is estimated at 40% of household consumption, increasing by 4% annually. The initial household and non-residential demand shown in Table 8.2.
6. Input: The inputs required for the model are rainfall values and hydrological parameters as shown in Table 8.3.

8.5.1.2 Results and Discussion

The ABM framework used to construct the three scenarios listed in Table 8.3. In the first scenario, household agents do not adapt their demand values. In Scenario 2, household agents respond to drought by reducing outdoor consumption. In the third scenario, the household agents cooperate and respond to drought stage by adopting with water-efficient appliances and reduced outdoor water consumption. For each scenario, the population reaches 325,000 from 100,000 household agents in 50 years. In each scenario, demand increases based on population growth and the rules that the consumer agent must comply with. Non-residential demand does not affect by agent actions and remains among the scenarios. For Scenarios 1 and 2, there is no stochasticity sources and result reported for one step of running the simulation model. For Scenario 3, agents select water-efficient appliances based on individual probability adoption and probability communicate with random household agents, with results reported for 30 stages of random execution.

Due to the volume change, release and estimated demand, household agents seeking to match their demand affect the dynamics of the reservoir. Especially the least amount of storage and release or annual outflow from the tank can compare between scenarios. In the first scenario, due to constant rainfall and increasing demand due to population growth, the lowest annual volume decreases each year. The lowest annual volume of the reservoir has decreased steadily over the past 20 years of simulation. The lowest annual release volume from the reservoir observed in summer, reduced in a 50-year simulation. For Scenario 2, the adoption of the drought steps reduces impact on reservoir volume slightly, although the volume of the lowest annual in the last several simulation years approaches zero. The decrease in demand for Scenario 3 reduces the impact of population growth on reservoir storage, and the lowest annual reserve exceeding 44 billion liters in 50 years of simulation.

For scenarios 2 and 3, the drought stages are applied in the first year, and as a result of increasing population and demand, the result is an increase in the simulation period. The average monthly demand for the 50 years of simulation is 17.67, 17.17 and 10.42 ± 0.003 billion liters per month for Scenarios 1, 2 and 3 respectively (Scenario 3 results as plus or minus averages the standard deviation of 30 simulations is reported). Demand is significantly reduced by the behaviors identified in Scenario 3 compared to Scenario 2. As a result, the adoption of water-efficient appliances over the 50-year period has led to less drought stage for scenario 3 compared to scenario 2. The drought stage of scenario 3 in many time steps has been reduced to drought stage 1 and 2 and consequently stages 1 and 2 occur more frequently in scenario 3 than in scenario 2. Stage 3 shows a very serious threat to water supply, as in scenario 2 in month 156 and in scenario 3 in month 204.

This analysis used to determine the impacts of water-efficient appliances and water scarcity response programs and the need to identify new sources of water. The model results are not significantly sensitive to initial values such as reservoir levels. Instead, they are sensitive to the time of monthly inflows and withdrows. This model can be used to assess the long-term effects of climate change and land use changes to balance supply and demand.

It is also used to illustrate how simple ABM is used in urban water supply design. The model involves interactions in the urban water system that are expected to affect the stability of the water stored in the reservoir. The interactions include advertising and innovation when consumers accept new appliances and the use of water demand restrictions. Agents are homogeneous in their characteristics but heterogeneous in their behaviors because they choose water-efficient appliances based on their personal relationship with other agents. The ABM framework examines the impact of interactions on reservoir storage and complete supply of water demand affecting drought stages that restrict water use. It also simulates water conservation and reuse programs based on the timing and use of new technologies.

Other sources of uncertainty can have used to estimate the effects of uncertainty on results. Rainfall and heterogeneity of agents can increase the complexity of the

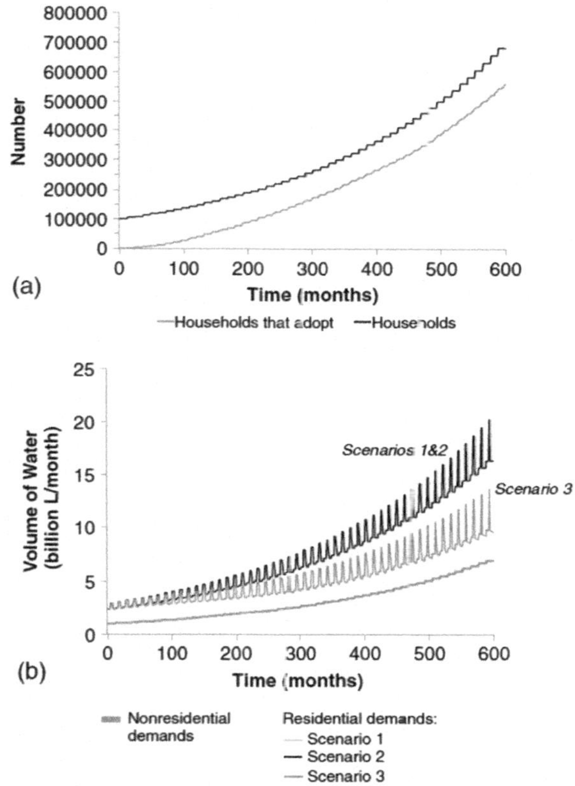

Fig. 8.2 **a** Growth in number of households and adoption of water efficient appliances for Scenario 3; **b** nonresidential and residential demands for Scenarios 1, 2, and 3 (nonresidential demands do not vary between scenarios); results are shown for one simulation of Scenarios 1 and 2 and average over 30 simulations of Scenario 3

results. The model performs a random multiple simulation to understand the wide range of demand, storage and drought processes expected to used to find alternative management strategies such as resource deployment and demand reduction programs. Realistic probabilistic representations, management strategies, and heterogeneity among consumers in urban water supply within an ABM framework may produce results that can guide water resources management (e.g. Giacomoni et al. 2013; Kanta and Zechman 2014) (Figs. 8.2, 8.3 and 8.4).

8.5.2 An Agent-Based Model to Manage Water Resources Conflicts

In this example, a new approach presented to simulate the process of encouraging those involved in conflict game to cooperate. This encouragement is applied through social institutions and organizations in the form of new incentives, penalties and regulations. For this reason, agent based model's framework for introducing the behaviors

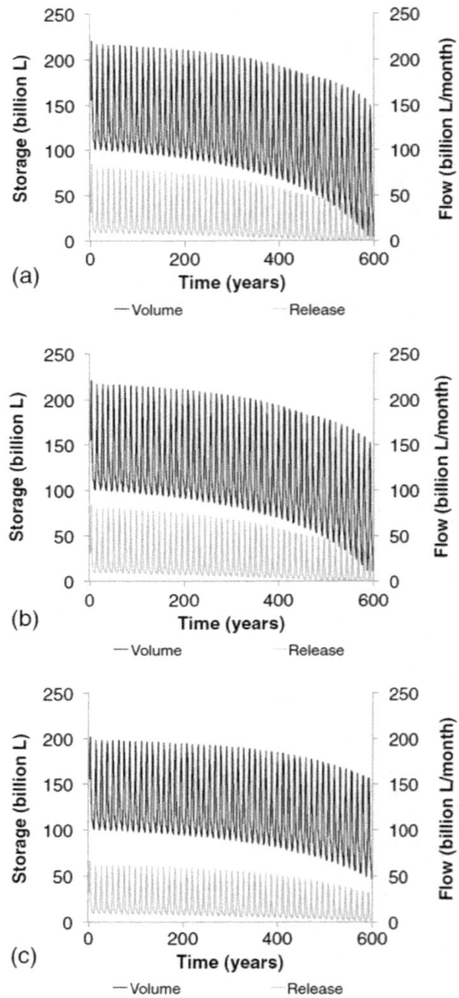

Fig. 8.3 Reservoir storage and release for **a** one simulation of Scenario 1; **b** one simulation of Scenario 2; **c** average over 30 simulations of Scenario 3

of different water users and their responses to the system's managers and stakeholders as well as their response to different simulation scenarios is introduced. The model simplifies the complexity of all conflicting views and interactions of competitors. The approach described is also a powerful tool in social network simulation that provides an opportunity to test new managerial scenarios and understand the consequences of decisions in a simple but reliable form without the need for the user to develop sophisticated new scenario formulas. On the other hand, it helps to assess the effectiveness of institutional and social policies to reduce conflict. The model should continuously associate with a watershed simulation model to monitor the impact of measures taken on the quality and quantity of water needs with a dynamic approach.

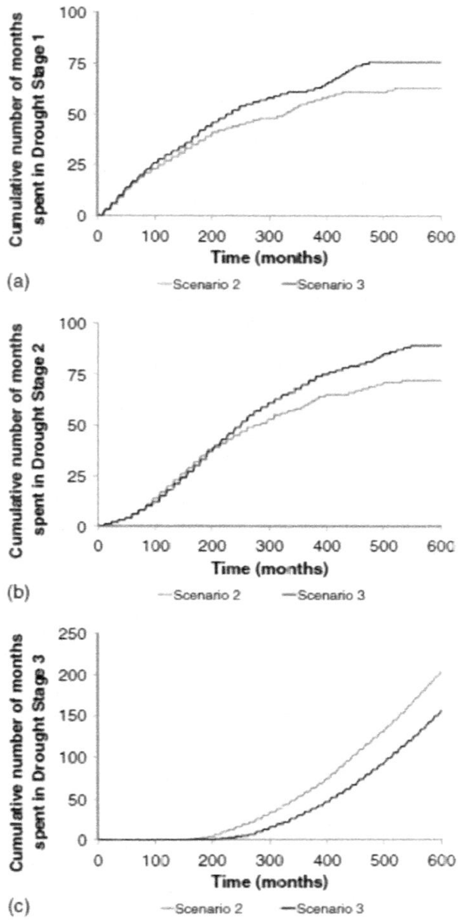

Fig. 8.4 Cumulative number of months spent in each drought stage(DS) for Scenarios 2 and 3: **a** DS 1; **b** DS 2; **c** DS 3; results shown are for one simulation of Scenario 2 and average over 30 simulations of Scenario 3

This model can then use to write rules tailored to time varying water needs and environmental concerns. This study was conducted for the San Joaquin Watershed in California but can apply to other basins.

The way water resources were managed, such as water transfer or non-transfer from the area for other users, and how to transport it for decades had created serious differences. The situation became more complex after new restrictions on water resource supply were imposed due to lay down new regulations to protect the region's ecosystem. New and innovative ideas were created to resolve the tensions, one of the most prominent being the California Delta Bay Program (CALFED) in 1995. However, this method has not achieved much success in practice. The Little Hoover Commission (2005), for example, stated that the program was "costly, non-transparent and undervalued" and did not perform well as expectedTherefore, basic changes to the policies put forward must be made to deal with tensions, resolve conflicts and protect the Delta from collapse. The nature of the conflict in the Delta is that the parties act individually and do not cooperate with each other. This behavioral strategy ultimately reduces successfully implement of the Pareto solution compare to when people cooperative together. At present, some parties have been willing to cooperate due to the delicate situation of the Delta and further environmental, social and political constraints. However, the delta's dire situation and the potential for its collapse imposes significant costs on the government and stakeholders. Therefore, the cooperation of the parties will be less harmful to them. The proposed ABM framework is designed with the aim of finding effective management scenarios to encourage tension parties to cooperative.

To create the ABM framework, the environment must take into account the entire Delta system and the areas that are hydraulically connected to the system. Agents are also water consumers, operators, stakeholders, and water-use sectors. Due to computational limitations, the system was simplified and this study was performed only in a part of the region (San Jose Watershed). Due to the importance of agricultural activities in the area and high water salinity in the river, stresses regarding water quality management are high. To management these conditions of simulation, the San Jose Basin is considered as the environment for ABM. In simple approach, the agents include a decision maker, the government and the demand agent (farmers and environmental sectors that need sufficient water flow in the river of acceptable quality). The agent of government as the active agent and the other two agents are reactive and interact with each other indirectly. The water demanding agent may collaborate to gain more benefits or not care about the environment and avoid non-cooperation. So this agent has two cooperative and non-cooperative modes. On the other hand, the governing units can be considered as decision makers (Fig. 8.5). The agent's behavior depends on his or her understanding of the system and problems that environment and other agents affected to them. Figure 8.6 shows the overall impact of the environment and agents on each other.

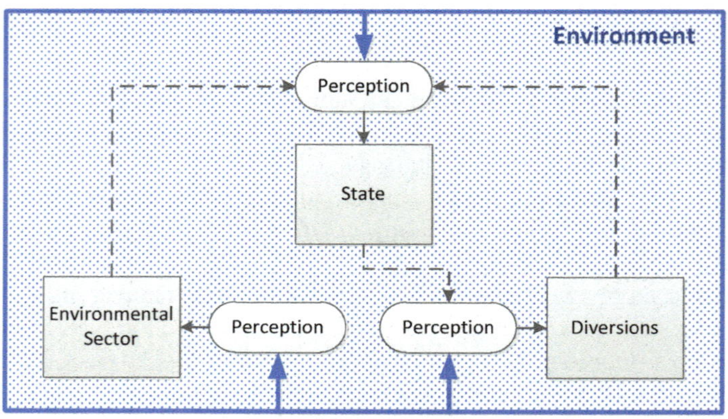

Fig. 8.5 The influence of the environment and other agents on each agent

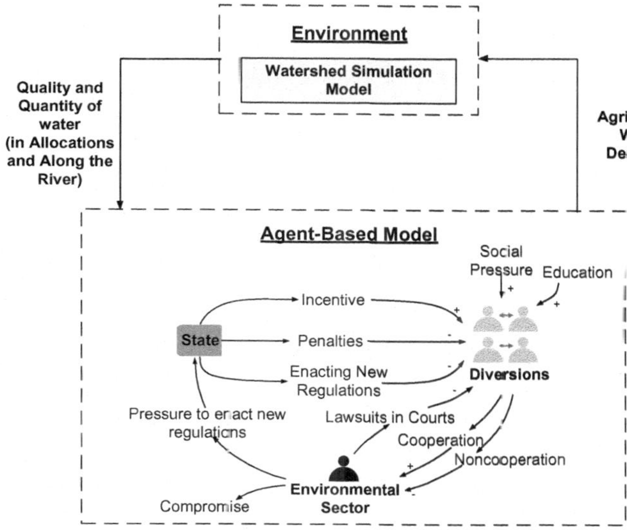

Fig. 8.6 Interactions between agents and agent-environment interactions

As can be seen, the environment affects all the factors as it determines the environment of water availability and system constraints. Demand water agent and the environment influence government perception by informing the government of their concerns, desires and the importance of their goals. The government also influence to understanding the demand agents by informing them about new laws and regulations, educational programs, approved incentives, and so on. Figure 8.7 illustrates these inter;actions.

In addition to environmental and government policies, social pressures and education are two other factors for water applicants to cooperate. Social pressure causes a change in agent's behavior. Figure 8.8 illustrates this social pressure. Here, "C" is syncnym for cooperative and "NC" is for non-cooperative.

As is evident, when most agents choose the way of interaction, the agent is likely to change his or her behavior in accordance with other neighbors. In addition to the social pressure, education can also stimulate agents to optimize their understanding by increasing knowledge of future conditions and changing attitudes.

Fig. 8.7 The influence of the social network on each agent

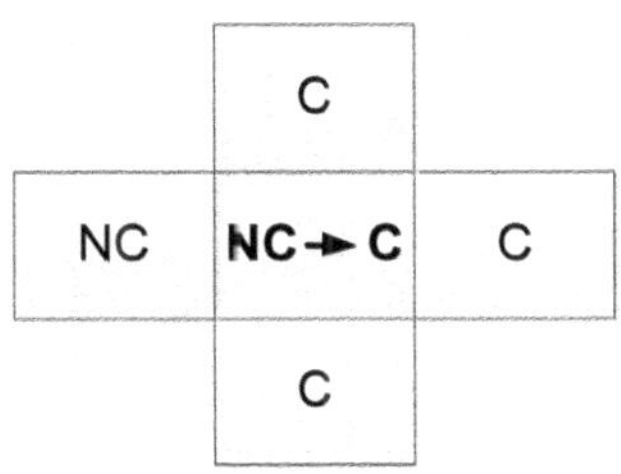

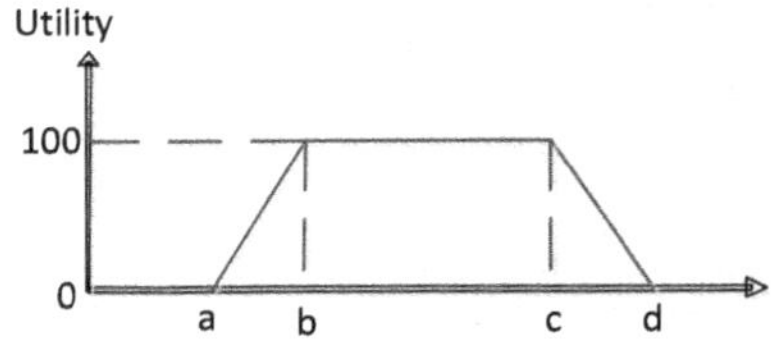

Fig. 8.8 The general form of a utility function

8.5.3 *The ABM Formulation*

The total amount of water available to allocate water applicants calculate by subtracting the minimum amount of environmental flow in the river from all inputs (including rainfall, upstream flow and branch inflow) that shown in Eq. 8.11. This value is divided by the total area of the study area and then multiplied by each individual land to determine the total amount of water existing to the applicants (Eq. 8.12). If the amount of water demanded by agent i is greater than the water available, the agent's behavior falls into the non-cooperative category. Otherwise the cooperative agent is defined (Eq. 8.13):

$$TAW = \sum_{in=1}^{N} Q_{in} - Q_{\min}; \qquad \forall\, y, m \tag{8.11}$$

$$AW_i = \left[TAW / \sum_{i=1}^{I} LA_i \right] \times LA_i; \qquad \forall\, y, m \tag{8.12}$$

$$\begin{cases} if\ AW_i < D_{\max,i} \quad => \quad i \to NC \\ if\ AW_i \geq D_{\max,i} \quad => \quad i \to C \end{cases} \tag{8.13}$$

where TAW is the total available water (cms); Q_{in} is the inflow to the river from the upstream and all tributaries (cms); Q_{min} is the minimum river water flow rate required for environmental purposes (cms); AW_i is the amount of available water for diversion i (cms); LA_i is the area of the land belong to diversion i (hectare); $D_{max,\,i}$ is the maximum water demand for water user i (cms); m is the number of months (from 1 to 12); and, y is the number of years in the time series.

After identifying cooperative and non-cooperative agents, the degree of the agent's willingness to change behavior is determined. The willingness of various agents, Ui, to change or not change behavior at the present time is determined by Eqs. 8.14–8.17. These formulas were developed by adopting Edwards et al.'s (2005) and Young's (1999) sociological dissemination model for the home sector.

$$U_i(C \to C) = a \times V_i(C) + F_m \tag{8.14}$$

$$U_i(C \to NC) = b \times V_i(NC) \tag{8.15}$$

$$U_i(NC \rightarrow C) = c \times V_i(C) + F_m \quad (8.16)$$

$$U_i(NC \rightarrow NC) = d \times V_i(NC) \quad (8.17)$$

where that agent ith is the cooperative behavior and the agent's willingness to maintain this type of behavior ($U_i(C \rightarrow C)$). The ith agent is cooperative behavior and the agent's decision to change behavior $U_i(C \rightarrow NC)$. V_i (C) and V_i (NC) are the neighborhoods of ith agent, which are characterized by cooperative and non-cooperative behavior, respectively. a, b, c and d are the model parameters. Edwards et al (2005) suggest that a, b = 0.7, and d = c = 0.3. F_m is a modified factor based on water availability, training and environmental pressure related to the environment and the government. In the above equations, the first part on the right represents social pressure and the second part (Eqs. 8.14 and 8.16) shows the pressure on the other side of factors and environment as well as the impact of education.

If enough water is available to allocate to consumers, then $F_m = F_m^*$, so:

$$F_m^* = \begin{cases} 1 - [a \times V_i(C)] & For \quad Eq.4 \\ 1 - [c \times V_i(C)] & For \quad Eq.6 \end{cases} \quad (8.18)$$

where the sum of the Eqs. 8.14 and 8.16 equals U_i = 1, then the consuming water is fully supplied and agent has cooperative behavior. Table 8.4 shows these modification factors.

According to the table, the water requesting agent obliged to cooperate if the environmental department files in a lawsuit or a new plan approved by government. In this case, the modification factor equal to F_m^*, in order to achieve 100% utility for the relevant agent to cooperation. In the event of the environmental sector being compromised, there will not operating pressure on agent for cooperative. The agent may only have affected by the social network (neighbors) that in this case the modification factor considered equal to zero. Assuming that incentives provided by the government cause the agent to cooperate, the amount of the modification factor corresponds to the amount of incentive approved. Modifying the value of the demand for factor ith, D_i^m, is calculated from the following relation:

Table 8.4 Modification factors for different state and environmental sector pressure

Category	Action	Modification factor
Legal	Filling a Lawsuit in a Court	$F_m = F_m^*$
	Environmental Sector Compromises	$F_m = 0$
Management	Providing Incentives by the State	Percent of the lost benefit
	Considering Penalties by the State	$F_m = f(D\max_i)$
	Education	$F_m = f(PV \mid future\ damages)$
Legislative	Enacting New Regulations	$F_m = F_m^*$

$$\begin{cases} D_i^m = \left(D\text{max}_{,i} - AW_i\right) \times (1 - U_i); & \forall y, m,\ D\text{max}_{,i} > AW_i \\ D_i^m = 0; & \forall y, m,\ D\text{max}_{,i} \leq AW_i \end{cases} \tag{8.19}$$

And the maximum demand for new agent ith, ND_i, is determined as follows:

$$ND\text{max}_{,i} = AW_i + D_i^m; \qquad \forall y, m \tag{20}$$

In water scarcity conditions, if the agent behaves cooperatively, $Ui = 1$, the agent's demand modification value becomes zero and the new demand equals water availability. Otherwise the demand modification value will be greater than zero and the agent will try to reach more water. However, the new demand is not the same as the initial water demand by the agent because it is influenced by society, environment and other agents.

8.5.3.1 Measures

It is necessary to evaluate the performance of the ABM model and the impact of the defined scenarios. The calculation may determine the level of water users' satisfaction due to the allocations and decisions made. To this end, the benefit function for different water users is defined as satisfaction measure. The general form of the function is shown in Fig. 11. The vertical axis represents zero to 100% of profit and the horizontal axis depends on the type of beneficiary for which the profit function is made. To the water applicant, the horizontal axis indicates the allocation range. For the environmental sector, two profit functions must construct. The first is the flow rate in the river and the second is the expected water quality in the river.

Using these utility functions, the performance of each water consumer can determined by the proportion of water allocated or the quality of the river because of existence all the scenarios. The final stage examines whether the existing conflicts have diminished. If there is no increase in the water consumption of the stakeholders/consumers, it can be claimed that the tension has decreased.

8.5.3.2 Discussion

The proposed model can help decision-makers in water management section to manage conflicts in complex water resources systems by considering laws based on water demand, environmental concerns, timing of flow. In this model, different management scenarios can be considered to determine the most effective scenarios to encourage different parties to cooperate. Therefore, new management situations can be evaluated without the direct involvement of agents and without the use of complex formulas.

The results of the model showed that factors such as social pressure, education and incentives are the most important scenarios that lead to reduced diversions. Reducing diversions by 20% can keep agricultural organizations fully satisfied (100%). incentives should be provided in the form of agricultural subsidies, but it should be noted that incentives should be based on available financial, legal, social and technical resources.

8.6 Summary

Water resources management in arid and semi-arid regions is very importance. Knowing the behaviors, interactions and reactions of social actors (water users) is a very effective help in the management discussion. To study these interactions and examine different management scenarios by changing the parameters, simulation and modeling is the best method. In the last two decades, the use of agent-based model for the study of social network has expanded and the application of this modeling method in the water sector has been considered, which needs to be further developed. In this article, we tried to introduce the method of ABM, the various cases of its efficiency in the water sector, as well as how to build it. To understand the application of this method, two different issues regarding the management of urban water resources and agriculture were explained so that researchers can use them to develop this method in line with the new concept of good governance.

Acknowledgements The authors thank Iran's National Science Foundation (INSF) for the support of this research.

Appendix A

According to references describing creating an ABM for a complex system, the steps can be divided into the following sections:

A.1 Designing Agent Behavior

An agent's behavior is the most important part of the evaluation tool. This behavior described by thresholds, rules, and nonlinear functions. The behavioral models used in ABM have answered specific research questions and a wide range of models have been used in various disciplines. Using empirical data, laws can be developed to represent real human behavior by taking one of several approaches (Janssen and Ostrom 2006). The collective of high quality observations can be used to select statistical distributions that describe the properties and decision factors. Controlled laboratory experiments are used to gather information to test hypotheses about alternative models of human decision making such as individual reasoning, identification, learning, and reminder. However, they may be of limited use in studies aimed at decision making in a particular context. To construct behaviors in the model techniques such as role-play, companion modeling can be used to form the stakeholders for particular community. The game information used to create ABM rules and its results can be evaluated by stakeholders. Finally, for a particular system, data from various sources including remote sensing, maps, census data, and field observations that describe the subsystems to be analyzed can be used.

A.2 Agents Relationships and Interactions

Agent interactions are modeled as effective behaviors that lead to new information in the form of rules. For example, the influence of interests on agent's behavior can demonstrated using a threshold. That is, the factor that has adapted to the new behavior is simulated after the agent has communicated with neighbors who have adopted the behavior. Assigning a range of values to the threshold for the agents causes heterogeneous behaviors within a population. The relationship between the environment and agent and between agents can limit the interactions. For example, an agent may exchange information with other agents that common-pool in space, or are otherwise networked, and ignore agents outside that space, and vice versa. It should be noted that agent interactions are heterogeneous because of the large number of agents, or the interactions may be to a limited number of agents, or the agent may even be secluded. From a social network perspective, communication is simulated in a heterogeneous way between agents. General network models include random network, two-dimensional (2D) grid network (cellular network), small-world networks, etc. The random network associates each factor with the number of randomly selected neighbors. The 2D grid selects the location of each agent and determines the neighbors based on the proximity to that location. The global small network focuses on either random and clustered or centralized connections between small sets of agents. Also, networks can be formed by social media such as blogs or social networks and modeled by ABM.

A.3 Agent Adaptations

Agents can be coded with their adaptive ability that represents different levels of learning. The agents can be divided into reactive and active groups. Reactive agents are those who have not updated their laws and continue to work in the same manner as before and passively respond to other factors and the environment. That group responds to new conditions by understanding their environment over a period. This group can be simulated using a set of simple or complex rules to examine how events and environments respond. The second group consists of those who are purposeful and do their best to achieve their personal goals. Sometimes optimization methods may be used to achieve the goal. Active agents change rules based on performance for more correct behavior (not just for better decision making) and to achieve a level of satisfaction with personal goals. The ABM is used for active agents in a process of discovering the effective way of decentralized decision-making.

Therefore, stakeholder engagement forms an important part of the evaluation. Most research on ABM suggests that this type of modeling can provide specific options in the evaluation process for this partnership. Some believe that this type of modeling can create a debate between agents and guide stakeholders toward a consensus on a particular topic. Some also acknowledge that they are very powerful in facilitating stakeholders' understanding of the results and for explaining the data obtained.

Appendix B: Communication Between Social and Ecological Sectors

As mentioned, the strength of ABM is the ability to consider the two components, social and ecological, and the relationship between them. In socio-ecological systems of water resources, users influence and interact with water resources through their behaviors and decisions. The ecological part of the model constitutes the environment and the factors related to it, and is defined according to the type of problem. For example, for water resources issues and their application in agriculture, the ecological model of the model depends on the type of problem, including groundwater simulation model, runoff rainfall, dam and river set, vegetation growth and evapotranspiration (Becu et al. 2003; Berger et al. 2007) or climate (van Oel et al. 2012), or combinations of all of them. Therefore, the relationship between social and ecological sectors in the model can be examined from two general perspectives: 1- how the ecological sector is applied; 2- the implementation and updating of the ecological sector.

Appendix C: Modeling the Location

The diversity and spatial distribution of the data, including its heterogeneity is another issue. Ecological variables such as elevation and distribution of precipitation, amount of water resources, soil type, temperature, slope, etc., vary from location to location. The presence of agents in different locations also makes a difference in decision-making, and interactions between agents and agents with their environment. Agents can also be categorized according to their differences in behavior patterns (decisions, attitudes, and characteristics) depending on the information available and the nature of the problem. For example, in examining the impact of upstream river users' decisions on downstream users (Becu et al. 2003) living along the river, it was found that they have different approaches and are influenced by each other's decisions. That is, upstream users of the river make freer decisions about water harvesting, while downstream users have to make decisions based on the behavior of upstream users. The second challenge in this section is the accuracy of location units. The spatial cells in ABM are the smallest spatial units in computational units of the model. The value of each parameter is homogeneous within the cell but varies from cell to cell. Depending on the conditions of the issues, each cell can include various parameters such as rainfall height, slope, soil type, and integration of cells can form a larger unit. For calculation, the size of computational cells can refer to common methods such as digital elevation maps (DEM) in the study area.

Appendix D: Selection of Parameters Involved in Modeling

Modelers differ in creating models and writing new applications. Agent-based models, by preparing the thinking about the potential effectiveness of the various options involved in modeling to obtain optimal results, can have a favorable perspective on the choice of these initial parameters. Modelers develop and create models, to select, change or modify each parameter and estimate their possible effect on the range of results obtained. By comparing parameters or a combination of parameters, they can see how different options will achieve different results (Chalabi and Lorenc 2013). In addition to examining and comparing parameters, ABM can assess the unpredictable and adverse effects of parameters that may occur over time and alert to decision makers about them (Morell et al. 2010). Therefore, ABM has the ability to realistically simulate various complex environmental changes and to investigate the reactions of many different agents. That is, it creates a complex bottom-up behavioral model between agents and the environment in both time and space. It provides a way to predict future outcomes that were previously out of the reach conventional (traditional) methods.

Appendix E: Modeling Program Theory

After selecting the parameters involved, it is the turn of the stakeholders' choice. Modelers determine the conceptual model of how key stakeholders work, the relationships between them, their performance, and ultimately the desired outcome that they want to achieve. This conceptual model is called "program theory" and called the program conceptualization, blueprint or design (Rossi et al. 2004). An explicit design provided by many scientists is the basis for evaluation, because this design clarifies the process from the questions to the evaluation and interpretation of the results. The first step is to examine the logic and probability of the theory. This is an important step in the "evaluationability assessment" because it turns out that the theory under consideration has at least the necessary criteria for evaluation, or not (Wholey 1979). ABM can be used to evaluate and build program theory. That is, the researcher can use ABM's proposed model to validate the different functions of his/her theory. It should be noted, however, that these functions, although built on the latest experience and research, can lead to problems in the future without evaluation.

Therefore, ABM can help to identify the relationship between program activities and projected outcomes by testing hypotheses. One can be sure whether these links are possible or not. By doing various model tests, researchers can determine whether or not this program leads to desirable results, and if the answer is positive, they can adjust the important elements in the program to achieve these desirable results.

Using of ABM also determines how changes in stakeholders and environmental factors can account for the differences between the new theory and the previous one, which may have been very strong. ABM does this by allowing agents to represent heterogeneous features of the proposed population and reflect the reality of the environment of agents. In this way, the researchers can find whether previous hypotheses apply in the new situation as well. Otherwise, it turns out that the previous proposed framework is most likely not similar to what is happening in reality. Therefore, an accurate theory is needed for effective evaluation. Therefore, ABM helps researchers build real program theory. It should be noted that the researcher can easily test the program with the ABM before applying it, because many parameters can be entered and varies, even in a simple program. Because it is impossible to consider a large number of parameters in the model, results also confront with some challenges. On the other hand, ABM is best suited to evaluate a particular part of the theory. Considering situations that are exclusively modeled, such as the interaction of the agents or of agents with their environment, it is very important for selecting the model or parts of the model to use in the evaluation. Therefore, ABM is very suitable for examining the elements of a program. Finally, the researchers make suggestions regarding the prioritization of the relationships or changes in the model that are important on the parameters involved in the modeling. These priorities have the greatest potential to achieve the expected results.

Appendix F: Design for Evaluation

Designers need to know what data is needed and how it should have collected. They should anticipate what data is needed in the future to run the program, although differences may occur in the actual process and during the implementation. That is, when designing approaches, the researcher must know what to expect from the program, how the program operates, and what results are sough. Researchers can incorporate ABM into the design process and predict the types of events that may occur during program execution. Then researchers can observe the potential power of the results for different values of the parameters involved and design them to be prepared to deal with the sensitive issues that they are suddenly faced with and possibly not have anticipated. Thus, one can have a better understanding of how parameters are measured. ABM helps researchers refine their programs and bring the design path closer to what is actually happening. Such a model can act as a serious warning of the consequences or challenges that may arise or how their plans influence the results. This modeling can be effective in the program and provide the researcher with the insight that there is a need to change the design or not. Using

ABM and building it based on observations, the researcher can identify the dynamics of program results and illuminate the environment in which unforeseen events occur. Most unforeseen events are hidden by stakeholders who are prejudiced about their program functions and, as a result, the consequences are unclear.

Appendix G: Setting Performance Goals

Researchers help the program managers determine the goals that they are looking for in the program. For example, an organization wants to set up representative offices. Each office determines how much it will serve customers, service a certain number of customers, and report a certain level of customer satisfaction as a key outcome of its work. The level of service and its reporting should be evaluated to determine to what extent the implementation of the program complies with the set criteria. Ideally, gathering evidence and actions based on this information to help one reach a certain level of the program will clearly help improve the situation. However, it is difficult to gather information about it, and establish clear, meaningful, and consensual performance standards before implementing the program (Rosi et al. 2004). In addition, the amount of information available or previous operations may be low, such that precise information was not collected from the conditions. ABM is capable of providing empirical documentation to enable researchers to achieve more meaningful performance goals. For example, a researcher could develop a model to show how many plaintiffs ask their case manager about the complaint before the outcome of the hearing is announced for the case. Therefore, the goal here is called "file's average". Comparing the results with reality helps the researcher to guess how successful the program will achieve its intended goals, and ABM is one of the tools that can used to compare results.

Appendix H: Interpretation of Evaluation Results

ABM is useful after evaluations because stakeholders can see the results of evaluations. Modeling can help to extend the limited results as well as to distinguish between failure of theory and failure in implementation. When evaluation is unique to the findings, researchers can extend the findings by generating dynamic models. Researchers can extend the findings by producing dynamic models, when the evaluation is finished with the findings. If they do not already have a model, they can create a model to understand the evaluation results. Also, they can examine the variations of different parts of the program with changing in parameters and compare the different findings with each other. Therefore, when experimental data is not produced, the ABM is useful for generating causal data. On the other hand, modeling can be a way to combine data and simulate more data. It is useful here because researchers examine limited evaluation findings; arrive at inconsistent findings, or address nonlinearities in the evaluation program. To explain nonlinear and contradictory findings, researchers can develop a model that represents program factors and can be used to determine the parameters that contribute to the findings. Program impact assessment includes questions such as: Are possible goals and programs targets of the program achieved with the existing program? On the other hand, is the process of changing the assumed cause and effect in program theory acceptable? In other words, can the program take into account the social impacts of the activities that each agent performs? The positive answer is evidence that the links in theory can actually occur.

When reviewing the results, researchers may return to program theory and develop a new model to examine whether the program has the effects it should have. This is possible with the help of ABM. ABM can be used to evaluate process theory factors including financial resources, management, services, and other factors that may influence program apply.

References

Arnold RT, Troost C, Berger T (2015) Quantifying the economic importance of irrigation water reuse in a Chilean watershed using an integrated agent-based model. Water Resour Res:648–669

Belaqziz S, Le page M, Aparicio CF, Kharrou MH, Khabba S, Fazziki A, Hennigan P, Jarlan L (2016) Simulating negotiations over limited water resources: a multi-agent system approach for irrigation systems facilitates the analysis of the decision-making process. Int J Interbehaviorism Behav Anal 4(2):116–135

Berger T (2001) Agent-based spatial models applied to agriculture: a simulation tool for technology diffusion, resource use changes and policy analysis. Agric Econ 25(2–3):245–260

Chalabi Z, Lorenc T (2013) Using agent-based models to inform evaluation of complex interventions: examples from the built environment. Prev Med 57(5):434–435

Darbandsari P, Kerachian R, Malakpour S, Malakpour-Estalaki S (2017) An agent-based behavioral simulation model for residential water demand management: the case-study of Tehran, Iran. Simul Model Pract Theory 78(1):51–72

Edwards M, Ferrand N, Goreaud F, Huet S (2005) The relevance of aggregating a water consumption model cannot be disconnected from the choice of information available on the resource. Simul Model Pract Theory 13(4):287–307

Feuillette S, Bousquet F, Le Goulven P (2003) SINUSE: a multi-agent model to negotiate water demand management on a free access water table. Environ Modell Softw 18(5):413–427

Giacomoni M, Kanta L, Zechman EM (2013) A complex adaptive systems approach to simulate the sustainability of water resources and urbanization. J Water Resour Plan Manag:554–564

Grimm V et al (2006) A standard protocol for describing individual-based and agent-based models. Ecol Modell 198(1–2):115–126

Hare M, Deadman P (2004) Further towards a taxonomy of agent-based simulation models in environ- mental management. Math Comput Simul 64(1):25–40

Israel N, Wolf-Branigin M (2011) Nonlinearity in social service evaluation: a primer on agent-based modeling. Social Work Res 35(1):20–24

Izquierdo LR, Polhill JG (2006) Is your model susceptible to floating-point errors? J Artif Soc Social Simul 9(4)

Kanta L, Zechman EM (2014) Complex adaptive systems framework to assess supply-side and demand-side management for urban water resources. J Water Resour Plan Manag:75–85

Klotz C, Hiessl H (2005) Analysis of system innovation in urban waterinfrastructure systems: an agent-based modelling approach. Water Sci Technol 5(2):135–144

Madani K, Dinar A (2012) Cooperative institutions for sustainable common pool resource management: application to groundwater. Water Resour Res 48(9)

Montalto F et al (2013) Decentralized green infrastructure: The importance of stakeholder behaviour in determining spatial and temporal outcomes. Struct Infrastruct Eng 9(12):1187–1205

Morell J, Hilscher R, Magura S, Ford J (2010) Integrating evaluation and agent-based modeling: rationale and an example for adopting evidence-based practices. J Multidiscipl Eval 6(14):32–57

Mulligan KB, Brown C, Yang YCE, Ahlfeld DP (2014) Assessing groundwater policy with coupled economic-groundwater hydrologic modeling. Water Resour Res 50(3):2257–2274

Ng TL, Eheart W, Cai X, Braden B (2011) An agent-based model of farmer decision-making and water quality impacts at the watershed scale under markets for carbon allowances and a second-generation biofuel crop. Water Resour Res 4(1):158–175

Ostrom E (2007) A diagnostic approach for going beyond panaceas. Proc Nat Acad Sci 104(39):15181–15187

Reeves H, Zellner M (2010) Linking MODFLOW with an agent-based land-usemodel to support decision making. Groundwater 48(5):639–660

Riegels N et al (2007) Systems analysis approach to the design of efficient water pricing policies under the EU water framework directive. J Water Resour Plan Manage:574–582

Schwarz N, Ernst A (2009) Agent-based model of the diffusion ofenvironmental innovations—an empirical approach. Technol Forecast Soc Change 76:497–511

Sokolowski AJ, Banks MC (2009) Modeling and simulation for analyzing global events, vol 1. Wiley, pp 8–16

Zellner M (2007) Generating policies for sustainable water use in complexscenarios: An integrated land-use and water-use model of Monroe County, Michigan. Environ Plann B: Plann Des 34:664–686

Chapter 9
Water Resource Management Aided by Game Theory

Icen Yoosefdoost, Taufik Abrão, and Maria Josefa Santos

Abstract Game theory is a theoretical framework for conceiving social situations among competing players. In some respects, game theory is the science of strategy, or at least the optimal decision-making of independent and competing actors in a strategic setting. Using game theory, real-world scenarios for such situations as pricing competition and product releases (and many more) can be laid out, and their outcomes predicted. In this chapter, after introduction and discussion about game theory, the main reason for applying game theory and the area of using this approach explained. In the next step, applying game theory in various fields, game-theoretical models, Game Classification are discussed. After that, the application of Game Theory in Water Resources Management and Useful definitions of applied GP in water resources conflict introduced. Three examples of game theory in water Allocation, water Costs, and groundwater conflicts are introduced and discussed in the final section. the result of these studies illustrates the noticeable performance of game theory approaches in water resources problems.

Keywords Game theory · Water resources management · Water allocation · Water costs · Groundwater

---

I. Yoosefdoost (✉)
Department of Sciences and Water Engineering, Faculty of Agriculture, University of Birjand, Birjand, Iran
e-mail: icenyd@birjand.ac.ir

T. Abrão
Electrical Engineering Department (DEEL), State University of Londrina (UEL), Londrina PR, Brazil
e-mail: taufik@uel.br

M. J. Santos
Chemistry Department (DEEL), State University of Londrina (UEL), Londrina PR, Brazil
e-mail: mjyabe@uel.br

O. Bozorg-Haddad (ed.), *Essential Tools for Water Resources Analysis, Planning, and Management*, Springer Water, https://doi.org/10.1007/978-981-33-4295-8_9

9.1 Introduction

9.1.1 Game Theory

Game theory (GT) is essentially the mathematical study of competition and cooperation. It illustrates how strategic interactions among players result in overall outcomes with respect to the references of those players. Such outcomes might not have been intended by any player (Snyder 2017).

Game theory can be used to predict how people behave, following their own interests, in conflicts. In a typical game, decision makers (players), with their own goals, try to outsmart one another by anticipating each other's decision. The game is resolved as a result of the players' decisions. GT analyses the strategies players use to maximize their payoffs. A solution to a game prescribes decisions the decision makers might make and describes the game's outcome. GT was established in 1944 with the publication of von Neumann and Morgenstern's "Theory of Games and Economic Behavior" book, which mainly dealt with quantitative game theory methods. After World War II, most scholars worked on developing quantitative game theory methods; and this trend still persists today (Hipel and Obeidi 2005).

Every baby realizes what games are. When a person overreacts, we sometimes tell, "it's just a game." Games are often not earnest. Mathematical games are different. It was the main goal of game theory GT from its beginnings in 1928 to be applied to serious situations in economics, politics, business, and other areas. Even we can analyze war by using mathematical game theory. Let's describe some ingredients of a mathematical game:

Rules: Mathematical games have strong rules. They determine what is allowed and what isn't. Although many real-world games allow for finding new moves or ways to act, games that can be analyzed mathematically have a strict set of possible steps, generally, all known in advance.

Outcomes and payoffs: people play games just for fun. Mathematical games may have several likely outcomes, each producing payoffs for the players. The payoffs possibly are monetary, or they may represent satisfied.

Uncertainty of the Outcome: In most cases, a mathematical game is "thrilling" because the result cannot be forecast in advance. As its rules are stable, this implies that a game must either have some accidental parameters or own more than a single player.

Decision making: A game with no resolve possibly is annoying, at least mentally. Running a 100 m race does not need mathematical proficiency; it needs just fast legs. However, most sports games also involve decisions, and can, therefore, at least partly be analyzed by game theory. No cheating in real-life games, fraud is possible. Cheating means not playing by the rules. If, when your chess rival is distracted, you take your queen and put it on a better square, you are cheating. Game theory doesn't even acknowledge the being of cheating. We will learn how to win without cheating.

9.2 Definition and Terms

9.2.1 Game Definition

Let's define the tuple:

$$G_T \triangleq \langle N, A, P, I, O \rangle$$

where:

G$_\mathbf{T}$—the game, which exist manly in two forms: Strategic (or Normal) games, denoted by G, and Extensive games, denoted by Γ;

N—set of players. N = {1, 2, ..., n} is a finite set. Every player is denoted by i; the other n − 1 players or i's opponents in some senses, denoted by −i; ∀ i, −i ∈ N;

A—the profile of action (or move) of the players. An action carried out by player is a variable of his decision, which is denoted by ai. The set of A_i, i = $\{a_i\}$ is player's action set, i.e. the entire set of actions available to him. The ordered set $a_i = \{a_i\}$, i ∈ {1, 2,..., n}, is an action combination for each of the n players in a game. In the action set, S is the strategy set (called strategies space) of the players. Strategy is the rule to choose actions. The strategy space of player *i*, denoted by S_i is the set of all the strategies which player *i* can choose from.

P—payoff (or utility). A payoff is the value of the outcome to the players. It refers to both actual payoff and expected payoff. Payoffs are based on benefits and costs of actions and outcomes of each player. $u_i(s_i, s_{-1}-)$ means player *i* 's payoff function, which is determined by the strategies chosen by himself and the other players;

I—information set. It is players' knowledge about another player, such as the characteristics, action profile, and payoff function in the game. If the payoff function of every player is a common knowledge among all the players, then it is complete information. Otherwise, it is called incomplete information. If the information is complete and perfect, it means that the players know well the former process of the game before he chooses his next move at each step. If the player who will choose his next move does not know the prior processing of the game at some steps, it is called complete but not perfect information;

O—outcomes of the game. An outcome is a strategy profile rusting from the action/moves combination chosen by all the players at the end of a game;

E—equilibrium or equilibria. In the equilibrium, each of the players can maximize his payoff. $s^* = \{s_1^*, s_2^*, \ldots s_n^*\}$ is a best strategy combination of the *n* players. For player *i*,s_i^* is the player's best response to the strategies specified for the *n-1* other players, i.e. $u_i\left[s_i^* - s_i^*\right] \geq u_i\left\{s_i - s_i^*\right\}$.

Generally speaking, the elements of game theory includes ***N***—Players, ***A***—Action, ***P***—Payoff, ***I***—Information, ***O***—Outcome, and ***E***—Equilibrium, i.e., NAPI-OE. NAPI are concertedly known as the rules of a game and *OE* are the game results. The main task of constructing a game model is to define the rules (NAPI) in mathematical language and get the solution from OE. The detailed GT can be referred by

(Friedman 1998; Gibbons 1992; Kreps 1990; Straffin and Philip 1993; Gardner 1995; Myerson 2013; Stahl 1999; Osborne 2004; Gintis 2001). Every player has different strategies; however, the optimal strategy for an individual player is to maximize his benefits by using the game rules; while the optimal strategy for the player of a society as whole is to maximize the common welfare of the society through the rules. GT models involve the following conditions and assumptions:

- Players in the game models are regarded as "intelligent and rational". Rational payer means that each player will choose an action or strategy which can maximize his expected utility given he thinks what action other players will choose. Intelligent player means that each player understands the situation, and he knows the fact that others are intelligent and rational;
- Each player considers not only his own knowledge and behavior but also others' during pursuing exogenous aims;
- Each player has more than one choice or sequence ("plays");
- All possible combinations of choices or plays result in a well-defined outcome: win or lose, or mutual gains and losses.
- The players are aware of the rules of the game and the options of other players, but they do not know the real decisions of other players in advance. Therefore, every player has to choose options based on his assumption of what other player will choose;
- Each player knows that his actions can affect the others, and the actions of others affect him;
- Each player makes the best possible move, and he knows that his opponent is also making the best possible move (Wei 2008).

9.3 Basic Principles and Logics

9.3.1 Game Theoretical Models

In the last years, game theoretical modelling is becoming an indispensable approach to analyze, understand, and solve many water problems around the world. Like other sciences, game GT itself is comprised of a collection of models. There are different methods to classify these models. In general, they are summarized as follows:

- Binding agreements: non-cooperative and cooperative games;
- Numbers of players: single player game (decision problem), two-persons game and multi-person game;
- Order of actions (moves): static and dynamic games;
- Elements of actions (moves) set: finite and infinite games;
- Sum of payoffs: zero-sum and non-zero-sum games;
- Information set: complete information and incomplete information games;
- Numbers of the same play in a game: one-shot game and repeated game (Wei 2008).

9.3.2 *Some More Definitions: Game, Play, Action*

The entire collection of rules describes a game. A play is a sample of the game. In specified situations, called positions, a player has done make a decision, called a **move** or an **action**. This is not the same as a strategy. A strategy is a plan that expresses to the player what move to choose in every feasible position. Rational behavior is almost supposed for all players. That is, players have priority, beliefs about the world (including the other players) and try to optimize their individual payoffs. Besides, players are aware that other players are trying to maximize their payoffs (Dinar and Hogarth 2015).

9.3.3 *Game Classification*

Game can be categorized according to several criteria, including:

Number of players

Usually there should be more than one player. However, you can play roulette alone—the casino doesn't count as player since it doesn't make any decisions. It collects or gives out money. Most references on GT do not treat one-player games, but (Dinar and Hogarth 2015) discuss one-player games provided they contain elements of randomness.

Simultaneous or sequential play

In a simultaneous game, each player has just one move, and all movements are made simultaneously. In a sequential game, it is forbidden two or more players move at the same time, and players possibly have to move multiple times. Some games are neither simultaneous nor sequential (Dinar and Hogarth 2015).

The game with random moves

Games possibly can contain random events which can influence its outcome. They are called random moves (Dinar and Hogarth 2015).

Players with perfect information

A sequential game has perfect information if every player, when about to move, knows all previous moves (Dinar and Hogarth 2015).

Players with complete information

This means that whole players are aware of the structure of the game. The order in which the players move, all thinkable moves in every position, and the payoffs for all outcomes. Actual world games generally do not have complete information. In our games, we consider complete information in most cases, since games of incomplete information are more difficult to analyze (Dinar and Hogarth 2015).

Zero-sum game

Zero-sum games possess the property that the sum of the payoffs to the players equals zero. A player can have a positive payoff just if the other has a negative payoff. Poker and chess can be great examples of zero-sum games. Actual-world games are scarcely zero-sum (Dinar and Hogarth 2015).

Permitted communication

Sometimes the relationship between the players is allowed before the game starts and between the moves and sometimes it is not (Dinar and Hogarth 2015).

The cooperative and non-cooperative game

Game theoretical models are usually divided broadly into two branches, either noncooperative game or cooperative game. It does not mean that these two branches are applied to analyze different kinds of games, but they are just two ways to view the same game (Gibbons 1992; Wei 2008).

Even if players converse, the question is whether the results of the negotiations can be performed. If not, a player can always move differently from what was promised in the talks. Then the relationship is named "cheap talk." A cooperative game is one where the outcome of the negotiations can be put into a contract and be performed. There should also be a method of distributing the payoff among the members of the coalition (Dinar and Hogarth 2015).

Non-cooperative game can be defined from the following aspects:

- modelling the situation of lacking binding agreements;
- what actions (moves) that players can take;
- how players interact with each other to maximize individual welfares;
- solutions concepts: Nash equilibrium, sub-game perfect Nash Equilibrium, Bayesian Nash Equilibrium and perfect Bayesian (sequential) Equilibrium;
- mainly stressing individual rationality, individual optimal strategies and payoff;
- the results may be efficient and maybe not (Wei 2008).

Cooperative game can be defined by:

- modelling the situation of binding agreements;
- what coalitions forms that players can use to maximize the collective welfare of all the players;
- how the available total value split in a satisfactory way;
- most popular solution concepts include: the stable set, equity-based rule, the core Shapley value, as well as the Nash bargaining solution;
- Stressing mainly collective rationality, efficiency and fairness;
- the results are usually social optimum (Wei 2008).

In summary, the non-cooperative and cooperative game theories are similar to the positive and normative approaches that economists use. In economics, the positive approach describes what the real world is, and it usually deals with analyzing and prediction. However, normative approach deals with what the world should be, and it focuses on the methods to change the world. Moreover, noncooperative game theory is a strategy-oriented game, and it studies what players expect to do and how they do it. On the other hand, cooperative game theory establishes what the players can achieve and how they can achieve it (Wei 2008).

9.3.4 Why Game Theory?

"Game theory is essentially the mathematical study of competition and cooperation. It illustrates how strategic interactions among players result in overall outcomes with respect to the preferences of those players." The objective is to predict how people behave, trying to achieve their goals while in conflict. It includes decision makers (players) trying to outsmart one another by anticipating each other's decision. The game is resolved as a byproduct of the players' decisions. The resolution of the 'game' leads to optimal decision making and describes the game's outcome. The application of GT in the industry could result in a revolutionary change in process efficiency and optimality, saving significant amounts of the essential scarce resource water.

9.3.4.1 GT Features

- Game theory creates a realistic simulation of stakeholders' interest-based behavior. The self-optimizing behavior of players and stakeholders often results in non-cooperative behaviors, although cooperative competition could be a win-win situation.
- The model can create planning, policy, and design insights that would be unavailable from other traditional systems and engineering methods.
- Another advantage of GT over traditional methods is its ability to simulate different aspects of the conflict, incorporate various characteristics of the problem, and predict the possible resolutions in absence of quantitative payoff information.
- Often non-cooperative GT methods can help resolve the conflict based on the qualitative knowledge about the players' payoffs. This enables to handle the socio-economic aspects of conflicts and planning, design, and policy problem when quantitative information is not readily available available (Madani 2010; Jhawar et al. 2018).

9.3.4.2 Challenges in Applying GT to Water Resource Management

- The complexity in the economics that comes with large-scale water resources projects are challenging; the industry impacts different strata of the society and varied geographical locations differently.

- The unpredictability of natural and climatic conditions makes creating forecasts even more challenging. However, the implications would be significant in extent and variety.
- The large number of decision variables involved, stochastic nature of the inputs, and multiple objectives makes this sector an obstacle course towards optimality using the given model (Datta 2005). Therefore, we can see the relevance, challenge, and important resource incorporating operations research into the management of water resources (Jhawar et al. 2018).

9.4 Importance and Necessity of Game Theory

GT was established in 1944 with the publication of von Neumann and Morgenstern's "Theory of Games and Economic Behavior" book, which mainly dealt with quantitative game theory methods. After World War II, most scholars worked on developing quantitative GT methods; this trend still persists today (Hipel and Obeidi 2005). Besides, over the years, GT applications have been developed for several water sectors. Many researchers have attempted water conflict resolution in different area of studies in a game-theoretic framework.

9.4.1 Application of Game Theory

9.4.1.1 Water Resource Management

Leoneti and Pires (2017a) conducted research which the main goal was to supply a review of the literature from the field of decision sciences to the area of *water resource management*.They discuss the application of multi-criteria methods, including Analytical Hierarchy Process, Measuring Attractiveness by a Categorical Based Evaluation Technique, Multi-attribute Utility Theory, Elimination and Choice Translating Reality, Preference Ranking Organization Method for Enrichment Evaluations, Technique for Order Preference by Similarity to Ideal Solution, and GT, containing cooperative and non-cooperative bargaining games. Their numerical results illustrate that, these techniques are useful for creation and comparison of scenarios, decrease the time needed to achieve a solution for complicate problems containing a large number of criteria and agents. It besides shows the benefits of creating greater transparency in the decision-making process, thus improve the potential for a solution acceptable to all the parties involved. Although still little explored, discussions of sanitation problems can and must be raised with the use of techniques and methods of decision sciences, while multi-criteria and game theory techniques are particularly suitable for this task.

9.4.1.2 Water Resource Conflict of Interest

Varouchakis et al. (2018) in one study tried to help people who live in the city of Chania, Greece, the resident have asked for a fairer tariff policy and represent the purpose to save water under the performance of stricter measures. A two-person zero-sum game was proposed, including a conflict of interest among the Municipal Enterprise for Water and Sewage of Chania (MEWS) (Player 1) and the city's near 108,000 residents (Player 2). Three scenarios for the gradual decrease for the fixed charges and the continuous growth for volumetric charges were developed, assuming various degrees of change in water use behavior by each consumption block of consumers. The payoff matrices, for each scenario, incorporated two clear cost strategies for Player 1, in terms of changing the current tariff policy, and four clear cost strategies for Player 2, regarding the change in consumers' behavior. The optimal decision for both players, derived from the identification of the equilibrium point, demonstrated that domestic water consumption might be reduced by up to 4.6% while maintaining the MEWS's profit. The proposed model can provide a guide for other similar applications.

Conflicts in Irrigation Area and Drainage Network

Gholami et al. (2017). In the study with the purpose of analyze and to provide solutions for solving the conflicts in Sefidrud irrigation and drainage network the GT approach was employed. For modeling and analysis of the conflict, Graph Model for Conflict Resolution (GMCR) were used. After determining players and options and inserting them into the model, 64 states were created in this conflict arena. Using non-cooperative solution concepts with regard to prioritize strategies by decision-makers, 4 situations were identified as equilibrium points. After the final analysis of the strongest points of equilibrium, status quo and the base status form one of the points of equilibrium. The other equilibrium point was situations that farmers take alternative irrigation tout for the favorable situation in the future. Therefore, it is essential to train farmers as a primary player and involve them in decisions and to form water user associations to improve the condition of their participation in the management of water resources. This would reduce the conflict using stability of the resources.

Irrigation Water Conflict in Southeastern Brazil

Getirana and de Fátima Malta (2010) apply GT approaches to conflict among irrigators among the Coqueiros Canal water users, located in the Campos dos Goytacazes municipality, in the northern region of the State of Rio de Janeiro, and a canal in Rio de Janeiro State in southeastern Brazil. The Graph Model for Conflict Resolution (GM-CR) tool, which is able to solve Non-Cooperative Games (NCG) based on graph theory has been applied to efficiently solve such water resource dispute. The authors developed six scenarios pertaining to the decision makers, and their options and strategies. Then they identified two possible roles for the managing institution: (a) the conflict resolution managing institution takes into the account the fact that it has no explicit preferences for any of the outcomes; (b) the managing institution explicitly demonstrates preference for those scenarios and solutions that provide more income taxes. The results suggest a solution to the conflict among the irrigators, with the demand for irrigation water affecting the priorities in attaining possible equilibria.

9.4.1.3 Relicensing Process with Bargaining Solutions

Madani (2011) developed a method based on Nash and Nash–Harsanyi bargaining solutions to illustrate the Federal Energy Regulatory Commission (FERC) relicensing process, in which owners of non-federal hydropower projects in the United States must negotiate their suitable operations, with other interest groups (mainly environmental). In this study, the connection of games to develop the possible solution range and the "strategic loss" concept are considered, and a FERC relicensing bargaining model is expanded for studying the bargaining stage (third stage) of the relicensing process. According to the suggested solution method, how the lack of incentive for cooperation results in a long delay in FERC relicensing in practice is explained. Further, the potential impacts of climate change on the FERC relicensing are express, and how climate change may provide an incentive for collaboration between the parties to hasten the relicensing is discussed. An "adaptive FERC license" framework is proposed, according to cooperative game theory, to increase the efficiency and competitivity of the system to future changes with no cost to the FERC in the meat uncertainty about future hydrological and ecological conditions.

9.4.1.4 Multiple-Reservoir Cooperation in Hydropower System in China

In a methodology namely Progress Optimality Algorithm based on Discrete Differential Dynamic Programming (POA-DDDP) and implemented by the Multidimensional Search Algorithm (PDMSA) combined to game theory is proposed to address the challenge of fairly allocate the incremental benefits of cooperation among all hydropower plants participants/players. In this study, The PDMSA expands to define optimal operation decisions, obtaining a multi-yearly average earning under all feasible coalitions of plants. Then, the collaboration benefit can be correctly calculated based on the differences of generation production revenue among various alliances. In addition, the game-theoretic Shapley method is used to discover the suitable share of each plan cooperator from overall cooperation benefits. The cooperative core based on a set of necessary conditions helps to choose a possible, stable allocation plan, while their stability is evaluated by the propensity to disrupt (PTD). The proposed methodology is used to a multiple-reservoir hydropower system on the Lancang River, which is one of 14 largest hydropower bases in China. This case illustrates that the method provides the most stable incremental allocation project by comparison with different generally used methods.

9.4.2 An Overview to Studies, Applying Game Theory Approach in Water-Resources Systems

Publications dealing with Game theory have covered several domains of water-resources systems. We can categories these studies into eleven typologies that mark an essential contribution to the literature in the last fifty years, such as **Urban water supply and sanitation** (Barreiro-Gomez et al. 2017; Leoneti and Pires 2017b; An et al. 2018; Zarei et al. 2019a; Goksu et al. 2019; Chhipi-Shrestha et al. 2019). **Irrigation** (Omidvar et al. 2016; Xing and Yuan 2017; Mukherjee 2017; Chhipi-Shrestha et al. 2019; Ristić and Madani 2019; Hone et al. 2020). **Hydro-electric power** (Gately 1974a, b; Anderson 2016; García Mazo et al. 2020). **Water pollution control** (Adhami and Sadeghi 2016; Guo 2016; Shi et al. 2016; Yong et al. 2017; Xu et al. 2017; William et al. 2017; Zhang et al. 2019; Hu et al. 2019). **Groundwater** (Huang et al. 2016; Gao et al. 2017; López-Corona et al. 2018; Tian and Wu 2019; Tian and Wu 2019; Ghadimi and Ketabchi 2019; Nazari and Ahmadi 2019; Nazari et al. 2020). **Allocation issues** (Xiao et al. 2016; Shi et al. 2016; Oftadeh et al. 2017; Yuan et al. 2017; Han et al. 2018; Degefu et al. 2016; Zarei et al. 2019a). **International/transboundary water** (Menga 2016; Li et al. 2016, 2019; Fu et al. 2018; Khachaturyan and Schoengold 2019; Zhang et al. 2019; Tayia 2019; Zeng et al. 2019; Janjua and Hassan 2020). **Water conflict and negotiations** (Mehrparvar et al. 2016; Zomorodian et al. 2017; Oftadeh et al. 2017; Zanjanian et al. 2018; Mogomotsi et al. 2019; Zeng et al. 2019; Yu et al. 2019). **Water and ecological systems** (Dinar et al. 2013; Hachoł et al. 2019). **Watershed management and regulation/river basin planning** (Girard et al. 2016; Hui et al. 2016; Jeong et al. 2018a; Rahmoun and Rahmoun 2019; Andik and Niksokhan 2020; Lee et al. 2020; Adhami et al. 2020; Janssen et al. 2020). **Multipurpose water projects** (Kahil et al. 2016; Jeong et al. 2018b; Alahdin et al. 2018; Alaghbandrad and Hammad 2018; Zarei et al. 2019b; Chhipi-Shrestha et al. 2019).

9.5 Methodology

9.5.1 Game Theory Approach in Water Resources Management

Game theory began as applied mathematics and microeconomic theory, but it serves here as a modelling approach to manage water resources. The questions arose in the game modelling of water resources management are as follows:

- What kind(s) of the game (games) can water resources management be modeled as? In other words, what kind(s) of the game (games) is (are) involved in water resources management? Can the rational choices of multi-stakeholders be translated into a mathematical or/and economic problems? Can the rational outcome be as the "solution" to the game?
- How to translate a case of water resources management into a game in mathematic or/and economic language? In details, what is (are) the player(s)? What are the strategies available to each player? What is payoff that each player can obtain from the combination of strategies chosen by the players? what methods can be used to solve for the Nash Equilibria of strategies?
- What is the strategy space? In which condition does a player use pure strategies or mixed strategies? How to choose dominated strategy (strategies)?
- What does it mean complete and incomplete information in the games of water resources management? What uncertainties or risks are there in a game of water resources management? How to predict them?
- How to value the problems and benefits in payoff terms? How to value the payoff and make right decisions? If the game is cooperative one, how to divide the joint payoff? (Wei 2008).

9.5.2 Types of Games

From the game theoretical point of view, there are full of games in human society and nature. Figure 9.1 depicts the nature and human society from a game point of view, and each interacting and interdependent group or/and individual can be modelled as game(s). For examples, the game can be between human and rain, rivers, lakes and animals, and between animals and animals, plants and plants, animals, plants and their habitats, human and human, and so on (Wei 2008).

The game components/elements for Fig. 9.1 could define as:

N (set of players), In this picture according to each game, could be different, for example in HH-G it can be several countries, people, company, etc. In **NN-G** it is different species of animals in front of each other, river and animal, rain, animal, and, ... In HN-G The players are human-rain, human-force, human-water, human-animal, etc.

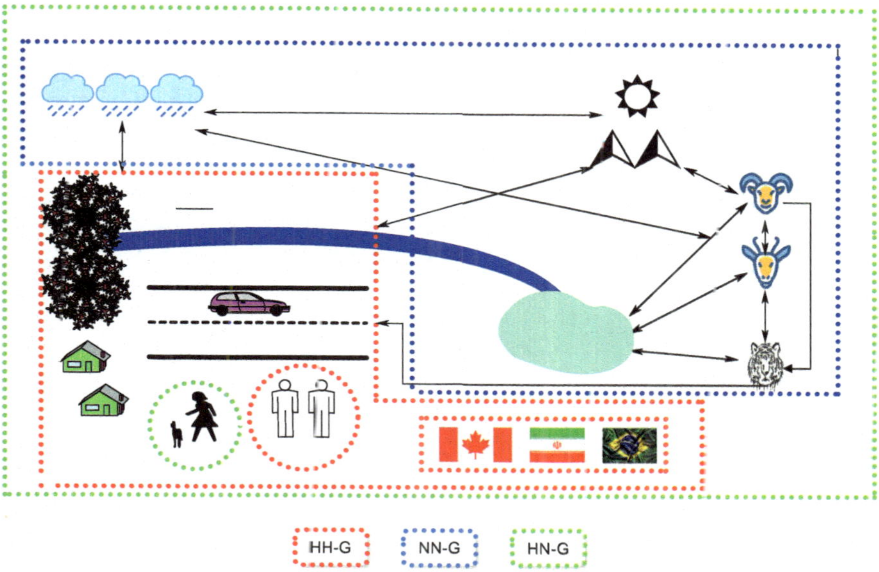

Fig. 9.1 Nature and human society from a game theoretical perspective (Wei 2008)

A (the profile of action (or move) of the players). For example, in HH-G, It can the way two countries treat each other, respectfully, aggressive, and so on. Or in HN-G, between human and groundwater resources, the action of humans is excessive, improper use of groundwater sources, land subsidence is nature's action. In NN-G, for example, hunting is an action for a predator, and camouflaging is an action of prey.

I (information set) is player knowledge about another player. For example, information which two countries have from each other, or The knowledge that prey and predator have from each other, in the subject of characteristics, action profile, and payoff function in the game.

O (outcomes of the game), In HH-G in two countries game, it can result in a loser country. Become a colony of another country or in NN-G; it can be a Successful predator to prey or escape prey from and reduce the number of one of them due to lack of food or hunting.

Vrieze (1995) classified environmental games in two ways: society's game and game of exhaustion, while (Kelly 2003) classified into games of skill, games of chance and games of strategy. In this study, the games involved in water or other nature resources management is classified into the following three kinds:

HH-G: Human and human games, the games played among human beings, including different countries, world regions, or areas within regions;

HN-G: Human and nature games, the games played between human beings and the nature;

NN-G: Nature and nature games, the games in nature itself (Wei 2008).

In definition, HH games are similar to society's game and games of strategy, and HN games are similar to the game of exhaustion and the combination of games of chance and games of skill. HN game is a close relative of decision theory. Parson and Wooldridge (2002) stated that decision theory could be considered to be the study of games against nature, where nature acts randomly. In the literature of game theory, nature usually is regarded as a pseudo player entering the game. Some people maybe do not believe that nature can be players because they cannot move. However, there are so many examples to show that nature really moves and strict back when humans use it improperly, such as pollution, the greenhouse effect, and so on. If so, the question is what their strategies and payoffs are since they are players. For the NN games, there are very few studies comparing with the former two kinds. Smith (1982) analyzed the NN games in his book Evolution and Theory of Games (Wei 2008).

9.5.3 A Game Theoretical Approach to Solve Conflicts

The question is how to construct a game model. Figure 9.2 depicts the process of using game theoretical approach to solve conflicts. Generally speaking, the process of game theoretical modeling approach can be divided into four steps (Wei 2008).

Step 1: Defining the game

Defining the players
Defining their payoff functions f
Defining their moves (strategies)
Defining information set

Step 2: Setting up game models

Non-cooperative game models
Cooperative game models

Step 3: Analyzing the game models

Getting the possible game outcomes
Comparing these outcomes

Step 4: Solving the game

Getting the equilibrium of non-cooperative games
Getting the trade-off point to share the benefit obtained from cooperative games.

This flow can be shortly summarized into the following questions:

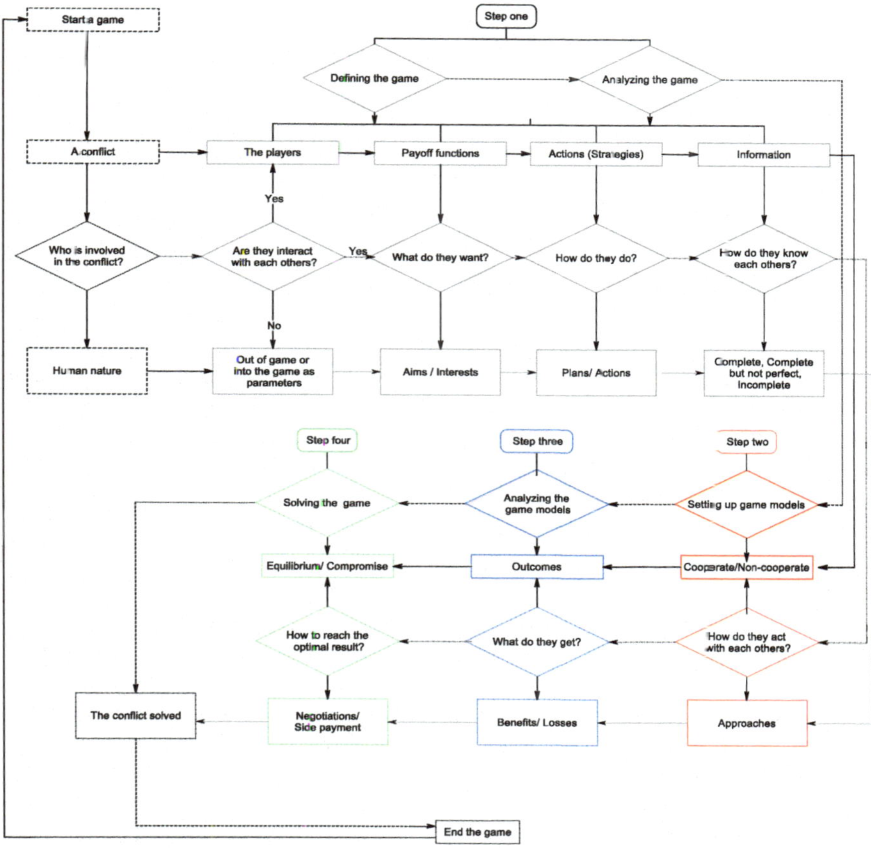

Fig. 9.2 General flow chart of game theoretic approach to solve conflicts (Wei 2008)

- Who involves in the conflict?
- What are their actions (strategies)?
- How to form the payoff function of each player?
- How does every player know the payoff function of others?
- Is the game one-time game, continuous game, finite game or infinite one?
- How to compute the equilibrium/equilibria of the game(s) in the case of a noncooperative game?
- Is every player better off if he cooperates with others?
- How to deal with the amount of benefit derived from cooperative games among the players? (Wei 2008).

9.5.4 Useful Definitions on GT Applied to Water Resources Conflicts

In a water conflict, several interest groups or persons can be modeled as decision-makers (players), where each decision-maker can make choices unilaterally, and the combined decisions of all players together determine the possible outcomes of the conflict. Instead of unilaterally moving, decision-makers also may decide to cooperate or form coalitions leading to Pareto-optimal outcomes. GT techniques create an efficient and *accurate language* for discussing *specific water conflicts.* A systematic study of a strategic water dispute provides insights about how the conflict can be better resolved and may suggest innovative solutions. Many researchers have attempted water conflict resolution using a game-theoretic framework. We provide a menu of several water resource allocation schemes available in the literature. In the next subsections, we provide a theory review behind each of these allocation schemes by visiting the sources provided for each resource allocation scheme. While we use cost allocation schemes, the reader can convert them very easily to benefit/profit allocation schemes as well. The examples are taken from (Dinar and Howitt 1997) and (Kreps 1990).

In Mehrparvar et al. (2016), cooperative game theory (CGT) approaches were used to water allocation in a river basin with attention to equity benefit shares between stakeholders. Firstly, to allocate water between competing users, an optimization model is developed based on the containing industrial, agricultural, and environmental users and their economic objectives. The model is elaborated to determine water shares for different likely coalitions among water users. Then, CGT approaches, including Shapely, Nucleolus, and Nash-Harsanyi methods, were used for reallocating net profits to the users as a solution to encourage them to participate in equitable cooperation. Then, the results from different game-theoretic approaches are evaluated by using the *stability index* and *voting* methods, such as *social choice* and *fallback bargaining*. This study was proceeded in the Zayandehrood River basin located in Iran, which struggles with water scarcity. The different CGT approaches applied to two predefined real-life scenarios in the basin under study, and their performance have been investigated. The results indicate the proper performance of both Nash-Harsanyi and Shapely methods for pessimist and optimistic scenarios, respectively. It is also found that the application of the proposed methodology effectively increases the users' benefits in the study region through optimal water allocation and reallocation of benefits.

9.5.5 Non-GT Cost Allocation Schemes Used in GT Studies

There are a wide variety of cost allocation schemes for joint operation of facilities proposed in the accounting and engineering literature (Biddle and Steinberg 1985) and (Alchian 1965), providing a comprehensive review, from which we use three main types:

- an engineering approach where the cost allocation is proportional to the physical use of the facility;
- marginal cost analysis based on economic efficiency principles;
- the separable cost remaining benefit (SCRB) principle, where the allocation of the fixed investment is based on an equitable division of the cost.

In the following subsection, the terms "player" and "user" are used interchangeably (Dinar and Hogarth 2015).

9.5.5.1 Resource Allocation Based on Pollution Generation

This water resource allocation scheme simply suggests that each user of the joint facility will be charged in proportion to the services the facility provides for this player (e.g., volume of pollution it generates that is treated in the joint facility). In summary, the cost allocated to user j is:

$$P_j = f^N . \frac{q_j}{\sum_{i \in N} q_i}$$

where, f^N is cost of the joint facility and q_j is the quantity of pollution generated by user j. This scheme allocates all of the joint cost among all N users (Dinar and Hogarth 2015).

9.5.5.2 Allocation Based on Marginal Cost

The allocation based on the marginal cost of the joint facility takes into account marginal quantities generated by each potential user. Since economies of scale in the joint cost function exist. the revenues generated by this allocation scheme will not cover the total cost. Therefore, an additional procedure is necessary to account for the remaining uncovered costs. Usually, this can be done using any *proportional rule*, such as pollution volume, or volume of production. The allocation in terms of joint cost for the j-th user is defined as:

$$b_j = \left\{ \frac{\partial f^N}{\partial q_j} + f^N \left[1 - \sum_{i \int N} \frac{\partial f^N}{\partial q_i} \right] \right\} \frac{q_j}{\sum_{i \int N} q_i}$$

where q_j is the quantity of pollution generated by j-th user; $\frac{\partial f^N}{\partial q_j}$ is the marginal cost associated with the use of user j; and $f^N[1 - \sum_{i \int N} \frac{\partial f^N}{\partial_{q_i}}]$ is the remaining uncovered cost, which is now included in the allocation scheme (Dinar and Hogarth 2015).

9.5.6 *Separable Cost Remaining Benefit (SCRB)*

The separable cost of user $J \in N$ is the incremental cost:

$$m_j = f^N - f^{N-\{j\}}.$$

The alternate cost for j is the cost $f^{\{j\}}$ it bears while acting alone, and the *remaining benefit* to j (after deducting the separable cost) is $r_j = f^{\{j\}} - m_j$. The SCRB assigns the joint cost according to:

$$k_j = m_j + \frac{r_j}{\sum_{i \in N} r_i} \left\{ f^N - \sum_{i \in N} m_i \right\}^+.$$

where the operator $\{x\}^+ = \max\{0, x\}$. In other words, each user pays their separable cost m_j, while the "non-separable costs" $f^N - \sum_{j \in N} m_j$ are then allocated in proportion to the remaining benefits, assuming that all remaining benefits r_j are nonnegative for each player (Dinar and Hogarth 2015).

9.5.6.1 Game Theory Cost Allocation Solutions

Given the initial conditions of voluntary collective action, and the prior establishment of independent resource management institutions among the users (a region, river basin, etc.), the problem of allocating the joint costs of a joint water facility (joint well, treatment facility, reservoirs, hydropower generation) is modeled as a game among the players. Based on the empirical situation, it can be assumed that institutional regulations facing the players are already in place and that the players agree to consider them. If a player chooses not to cooperate (not to participate in the investment and the operation of a joint facility), it faces a certain outcome resulting from the operation of a private facility or alternative measures needed to meet the regulations. If the players choose to cooperate, they may benefit from economies of scale embodied in the larger capacity of the joint facility with lower average treatment costs compared to the cost in private actions. Some players may cooperate while others may choose not to cooperate, depending on the degree to which they can reduce their cost under cooperation. As a result, the larger the economies of scale, the bigger the incentive for cooperation. the following is based (Martin Shubik 1982a, b, c; Shubik 1982a; Shapley 1952). Let N be the set of all players in the region, S (S $\subseteq$ N), the set of all feasible coalitions in the game, and s (s ∈ S) a feasible coalition in the game. The non-cooperative coalitions are {j}, j = 1, 2, ..., n, and the grand coalition is {N}.

Assuming that the players' objective is to minimize their cost, let f^s be the cost of coalition s, and $f^{\{j\}}$ be the cost of the j-th member in non-cooperation. A necessary condition for *regional cooperation* is that the joint cost will be less than the sum of the individual costs:

$$f^s \leq \sum_{j \in S} f^{\{j\}}, \forall s \in S \subseteq \mathrm{N}.$$

The *joint savings* that are allocated among the players are defined simply by

$$\sum_{j \in S} f^{\{j\}} - f^{\{s\}} \geq 0, \forall s \in S \subseteq \mathrm{N}.$$

The above inequality can be interpreted as a cooperative game, with side payments, and can be described in terms of a characteristic function. The value of a characteristic function for any coalition expresses the coalition expenses, or profit, in the case of a benefit game:

$$\vartheta(s) = f^s, \forall_s \in s \subseteq N :$$

for details, see (Owen, n.d.).

We will consider four GT allocation schemes that have been widely used in water resources: the Core, the Shapley Value, the Nucleolus, and the Nash–Harsanyi allocation schemes (Dinar and Hogarth 2015).

9.5.6.2 The Core Allocation Scheme

The Core of an n player-cooperative game in the characteristic function form is a set of game allocation increasing that is not dominated by any other allocation set. The Core game theory (CGT) provides a locus for the maximum (or minimum in terms of cost) allocation each player may request. In this respect, it is an overall solution for several allocation schemes that are contained within the Core. The CGT scheme fulfills requirements for individual and group rationality, and for joint efficiency (Martin. Shubik 1982b).

CGT scheme is conducted under the assumption that the players in the game are economically rational. This means that the decision of each player to join a given coalition is voluntary, and it is based on the minimal cost they bear by joining that coalition.[1] Let ω_j be the j-th Core player allocation for the cost from the game. In case of a cost allocation game, the CGT equations can be defined as:

---

[1] In a benefit game it is the incremental benefit that such players gain.

$$\omega_i \leq \vartheta(\{j\}), \forall_j \in N,$$

$$\sum_{j \in s} \omega_i \leq \vartheta(\{s\}), \forall_s \in S,$$

$$\sum_{j \in N} \omega_i = \vartheta(N).$$

The first inequality in the CGT resource allocation scheme fulfills the conditions for individual rationality, i.e., the cooperative solution for each player is preferred to the non-cooperation case. The second inequality fulfills the group rationality conditions, meaning that the cooperative allocation to any combination of players is preferred regarding any allocation in any sub-coalition the player could establish. The third inequality fulfills the efficiency condition, which is the joint cost to be fully covered by the grand coalition participants. The system of these three inequalities has more than one allocation solution. A method of calculating the extreme points of the Core (Shapley 1971) provides the incremental contributions of each player when joining any existing coalition, and assigns these contributions to that player. Thus, having a non-empty Core allocation for a cooperative game provides the necessary condition for a solution that will be acceptable to the players (Dinar and Hogarth 2015).

9.5.6.3 The Nucleolus Allocation Scheme

The Nucleolus (Schmeidler 1969a) is a single point solution that always exists (if the Core is non-empty) and minimizes the dissatisfaction of the most dissatisfied coalition. To obtain the Nucleolus, we define the ε-core of the game to be the set of allocations that would be in the Core if each coalition given a subsidy at the same level of ε. By varying ε one finds the smallest non-empty ε-core, namely the *least* Core. The least Core is the intersection of all ε-cores. The least Core for a cost allocation game satisfies:

$$\text{minimize } \varepsilon$$

$$s.t. \sum_{j \in N} \omega_j \leq v(s) + \varepsilon, \forall s \subseteq S$$

$$\sum_{j \in N} \omega_j = v(N),$$

$$\varepsilon \lesseqgtr 0.$$

The solution to the minimization problem above may provide the Nucleolus (as a single solution) but it may also provide several individual cost allocations ω_j for the same value of ε for each coalition s. In this case, we define the *excess function* e(ε, s) for each s, that measures how much less a coalition costs to act alone, and in a lexicographical process (Schmeidler 1969b) obtain the Nucleolus, for which the value of the smallest excess e(ε, s) is as large as possible. The interpretation of ε is interesting. It can be used as a tax or a subsidy to change the size of the Core. If the Core is empty, then e ($\varepsilon < 0$) is an "organizational fee" for the players in sub-coalitions, causing them to prefer the grand coalition. If the Core is too big, ε might reduce it ($\varepsilon > 0$) by subsidizing sub-coalitions. The Nucleolus is always in the Core if it exists (Dinar and Hogarth 2015).

9.5.6.4 The Shapley Value Allocation Scheme

The Shapley Value (Shapley 1952) resource allocation scheme allocates θ_j to each player based on the weighted average of their contributions to all possible coalitions and sequences. In the calculation of the Shapley Value, an equal probability is assigned to the formation of any coalition of the same size, assuming all possible sequences of formation. The Shapley value can be calculated as (Dinar and Hogarth 2015):

$$\theta_j = \sum_{s \subseteq S j \in s} \frac{(n - |s|)!(|s| - 1)!}{n!} [v(s) - v(s - \{j\})], \forall j \in N,$$

where n is the number of players in the game, $|s|$ is the number of members in coalition s, i.e., the cardinality of the subset s; the function $v(.)$ is a characteristic function. It mean that, if s is a coalition of players, then $v(s)$ representing the worth of coalition s, describes the total expected sum of payoffs of the members of S; indeed, the payoffs maximization can be obtained by cooperation (Dinar and Hogarth 2015).

9.5.6.5 The Nash–Harsanyi (N–H) Allocation Scheme

The N–H Solution (Harsanyi 1958; Dinar and Hogarth 2015) for an n-person bargaining game is a modification of the 2-player Nash Solution (Nash 1953). This solution concept maximizes the product of the grand coalition members' additional utilities (income or savings) from cooperation compared to the non-cooperation case, subject to Core conditions, by equating the utility gains of all players. The N–H solution satisfies the Nash axioms (Nash 1953); it is unique and it is contained in the Core (if it exists). The solution might provide unfair allocations if there are big utility differences between the players, e.g., very rich player and very poor player.

The N–H solution for the j-th player, h_j, is calculated as:

$$\text{maximize} \prod_{j \in N} (f^j - h_j)$$

$$s.t.h_j = f^N, \forall_j \subseteq N,$$

$$\sum_{j \epsilon s} h_j \leq f^s, \forall_s \subseteq S,$$

$$\sum_{j \epsilon N} h_j = f^N$$

where h_j is the N–H allocation that satisfies efficiency and individual rationality conditions.

The fulfillment of the Core conditions for an allocation scheme is a necessary condition for its acceptability by the players. Thus, solutions not included in the Core are also not stable. Although an allocation scheme may fulfill the Core requirements for the regional game, it still may not be accepted by some players that might view it as relatively unfair compared to another allocation. Allocations that are viewed as unfair by some players are less stable. Some players might threaten to leave the grand coalition and form sub-coalitions because of their critical situation in the grand coalition. The consistency of any solution is essential given the existence of constants investments, and a more fix solution might be preferred even if it is harder to perform. We do not discuss coalitional stability here. The reader is referred for more reading to accessing (Shapley and Shubik 1954) and (Loehman et al. 1979), who used a measure of power in voting games. This power index is also used in (Williams 1988). Another measure of stability was introduced in (Dermot Gately 1974b) as the "propensity to disrupt" the grand coalition and was modified and applied considering N > 3 by (Straffin and Heaney 1981) to the case of the Tennessee Valley (Dinar and Hogarth 2015).

9.5.7 A Strategy for Water Resources Management Using Game Theory

For water resources management using GT approach on a river basin scale, mainly includes three important steps (Fig. 9.3), as follows:

(1) First important step is to decompose the river system and define the conflicting areas and/or bodies. After the players are defined, their moves (or action) and strategies, as well as their information set, and their payoff function can be defined.
(2) In the second step is defined how each player optimizes water quantity in order to maximize his payoff. Rather, this step includes the socio-macroeconomic

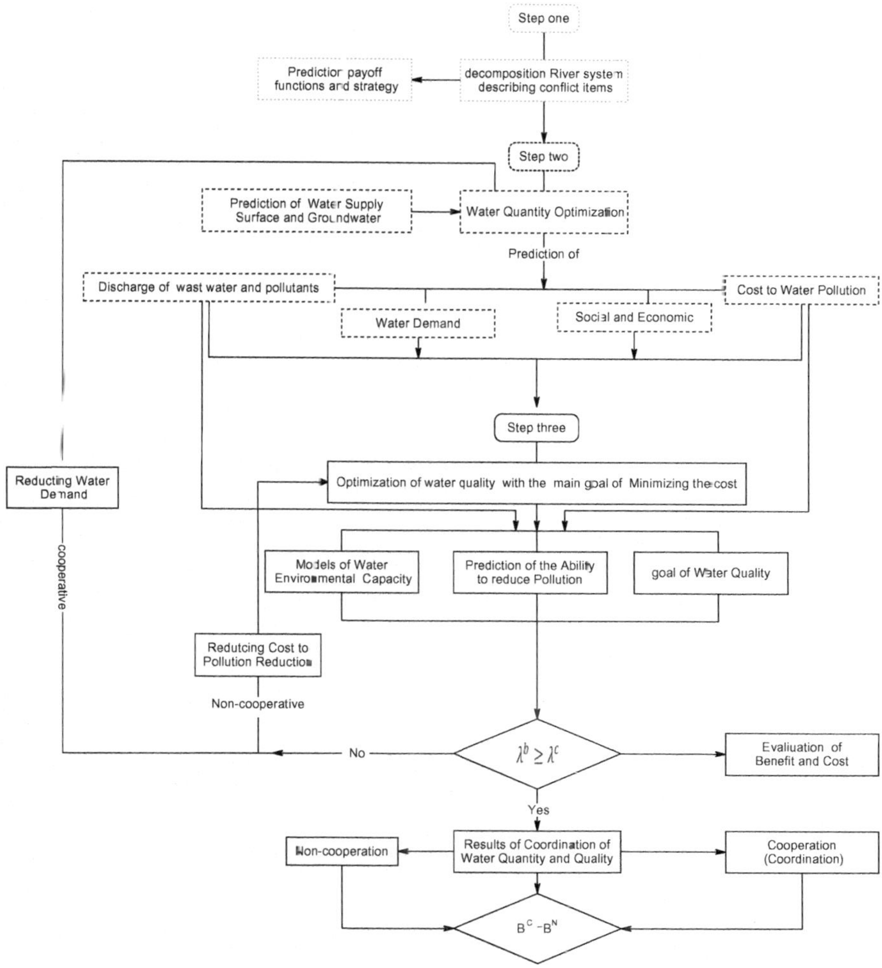

Fig. 9.3 A strategy for water resources management using game theory (Wei 2008)

predictions,[2] water supply and water demand predictions for different players, wasting water and pollutants predictions discarded by different players, as well as the cost of each player investment to treat his sewage. Step 2 is the benefiting process in which each player usually maximizes the output values per unit water (Wei 2008).

(3) Third step is to optimize water quality so that every player can maximize his payoff. This step consists of setting up models of pollutant capability in different

[2]Such as population, GDP, output values of agriculture and industry and the net incomes of household.

river sections, predicting each player's ability to reduce wasting water discharge and treat water pollution, while setting a target for water quality or water quality standard. In this step each player decides if they impose cost to reduce waste. The rational players will make planning by calculating the benefits and costs. From an economic point of view, waste production or pollution is public good or bad. In the non-cooperative situation, each player usually cut the waste treatment cost because he can freeride on other players' achievement of waste reduction. If all the players choose the strategies of free riding, equilibrium of prisoner dilemma will be reached. In the cooperative situation, the players will maximize their welfare by efficient water use (Wei 2008).

9.6 Practical Examples

9.6.1 Cooperative Water Allocation: A Cooperative GT Approach

Water allocation is essentially a practice in allocating available water to different demanding users. Water allocations merely based on a water rights approach, always do not make efficient use of water for the whole river basin. Meanwhile, an economy. Efficient water allocation plan cannot be well implemented if the involved participants or stakeholders do not regard it as being fair. in the study which is done by (Wang 2003), an equitable and efficient cooperative allocation approach had been proposed to solve water allocation problems in two steps. Water rights are initially allocated to water stakeholders and users based on existing water rights systems or agreements, and then the water had been reallocated to achieve efficient use of water through water transfers. The associated net benefit reallocation had been carried out by the application of cooperative game theory. The integrated cooperative water allocation modeling approach had been designed to promote and guide equitable water transfers and cooperation of relevant stakeholders to achieve optimally economic and environmental values of water, subject to hydrological and other constraints. A cooperative game theoretic approach which is proposed to solve water allocation problems is in two steps (Wang 2003):

(1) **initial water rights allocation** to water consumer or stakeholders based on existing water rights systems or agreements
(2) **reallocation of water to achieve** efficient consumption of water by water transfers.

A rational example is utilized to show the effectiveness and potential benefit of this approach.

9.6.1.1 Initial Water Rights Allocation

Generalized transboundary water allocation principles for sharing the water resources of international river basins between countries include:

(1) absolute sovereignty,
(2) absolute riverine integrity,
(3) limited territorial sovereignty
(4) economic criteria (Dinar and Wolf 1994; Wang 2003).

The seemingly fair and simple principles or guidelines of reasonable and equitable use are difficult to be applied in practice, especially for an inter-country river basin. Measurable criteria and models for water allocation need to be constructed and used to achieve fair apportionment of water (Seyam et al. 2000; Van Der Zaag et al. 2002).

In a water allocation problem, resource users have heterogeneities arising from physical resource characteristics, users' technologies and skill levels, and institutional arrangements. An institution can cause heterogeneities in pricing, property rights and political power (Schlager and Resource, n.d.).The water rights are allocated according to a legal intra-country water rights system and water policies or inter-country agreements before moving to the second gaming stage of the cooperative water allocation model (Wang 2003).

9.6.1.2 Cooperative Water Allocation Game

Recall that $N = \{1, 2, \ldots, n\}$ is the set of water stakeholders or *players* competing for water allocations in the concerned river basin or sub-watershed, and iN a typical stakeholder. A group of stakeholders SN entering a cooperative agreement and working together is called a coalition. N itself is called the grand coalition, the coalition consisting of all stakeholders (Wang 2003).

A coalition structure is a partition $\pi = \{s_1, s_2, \ldots, s_m\}$ of the n stakeholders, in which $\cup_i^m s_i = N$ and for all i $\neq$ j, $S_i - S_j = \emptyset$. For a game with n players, 2^n coalitions are possible, or $2^n - 1$ if the null coalition is excluded. The expression $v(S)$ is used to represent the aggregate payoff to the members of coalition S, while the *payoffs* to individual stakeholders acting in isolation are represented as $v(\{1\}), v(\{2\}), \ldots, v(\{n\})$. In a cooperative water allocation game, the generic notations of *payoffs* $v(\{i\})$ and $v(S)$ are interpreted specifically as the net benefits by the following definitions. The payoff $v(\{i\})$ of a stakeholder i is the maximum total net benefit $NB(i)$ that stakeholder i can gain based on its water rights over the entire planning period, subject to not decreasing the water flows and not increasing the pollutant concentrations in the flows to other stakeholders. Thus, the *payoff* $v(\{i\})$ is normally greater than the total net benefit $NB(i)$ gained with the initial water rights since there is additional value for the internal cooperation among the uses and users within stakeholder i (Wang 2003). Thus, the *payoff* $v(\{i\})$ optimization problem can be formulated as the maximization problem of the *total net benefit* (NB) subject to the water balance and hydrological constraints:

$$\vartheta(\{i\}) = \text{maximize} NB(i) = \sum_{t\in T} NB_{i,t} = \sum_{t\in T}\sum_{j\in U_i} NB_{i,j,t}$$

$$\text{s.t.} Q(k, j, t) \geq Q_R(K, j, t), \forall k \in V, \forall j \in U \; and \; j \notin U_i$$

$$S(j, t) \geq S_R(j, t), \forall k \in V, \forall j \in RES \, and \; j \notin U_i$$

$$C_p(k, j, t) \geq C_{PR}(K, j, t), \forall k \in V, \forall j \in U \; and \; j \notin U_i$$

$$C_p(j, t) \geq C_{PR}(j, t), \forall k \in V, \forall j \in RES \, and \; j \notin U_i$$

where, RES is the set of reservoirs.

Moreover, the payoff $v(S)$ of a coalition S is the maximum total net benefit $NB(S)$ that coalition S can gain based on coalition members' water rights over the entire planning period, subject to not decreasing the water flows and not increasing the pollutant concentrations in the flows to other stakeholders not taking part in coalition S (Wang 2003). This total net benefit maximization from coalition S subject to the water balance and hydrological constraints can be formulated as:

$$\vartheta(S) = \text{maximize NB(S)} = \sum_{i\in T}\sum_{i\in s} NB_{i,t} = \sum_{t\in T}\sum_{i\in s}\sum_{j\in U_i} NB_{i,j,t}$$

$$\text{s.t.} Q(k, j, t) \geq Q_R(K, j, t), \forall k \in V, \forall j \in U \; and \; j \notin U_s$$

$$S(j, t) \geq S_R(j, t), \forall k \in V, \forall j \in RES \, and \; j \notin U_s$$

$$C_p(k, j, t) \geq C_{PR}(K, j, t), \forall k \in V, \forall j \in U \; and \; j \notin U_{is}$$

$$C_p(j, t) \geq C_{PR}(j, t), \forall k \in V, \forall j \in RES \, and \; j \notin U_{is}$$

where, $U_S = \underset{i\in S}{-} U_i$, and $NB_{i,j,t}$ is the net benefit function of stakeholder i' s water demand node j during time step t, given by:

$$NB_{i,j,t} = f_{i,j,t} Q(k_1, j, t), C_P(k_1, j, t), S(j, t), C_P(j, t),$$
$$Q(j, k_2, t), C(j, k_2, t), \; (k_1, j) \in A, (j, k_2)$$

The net benefit function and cost function for demand node j is determined by:

$$f_{i,j,t}(.) = B_{i.j,t}(.) - C_{i.j,t}(.),$$

where $B_{i,j,t}(.)$, and $C_{i,j,t}(.)$ is the benefit function and cost function for demand node j, respectively.

Notice that $f_{i,j,t}(.)$ can be estimated from historical data statistics and simulation or obtained through optimization with control variables such as use type, area, user's technology and skill level, price, and other economic and policy factors. Note that in the latter case, $Q(k_1, j, t)$, $C_P(k_1, j, t)$, $S(j, t)$, $C_P(j, t)$, $Q(j, k_2, t)$, and $C(j, k_2, t)$ are the control variables deployed in searching for $v(S)$, as well as parameters in searching for the optimal value of $f_{i,j,t}(.)$ (Wang 2003).

A "*solution*" to a game is a vector of the payoffs received by each stakeholder. This payoff or reward vector after a trade can be written as $x = \{x_1, x_2, \ldots, x_n\}$. This trade process to achieve a cooperative water allocation under certain water balance and hydrological constraints is essentially a cooperative water allocation game. The payoff vector is called an imputation to the cooperative game, and meets the conditions of individual rationality, group rationality and joint efficiency (Young et al. 1982; Tisdell and Harrison 1992; Wang 2003), i.e.:

Individual rationality: $x_i \geq v(\{i\})$

Group rationality: $\sum_{i \in S} x_i \geq \vartheta(S)$

Joint efficiency: $\sum_{i \in N} x_i = \vartheta(N)$

Let $x(S) = \sum_{i \in S} x_i$, then the above three conditions can be reduced to:

Individual and group rationality: $x(S)(S)$, for all $S \subset N$

Joint efficiency: $x(N) = v(N)$

The set of reward payoff vectors that satisfy the conditions of individual rationality, group rationality and joint efficiency forms the *core of a cooperative game*. The core of a cooperative game may not always exist. If it exists, there is no guarantee that it has a unique feasible solution. Core-based and non-core-based resource allocation concepts may be applied to reduce it to a unique one (Dinar et al. 1986). Nucleolus and related solutions are listed in Table 9.1. The nucleolus minimizes the maximum excess $e(S, X) = v(S) - x(S)$ of any coalition S lexicographically (Schmeidler 1969a; Wang 2003):

Table 9.1 Nucleolus and related solutions (Wang 2003)

Solution concepts	Net benefit excess	Individual and group rationalities
Nucleolus	$e = v(S) - x(S)$	*min e* *Subject to* $x(S) + e \geq e = v(S)$ *for all* $S \subset N$
Nucleolus	$e_w = {(v(S) - x(S))}/{\lvert S \rvert}$	*min* e_w *Subject to* $x(S) + e_w \lvert S \rvert \geq v(S)$ *for all* $S \subset N$
Proportion nucleolus	$e_P = {(v(S) - x(S))}/{v(S)}$	*min* e_P *Subject to* $x(S) + e_P v(S) \geq v(S)$ *for all* $S \subset N$
Normalized nucleolus	$e_n = {(v(S) - x(S))}/{x(S)}$	*min* e_n *Subject to* $x(S) + (1 + e_n) \geq v(S)$ *for all* $S \subset N$

$$\text{minimize } v(S) - \sum\nolimits_{i \in S} x_i$$

$$\text{s.t.} : x(S) + e \geq v(S) \text{ for all } S \subset N$$

$$x(N) = v(S)$$

Application of this optimizing algorithm narrows the core solution space. Successive applications of the algorithm involve setting aside coalitions for which $e(S, X)$ equals the *critical excess* e_{crit} value found at each step and running the optimization program for remaining coalitions. Each iteration further constrains the solution space until a unique point is ultimately reached. The excess e can be interpreted as *subsidies* ($e \geq 0$) or *tax* ($e < 0$) to the water stakeholders. The weak nucleolus concept (Young et al. 1982) replaces the excess e by the *average excess* e_{avg}.

Proportional nucleolus (Young et al. 1982) replaces excess e by the ratio of excess to net benefit of coalition S; while the normalized nucleolus replaces excess e with the ratio of excess to imputation of coalition S. The nucleolus and related variation approaches can reduce or expand the core to obtain a unique solution in both cases of large core and empty core (Dinar et al. 1986). Applying the Shapley value solution concept, each stakeholder's reward or value to the game should be equal a weighted average of the contributions the stakeholder makes to each coalition of which he or she is a member. The weighting depends on the number of total stakeholders and the number of stakeholders in each coalition. The Shapley value gives the *payoff* to the i-th stakeholder such that (Shapley 1971; Wang 2003):

$$\begin{aligned} \mathrm{X}_i &= \sum_{\substack{S \subseteq N \\ i \in s}} \frac{((|S|-1)!(|N|-|S|)!}{|N|!} [\mathrm{V}(S) - \mathrm{V}(\mathrm{S}) - \{i\}] \\ &= \sum_{\mathrm{s}=1}^{\mathrm{n}} \frac{((\mathrm{s}-1)!(\mathrm{n}-\mathrm{s})!)}{\mathrm{n}!} [\mathrm{v}(\mathrm{s}) - \mathrm{v}(\mathrm{s} - \{i\}], \text{ for all } i \in N \end{aligned}$$

where $|S|$ is the cardinality of coalition S.

By using the above description, a two-step cooperative water allocation approach can be formulated, which consists of an initial water rights allocation and a cooperative water reallocation game. Water rights are initially allocated based on existing water rights systems or agreements, while the cooperative water reallocation game is formulated by using net benefits as a stakeholder's payoff. The cooperative water reallocation game can be solved by solution concepts such as the nucleolus, weak nucleolus, proportional nucleolus, normalized nucleolus and Shapley value. Since the model performs initial water rights allocation and subsequent reallocation based on existing water rights systems or agreements, and it utilizes the node-link river basin network, water balance and hydrological constraints, with a time step length of Δt during a planning period, the model realistically takes into account knowledge and sub-models from hydrology, economics and cooperative GT. This makes it possible to reach fair and efficient water allocation among competing uses with multiple stakeholders in an operational way. The methodology can be applied to an entire river basin or a sub-watershed (Wang 2003).

9.6.2 Water Costs Allocation in Complex Systems Using a Cooperative GT

Sechi et al. (2013) present a methodology to allocate water service charges in a water resource system between several users that attempts to fulfill the WFD requirements. The method is according to Cooperative Game Theory (CGT) techniques, while the related characteristic function definition deploys a mathematical optimization approach. The CGT provides the facilities that are essential to analyze condition that needs a cost-sharing rule. The CGT approach can describe efficient and fair solutions that supply the appropriate incentives between the parties involved. So, the water system value allocation has been costed as a game in which it is essential to determine the right payoff for each player, in this case water costumers. To use the CGT principles in a water resources system, the specific function needs to be defined and evaluated using enough modeling approach; in this study, it is evaluated using the WARGI optimization model. The so-called "core" represents the game-solution set. It represents the area of the possible cost allocation values from which the borders on the cost values for each player can be provided. Within the core lie, all of the allocations must satisfy the principles of equity, fairness, justice, performance, and that guarantee cost recovery. The core of a cooperative game as a reliable support in water resource management to attain the economic analysis required with WFD. This methodology was applied to a multi-reservoir and multi-demand water system in Sardinia, Italy (Sechi et al. 2013).

9.6.2.1 Cost Allocation Problem and Cooperative Game Theory

The CGT belongs to the Game Theory (GT) scientific area developed in the first half of the last century (Von Neumann and Morgenstern 1953). In GT, conflict situations are analyzed and competitive and/or cooperative solutions among participants sought. In the literature, many cost allocation problems have been analyzed using CGT principles; however, the approaches vary significantly in the different research fields. The CGT principles have also been applied in studies related to water resources (Authority 1938; Young et al. 1980a; Lippai and Heaney 2000; Andreu et al. 2009). One of the most important aspects of the methodology is the definition of the game's characteristic function (CF), which is the focus of this study. To define the CF of the cooperative game for a water system the following definitions, which were described extensively by (Barile and Stoner 1994) are required. $N = (1, 2, .., n)$ refers to a set of players that are participating in the game. Each subset $S \subseteq N$ is defined as a "coalition", while the Grand Coalition (GC) occurs when $S = N$. The players can represent real subjects, such as the users of a water system, or members of a more abstract set, such as the sector of a company, or different planning alternatives that can be realized together or separately. The *stand-alone cost*, given as $c(i)$, represents the cost that is connected to the i-th user when the user is considered independent of the other players. The cost linked to the coalition S, i.e., the cost commonly sustained by all of the users that belong to S is represented as $c(S)$. Besides, the cost associated with the GC, i.e., the common cost sustained by all participants of the game (all users in the water system) is represented as $c(N)$. Finally, the cost linked to an empty coalition is zero by convention,$(c(\Phi) = 0$ (Sechi et al. 2013).

As defined previously, an allocation is a vector $x = [x_1, x_2, \ldots x_n]$, where x_i is the amount charged to the i-th player. The cost associated with a generic coalition, which can be formed by either a player, a partial coalition, or even all participants of the game, must represent the lowest cost of serving the coalition in the most efficient way, i.e., the minimum cost necessary to satisfy all of the players in such coalition. Moreover, the discrete function that is formed by the costs of every coalition is called the *Characteristic Function* (CF), which is the key element setting a cooperative game (Sechi et al. 2013).

If for every pair of disjoint coalitions *S*' and *S*",

$$c(S' \cup S'') \leq c(S') + c(S'')$$

then, the CF and the related game are sub-additives. In this case, the players cooperate because the unions of the two groups of players will determine a cost that is lower than the sum of the autonomous costs. Because a game with a sub-additive CF will be characterized by economies of scale, the GC should be the most efficient alternative. This is the case when it is economically more convenient to realise a common project than independent projects. For players that cooperate and collectively accomplish a project, the principle of stand-alone cost test, commonly called the *rationality principle*, must be guaranteed (Dinar and Hogarth 2015). However, it can be extended to each player and thus, it is also referred to as *individual rationality*. The principle of *individual rationality* can refer to an individual player or to a coalition, satisfying (Sechi et al. 2013):

$$x_i \leq c(i) \forall i \in N$$

$$\sum_{i \in s} x_i \leq c(S) \forall \mathrm{S} \subseteq N$$

where x_i is the amount of total game cost that is assigned to a given player. According to this principle, no player or group of players that forms a coalition would accept a cost assignment that is higher than the cost that it/they would sustain when participating autonomously, i.e., greater than its/their own opportunity cost.

Another principle is the so-called *marginality principle* or incremental cost test. In general, the incremental cost or the marginal cost of a coalition S is defined as $c_m = c(N) - c(N - S)$. According to the marginality principle, the following inequality:

$$\sum_{i \in s} x_i \geq c(N) - c(N - S) \forall S \subseteq N$$

must be verified (Sechi et al. 2013). Each player or set of players will have to sustain at least its/their own marginal cost c_m when joining a coalition. Otherwise, the coalition of pre-existing players will be inefficient because it has to finance the entry of the new player or set of players. The *rationality principle* produces an incentive for the players to cooperate voluntarily, while the *marginality principle* supplies the equity conditions in such game (Peter et al. 1994).

Furthermore, the cooperative-based games support the solutions that include all players, and therefore, the majority of the CGT solving methods are able to divide completely the cost among all of the game participants. Taking into account these aspects, a generic solution is defined by a vector $x = [x_1, x_2, \ldots, x_n]$, such that (Sechi et al. 2013):

$$\sum_{i \in N} x_i = c(N)$$

where x_i is the payoff assigned to the ith player. This generic solution satisfies the *efficiency principle;* besides, in such configuration, the *marginality* and *rationality* principles are equivalent. In terms of *individual rationality*, the amount i saves by cooperating rather than going alone is given by $v_i = c(i) - x_i$. As a result, one can define the *group rationality* saving:

$$v(S) = c(S) - \sum_{i \in s} x_i \forall S \subseteq N$$

The main application problems come out evaluating the *characteristic function*: each potential coalition must be defined and assessed. Therefore, the number of players affects the complexity of the problem; hence, for n players, there are $2^n - 1$ coalitions that must be analyzed (Sechi et al. 2013).

The *game solutions* can be grouped into two branches:

set-theoretical solutions: identify a set of vectors that shares the value of the game among all players, as the core;

point solutions: define only one division and are more similar to the classic idea of a unique problem solution, as the Shapley value or the so-called nucleolus (Schmeidler 1969; Sechi et al. 2013).

The adopted game solution should guarantee an acceptable cost allocation considering the particular characteristics belonging to the water system and requirements given by the water authorities and other decision makers. In (Sechi et al. 2013), the core of the solutions is used under de assumption that the core is a closed, compact, convex subset in R^N. Unfortunately, it may be empty, even if c is sub-additive. Moreover, inside the core, there are several cost allocations, which respect the efficiency and equity principles while incentivize cooperation among the players. Consequently, the decision maker is provided with an admissible, potentially easily acceptable range of alternatives for defining water rates. These aspects will be examined in the following revisited case study (Sechi et al. 2013).

9.6.2.2 Water Cost Allocation Methodology

The CGT approach is particularly appropriate for water services, in which it is important to define the agreements, encourage cooperation among the stakeholders and achieve more efficient solutions by determining a fair cost allocation. The proposed methodology, which uses CGT techniques to allocate the costs in a complex water resources system, consists of the following steps, further detailed in (Andreu et al. 2009; Andreu et al. 2009; Sechi et al. 2013):

(1) *Water system analysis*: functional definition of the water system and evaluation of its different aspects, e.g., hydrologic, hydraulic, infrastructural and economic;
(2) *Cooperative game definition*: identification of players and coalitions, analysis of priorities and so on;
(3) *Characteristic function* (CF) *evaluation*: set up of the optimization model, calculation of the minimum cost associated with each coalition potential solution, and game's CF evaluation;
(4) *Game solution*: application of the CGT methods necessary to share the costs among the players.

First step: the hydrologic, hydraulic and infrastructural aspects of the description and the characterization of the water system must be identified in this step. During this phase, the costs that characterize the water system, which are to be shared among the water users must be defined. In fact, the majority of European water systems are almost entirely equipped and new important works are rarely expected. Therefore, it is mainly the management costs of the existing infrastructure that need to be allocated. In this case, the ordinary and supplementary maintenance costs, the adaptation and substitution infrastructure costs and the energy costs of the pumping stations need to be considered (Sechi et al. 2013).

Second step: identifies the players and which cooperative game is set up. The players can represent individual users, sets of users or more abstract memberships, such as sectors of water services, agricultural associations and city services. In the following real case application, the elements of the players' set consist of users belonging to a unique macro-demand having the same interests and priorities of irrigational, industrial and urban municipal demands.

Third step: the CF of the game is defined. According to its definition, the CF consists of a set of minimum costs associated with all of the possible coalitions. The need to value each coalition's minimum costs is a key feature of CGT. Even if the costs of GC are to be shared, each coalition needs to be valued in order to estimate the parameters for efficient cost sharing among the players. Nevertheless, in CGT applications, the coalition's minimum cost is defined as the sum of the management costs in the "minimum" infrastructures set necessary to completely satisfy the water request of the players included in that coalition. This modelling method has significant differences from that described by Deidda et al. (2009). Actually, the approach considered herein specifically refers to water system management and the CF evaluation is reached using the optimization modelling tool WARGI-DSS. The WARGI. tool allows apply Linear (LP) and Quadratic (QP) Programming models, to obtain the optimum system infrastructures set definition and the optimum system performance to be achieved for each coalition. Indeed, the optimization model can be easily built using the WARGI graphical interface and solved using Cplex optimizer tool. Depending on the system infrastructure, on the system sources, as well as on the demand characterization, the number and the typology of the potential infrastructures should be varied for each coalition game. Consequently, we can evaluate the management costs of the entire system referring to the optimal flows assessment given by the WARGI optimization tool and the solution values that are associated to a specific coalition solution. In this manner, the *least cost* (optimal system assessment) of each coalition can be defined, and the CF of the cooperative game can be established (Sechi et al. 2013).

Last Phase: In the game solution, the water system costs are allocated among the players using CGT techniques. In this way, CGT gives an admissible range providing an easily acceptable tool to the decision maker for defining water costs allocation (Sechi et al. 2013).

9.6.2.3 Result

The core solutions-set of the game is graphically represented in Fig. 9.4. In the triangle, the heights are proportional to the cost of the Grand Coalition and each internal point represents a possible cost allocation between the macro-users defined as players in the game. Every side represents a player and the distance between the side and the point inside the triangle provides the cost amount that is assigned to the player. The barycenter is the point at which the costs are equally shared, whereas the vertices correspond to the situation in which the total cost is assigned to one user. The dashed lines represent the maximum and the minimum costs that are sustainable by each player according to below equation and the painted area represents the core solution of the game. The analytical formulation for the core deploying GC can be defined by the following cost boundaries:

$$civil + Irrigation + Industrial = 293.35$$

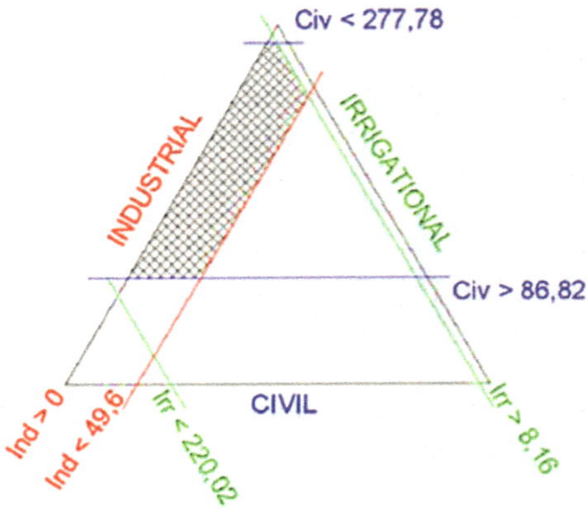

Fig. 9.4 Core of the game (Sechi et al. 2013)

$$86.82 \leq Civil \leq 277.78$$

$$8.16 \leq Irrigation \leq 220.02$$

$$0.00 \leq Industrial \leq 49.60$$

Each cost allocation that verifies the above boundaries also satisfies the rationality and marginality principles and guarantees a total cost recovery (Sechi et al. 2013).

The results presented in (Sechi et al. 2013) demonstrated that the evaluation of the CGT core of solutions represents the *set of admissible cost allocation* and supplies the boundary values for each player. Inside the core, each allocation satisfies the *marginality* and *rationality* principles; hence, the stakeholders should recognize equity and fairness. Moreover, the total cost recovery can be realized (Sechi et al. 2013).

9.6.3 GT Application for Groundwater Conflicts Resolution

In another study carried out by, game theory was applied to a multi-objective conflict problem for the Alto Rio Lerma Irrigation District, located in the state of Guanajuato, in Mexico, where economic benefits from agricultural production should be balanced with associated negative environmental impacts. The short period of rainfall in this area, combined with high groundwater withdrawals from irrigation wells, has produced severe aquifer overdraft. In addition, current agricultural practices of applying high loads of fertilizers and pesticides have contaminated regions of the aquifer. The net economic benefit to this agricultural region in the short-term lies with increasing crop yields, which require large pumping extractions for irrigation, as well as high chemical loading. In the longer term, this can produce economic loss due to higher pumping costs, i.e., higher lift requirements, or even loss of the aquifer as a viable source of water. Negative environmental impacts include continued diminishment of groundwater quality, and declining groundwater levels in the basin, which can damage surface water systems that support environmental habitats. The two primary stakeholders or players, the farmers in the irrigation district and the community at large, must find an optimal balance between positive economic benefits and negative environmental impacts. In GT was applied to find the optimal solution between the two conflicting objectives among twelve alternative groundwater extraction scenarios. Different attributes were used to quantify the benefits and costs of the two objectives; hence, following the Pareto frontier generation (trade-off curve), four conflict resolution methods have been identified and applied accordingly.

Step 1: water management problem

Modeling real water management problems, possible groundwater extraction scenarios have been proposed. The environmental and economic impacts of each groundwater scenario were measured by using the Groundwater Lading Effects of Agricultural Management Systems (GLEAMS), and then the identified water resource optimization problems solved by linear programming. Finally, conflict resolution methodology is applied to identify compromise solutions, which balances the economic and environmental concerns of the region.

Step 2: attributes estimation

Different groundwater extraction scenarios were proposed. For each groundwater extraction scenario, we have conflicting economic and environmental objectives. The economic attributes are the net income generated in the linear program and the pumping cost described below. The environmental attributes include nutrients and pesticides associated with irrigation runoff and percolation and a measure of groundwater depletion. Figure 9.5 shows the general hierarchy of the criteria. Table 9.2

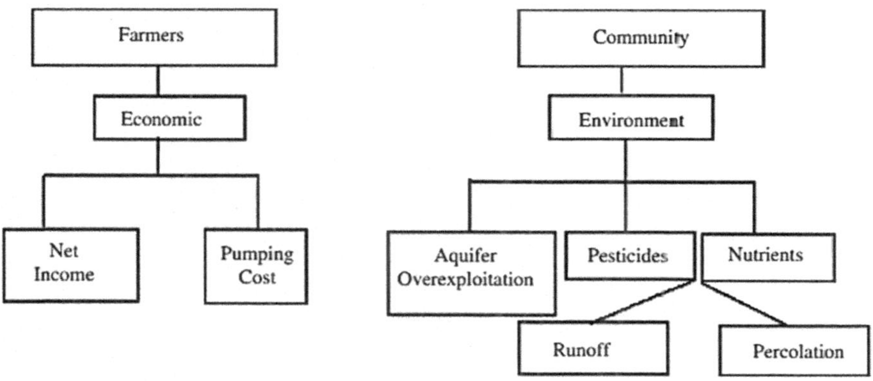

Fig. 9.5 Conflict resolution scheme

Table 9.2 List of attributes

Attribute	Estimation procedure
Economics Net benefits (106 $)	Generated by the linear program for each alternative
Pumping cost (106 $)	Calculated separately using data from ARLID and subtracted as production cost
Environmental Nitrogen in runoff (103 kg) Nitrate in percolation (10 crop 3 kg) Pesticides in runoff (103 g) Pesticides in percolation (103 g)	Output from GLEAMS for each crop
Aquifer overexploitation	Evaluated for each groundwater supply

presents a list of the alternatives and attributes, and show their estimation procedure, respectively.

Net Income: The farmers' net benefit for each groundwater extraction scenario was estimated by assuming that the farmers utilized crops selection optimally according to market prices and water availability. This can be formulated as a linear programming problem, where the total net benefit maximization can be defined as:

$$\text{maximize } NB = \sum_{j=1}^{n} [Y_j P_{cj} - C_j] A_{cj}$$

where the NB is defined as $(profit - cost\$)$, the A_{cj} holds for the j-th crop area (ha), Y_j is the yield of the j-th crop j (ton/ha), P_{cj} the price of crop j($/ton), the term C_j represents the production cost of crop j ($/ha), which also include pumping cost, and n is the number of crops. Regarding the constraints of the aforementioned net benefit maximization linear programming.

Pumping cost: The average pumping cost in month k is given as Cp Qk, where:

$$C_P = \left[\frac{h \times \mathrm{a}}{E_p}\right] C_e + C_{r,}$$

where a is the energy required to lift 1 m^3 of water to 1 m height (kWh/m^4), the variable h is the total head (m), C_e represents the average annual energy cost ($/kWh), E_p the pump efficiency, and C_r the repair cost ($/ha m).

Environmental attributes estimation: The environmental objective can be computed as a weighted sum of nitrates and pesticides in runoff and percolation, as well as aquifer overexploitation, which depend on crop volumes and water usage:

$$Env = Z \times 0.25 + P \times 0.25 + AO \times 0.50$$

where Z is the normalized measure of nitrates in runoff and percolation; P the normalized measure of pesticides in runoff and percolation and AO the aquifer overexploitation coefficient.

Step 3: conflict resolution methodology

Considering two conflicting objectives, the *feasible payoff* and the *worst payoff*, both conflicting objectives can be normalized in a way that zero value corresponds to the worst case and unit value to the best outcome. Hence, both objectives are now maximized. This conflict is mathematically defined by a pair (S, d), where $S \subseteq R^2$ is the *feasible payoff* set an $d \in R^2$ has the *worst* possible *payoff* values in its components. This vector is also known as the "nadir". The players want to increase their payoff values from these minimal values as much as possible. In the case of normalized objectives, $d_1 = d_2 = 0$. It is assumed that the Pareto frontier is given by the graph of a strictly decreasing concave function $g(.)$ defined in interval $\left[d_1, f_1^*\right]$, where $g\left(f_1^*\right) = d_2$, as depicted in Fig. 9.6. Herein, we also use the notation $\left(f_2^*\right) = g(d_1)$.

In many applications vector d is selected as the *current payoff vector* (called the "status quo" point), or the "*disagreement payoff*" vector, the components of which give the payoffs of the players in the case when they are unstable to reach an agreement. In such cases, the *feasible payoff* set S is restricted to the set.

$$S_+ = \{f = (f_1, f_2)/f \in S, f \geq d\},$$

since no rational player accepts an agreement which is worse than the outcome without an agreement or worse than the current situation. If vector d is selected as the nadir, then $S_+ = S$. The Nash-solution selects the unique point of the Pareto frontier, which maximizes the product of the gains from the disagreement payoff values. That is, the Nash solution is the unique solution of the following optimization problem (Optz-I):

Fig. 9.6 The area monotonic solution

$$\text{maximize}\,(f_1 - d_1)(f_2 - d_2)$$

$$\text{s.t.} d_1 \ll f_1 \ll f_1^*$$

$$f_2 = g(f_1).$$

Notice that at $f_1 = d_1$, and also at $f_1 = f_1^*$, the objective function is zero, and it is positive for all $f_1 \in (d_1, f_1^*)$. Therefore, the optimum is interior. The second constraint allows us to solve a single-dimensional problem:

$$\text{maximize}\,(f_1 - d_1)(g(f_1) - d_2)$$

$$\text{s.t.} d_1 \ll f_1 \ll f_1^*,$$

where a simple one-dimensional search algorithm can be used, or a single-dimensional monotonic equation can be solved based upon the first-order condition.

Four conflict resolution methods have been discussed in (Raquel et al. 2007b). These optimization methods are briefly revisited in the next subsections.

Method 1: Non-symmetric Nash Solution

The non-symmetric Nash solution is the unique optimal solution of the problem

$$\text{maximize}\,(f_1 - d_1)^{w1}(f_2 - d_2)^{w2}$$

$$\text{s.t.} d_1 \ll f_1 \ll f_1^*$$

$$f_2 = g(f_1),$$

where w_1 and w_2 are the powers of the two players, or the importance factors of their objectives. Clearly, it is a straightforward generalization of the previous formulation (Optz-I) but with unequal weights.

Method 2: Kalai–Smorodinsky Formulation

Method 2 uses the Kalai–Smorodinsky formulation, described as follows. Consider the linear segment between the disagreement point (d_1, d_2) and the "ideal" point $\in (f_1^*, f_2^*)$; then the solution is the unique intercept of this segment with the Pareto frontier. Hence, we have to compute the unique solution of equation

$$d_2 + \left\{\left(f_2^* - d_2/f_1^* - d_1\right)\right\}(f_1 - d_1) - g(f_1) = 0$$

in the interval $[d_1, f_1^*]$. If both objectives are normalized, then $(d_1 = d_2 = 0)$, and $(f_1^* = f_2^* = 1)$; so, along the linear segment connecting the disagreement and ideal points, the two objective $\bar{f}_1$ and $\bar{f}_2$ increase at the same rate. If the objectives have different importance weights, then the more important objective has to be improved more rapidly. This idea leads to the nonsymmetric Kalai–Smorodinsky solution that computes the unique intercept between the Pareto frontier and the straight line

$$\bar{g}(\bar{f}_1) = (w_1/w_2)\bar{f}_1,$$

where the two coordinate directions are the normalized objective functions.

Method 3: Area monotonic solution

The area monotonic solution is based on a linear segment starting at the disagreement point that divides S_+ into two subsets of equal area. If the conflict is not symmetric, meaning that $w_1 \neq w_2$, then we might define the non-symmetric area monotonic solution by requiring that the ratio of the areas of the two subsets be w_1/w_2. Hence, the first coordinate of the solution is the root of the nonlinear equation.

$$w_2\left[\int_{d_1}^{x} g(t)dt - 1/2(x - d_1)(g(x) + d_2)\right] = w_1\left[\int_{x}^{f_1^*} g(t)dt - (f_1^* - d_1)(g(x) + d_2)\right]$$

in the interval (d_1, f_1^*), as it is illustrated in Fig. 9.6.

Method 4: Equal loss solution

The equal loss solution was also originally developed for the symmetric case, when both payoffs are relaxed simultaneously with equal speed until an agreement is reached. If $w_1 \neq w_2$, then we may generalize this concept by requiring that the more important objective is relaxed slower than the other by assuming that the ratio of the relaxation speeds be equal to w_1/w_2. Therefore, one can determine a point $(x, g(x))$ on the Pareto frontier such that (Raquel et al. 2007b):

$$(f_1^* - x)w_1 = (f_2^* - g(x))w_2.$$

Notice that similarly to the other three methods, this is also a nonlinear equation for the single unknown x, which can be easily solved by using standard optimization methodology.

9.7 Summary

The results for the conflict resolution methods discussed in (Raquel et al. 2007b) demonstrate that the implemented linear programming model for each groundwater extraction scenario has produced less water extraction corresponds to less net income. The last result shows that farmers have the option of growing the most profitable crops if they know the future prices for the next season. The results obtained from the linear model also suggest leaving some land idle (without crops and irrigation) in order to maximize net income when water availability decreases.

The computed compromise solutions between the economic and environmental objectives applying different formulation methods in (Raquel et al. 2007b) have evidenced that the Kalai–Smorodinsky solution generated more uniformly distributed points along the Pareto frontier, demonstrating efficiency and effectiveness of the Method 2 in providing a multitude of solutions in the Pareto frontier. As expected, the more weight given to protecting the environment, the lower the optimal groundwater extraction volume for agricultural irrigation, with all four formulations exhibiting such basic feature.

The results of net income obtained in the four methods with different weights show that when applying Method 4, the net income increases linearly with increasing economic weight, while the remaining methods exhibit nonlinear behavior. When economic benefit is considered as the only objective, the optimal groundwater withdrawal attains its maximum level. At the other extreme, when only environment is considered, the optimal groundwater scenario is to extract the minimum volume of groundwater via the irrigation wells. When environment and economics are assigned equal importance one can provide water resource allocation solutions considering equilibrium among environment protection and the economic activities viability.

In conclusion, the results of the above and other studies indicated the noticeable performances of Game Theory approaches in solving water resources management issues. Therefore, we can consider this method as a strong tool in solving water management problems.

Acknowledgements The authors thank Iran's National Science Foundation (INSF) for the support of this research.

References

Adhami M, Sadeghi SH (2016) Sub-watershed prioritization based on sediment yield using game theory. J Hydrol 541:977–987

Adhami M, Sadeghi SH, Duttmann R, Sheikhmohammady M (2020) Best soil comanagement practices for two watersheds in Germany and Iran using game theory-based approaches. Sci Total Environ 698. doi:https://doi.org/10.1016/j.scitotenv.2019.134265

Alaghbandrad A, Hammad A (2018) FPP cost-sharing of multi-purpose utility tunnels. In: Lecture notes in computer science (Including Subseries Lecture notes in artificial intelligence and lecture notes in bioinformatics). Springer, pp 554–567. https://doi.org/10.1007/978-3-319-91635-4_29

Alahdin S, Ghafouri HR, Haghighi A (2018) Derivation of conflict resolution rule-curves in multi-purpose multi-reservoir system for inter-basin water transfer during drought. Environ Resour Res 6:51–68

Alchian AA (1965) The basis of some recent advances in the theory of management of the firm. J Ind Econ 14:30. https://doi.org/10.2307/2097649

An X, Li H, Wang L, Wang Z, Ding J, Cao Y (2018) Compensation mechanism for urban water environment treatment PPP project in China. J Clean Prod 201:246–253. https://doi.org/10.1016/j.jclepro.2018.08.003

Anderson JW (2016) Playing the dam game: how a proposed hydroelectric dam could trigger Central Asia's First Water War (thesis)

Andik B, Niksokhan MH (2020) Waste load allocation under uncertainty using game theory approach and simulation-optimization process. J Hydroinf. https://doi.org/10.2166/hydro.2020.181

Andreu J, Ángel Pérez-Martín M, Sechi GM, Zucca R, Andreu J, Pérez M, Sechi G, Zucca R, Zuddas P (2009) A cooperative game theory approach to water pricing in a complex water resource system. In: 18th World IMACS/MODSIM congress

Authority T (1938) Allocation of investment in Norris, Wheeler, and Wilson Projects, in: US House of Representatives Document, Congress Third Session. US Government Printing Office Washington

Barreiro-Gomez J, Riaño-Briceño G, Ocampo-Martínez C, Quijano N (2017) Data-driven evolutionary-game-based control for drinking-water networks. In: Advances in industrial control. Springer International Publishing, pp 363–383. doi:https://doi.org/10.1007/978-3-319-50751-4_18

Biddle GC, Steinberg R (1985) Common cost allocation in the firm. In: H. Young P (ed) Cost Alloc. methods, Princ Appl. North-holl

Chhipi-Shrestha G, Rodriguez M, Sadiq R (2019) Selection of sustainable municipal water reuse applications by multi-stakeholders using game theory. Sci Total Environ 650:2512–2526

Degefu DM, He W, Yuan L, Zhao JH (2016) Water allocation in transboundary river basins under water scarcity: a cooperative bargaining approach. Water Resour Manag 30:4451–4466

Dinar A, Dinar S, McCaffrey S (2007) Bridges over water- understanding transboundary water conflict, negotiation and cooperation

Dinar A, Dinar S, McCaffrey S, McKinney D (2013) Bridges over water: understanding transboundary water conflict, negotiationand cooperation second [WWW Document]. World Sci. Publ. Co. https://books.google.com/books?hl=en&lr=&id=3Ns7DQAAQBAJ&oi=fnd&pg=PR5&dq=Bridges+Over+Water+Understanding+Transboundary+Water+Conflict&ots=gX8VaAcFyl&sig=MQLgFSL8lkpTdjI1mQGTnu_9j-M#v=onepage&q=Bridges Over Water Understanding Transboundary Water Conflict&f=false. Accessed 27 April 2020

Dinar A, Hogarth M (2015) Game theory and water resources critical review of its contributions, progress and remaining challenges. Found Trends Microecon. https://doi.org/10.1561/0700000066

Dinar A, Howitt RE (1997) Mechanisms for allocation of environmental control cost: empirical tests of acceptability and stability. J Environ Manag 49:183–203. https://doi.org/10.1006/jema.1995.0088

Dinar A, Wolf A (1994) Economic potential and political considerations of regional water trade: the Western Middle East example. Resour Energy Econ 16:335–356. https://doi.org/10.1016/0928-7655(94)90025-6

Dinar A, Yaron D, Kannai Y (1986) Sharing regional cooperative gains from reusing effluent for irrigation. Water Resour Res 22:339–344. https://doi.org/10.1029/WR022i003p00339

Friedman D (1998) On economic applications of evolutionary game theory. J Evol Econ 8:15–43. https://doi.org/10.1007/s001910050054

Fu J, Zhong P-A, Zhu F, Chen J, Wu Y, Xu B (2018) Water resources allocation in transboundary river based on asymmetric Nash–Harsanyi Leader–Follower game model. Water 10:270
Gao F, Lin HH, Deng HW (2017) On the construction and application of the groundwater environment health assessment model based on the game theory-fuzzy matter element. J Saf Environ 17:1543–1547
García Mazo, C.M., Olaya, Y., Botero Botero, S., 2020. Investment in renewable energy considering game theory and wind-hydro diversification. Energy Strateg Rev. https://doi.org/10.1016/j.esr.2020.100447
Gardner R (1995) Games for business and economics. Wiley, New York
Gately D (1974a) Sharing the gains from regional cooperation: a game theoretic application to planning investment in electric power. Int Econ Rev (Philadelphia) 15:195–208
Gately D (1974b) Sharing the gains from regional cooperation: a game theoretic application to planning investment in electric power. Int Econ Rev (Philadelphia). 15:195. https://doi.org/10.2307/2526099
Getirana ACV, de Fátima Malta V (2010) Investigating Strategies of an Irrigation Conflict. Water Resour Manag 24:2893–2916. https://doi.org/10.1007/s11269-010-9586-z
Ghadimi S, Ketabchi H (2019) Possibility of cooperative management in groundwater resources using an evolutionary hydro-economic simulation-optimization model. J Hydrol 578. https://doi.org/10.1016/j.jhydrol.2019.124094
Gholami M., Shahnazari A, Mortezapour MR, Shahnourian MM (2017) Conflict resolution of sefidrud irrigation and drainage network using games theory. Iran-Water Resour Res:101–111
Gibbons R (1992) Game theory for applied economists. Princeton University Press
Gintis H (2001) Game theory evolving: a problem-centered introduction to modeling strategic. Princeton University Press, Princeton, New Jersey. https://doi.org/10.2307/41794873
Girard C, Rinaudo J-D, Pulido-Velazquez M (2016) Sharing the cost of river basin adaptation portfolios to climate change: Insights from social justice and cooperative game theory. Water Resour Res 52:7945–7962. https://doi.org/10.1002/2016WR018757
Goksu A, Bakalian A, Kingdom B, Saltiel G, Mumssen Y, Soppe G, Kolker J, Delmon V (2019) Reform and finance for the urban water supply and sanitation sector. Reform and finance for the urban water supply and sanitation sector. World Bank, New York. https://doi.org/10.1596/32244
Guo S (2016) Environmental options of local governments for regional air pollution joint control: application of evolutionary game theory. Econ Polit Stud 4:238–257
HP Young , Okada N, Hashimato T (1982) Cost allocation in water resources development. Water Resour Res 18:463–475
Hachoł J, Bondar-Nowakowska E, Hachaj PS (2019) Application of game theory against nature in the assessment of technical solutions used in river regulation in the context of aquatic plant protection. Sustainability 11:X. https://doi.org/10.3390/su11051260
Han Q, Tan G, Fu X, Mei Y, Yang Z (2018) Water resource optimal allocation based on multi-agent game theory of HanJiang river basin. Water 10:1184
Harsanyi JC (1958) A bargaining model for the cooperative n-person game. (Book, 1958) [WorldCat.org]
Hipel KW, Obeidi A (2005) Trade versus the environment: Strategic settlement from a systems engineering perspective. Syst Eng 8:211–233. https://doi.org/10.1002/sys.20031
Hone S, Crase L, Burton M, Cooper B, Gandhi VP, Ashfaq M, Lashari B, Ahmad B (2020) Farmer cooperation in participatory irrigation in south Asia: insights from game theory. Water (Switzerland) 12:1329. https://doi.org/10.3390/W12051329
Hu D, Liu H, Chen X, Chen Y (2019) Research on the ecological compensation standard of the basin pollution control project based on evolutionary game theory and by taking Xiangjiang River as a case. Front Eng Manag 6:575–583
Huang Y (2016) Multi-agent system for future groundwater depletion scenarios using game theory
Huang Y, Janovsky P, Das S, Welch SM, DeLoach S (2016) Multi-agent system for groundwater depletion using game theory. arXiv Prepr. arXiv1607.02376

Hui R, Lund JR, Madani K (2016) Game theory and risk-based leveed river system planning with noncooperation. Water Resour Res 52:119–134. https://doi.org/10.1002/2015WR017707
Janjua S, Hassan I (2020) Transboundary water allocation in critical scarcity conditions: a stochastic Bankruptcy approach. J Water Supply Res Technol 69:224–237
Janssen S, Vreugdenhil H, Hermans L, Slinger J (2020) On the nature based flood defence dilemma and its resolution: a game theory based analysis. Sci Total Environ 705:135359. https://doi.org/10.1016/j.scitotenv.2019.135359
Jeong H, Ha S, Kim H (2018a) Flood protection cost allocation using cooperative game theory for adapting infrastructure to climate change. J Water Resour Plan Manag 144:04018010. https://doi.org/10.1061/(ASCE)WR.1943-5452.0000920
Jeong H, Seo H, Kim H (2018b) Game theory-based analysis of decision making for coastal adaptation under multilateral participation. J Manag Eng 34:04018034. https://doi.org/10.1061/(ASCE)ME.1943-5479.0000637
Jhawar S, Agarwaal S, Oberoi T, Sharma T, Thakkar A (2018) Application of game theory in water resource management. Int J Adv Res Dev
Kahil MT, Dinar A, Albiac J (2016) Cooperative water management and ecosystem protection under scarcity and drought in arid and semiarid regions. Water Resour Econ 13:60–74. https://doi.org/10.1016/j.wre.2015.10.001
Kelly A (2003). Decision making using game theory: an introduction for managers. Cambridge University Press
Khachaturyan M, Schoengold K (2019) Applying interconnected game theory to analyze transboundary waters: a case study of the Kura-Araks Basin. Water Econ Policy 5:1850016
Kreps DM (1990) Game theory and economic modelling. Oxford University Press
Lee JH, Yamaguchi R, Yokomizo H, Nakamaru M (2020) Preservation of the value of rice paddy fields: investigating how to prevent farmers from abandoning the fields by means of evolutionary game theory. J Theor Biol 495:110247. https://doi.org/10.1016/j.jtbi.2020.110247
Leoneti AB, Pires EC (2017a) Decision sciences in the management of water resources: Multicriteria methods and game theory applied to the field of sanitation. J Water Sanit Hyg Dev. https://doi.org/10.2166/washdev.2017.171
Leoneti AB, Pires EC (2017b) Decision sciences in the management of water resources: Multicriteria methods and game theory applied to the field of sanitation. J Water Sanit Hyg Dev. https://doi.org/10.2166/washdev.2017.171
Li B, Tan G, Chen G (2016) Generalized uncooperative planar game theory model for water distribution in transboundary rivers. Water Resour Manag 30:225–241
Li D, Zhao J, Govindaraju RS (2019) Water benefits sharing under transboundary cooperation in the Lancang-Mekong River Basin. J Hydrol 577:X. https://doi.org/10.1016/j.jhydrol.2019.123989
Lippai I, Heaney JP (2000) Efficient and equitable impact fees for urban water systems. J Water Resour Plan Manag 126:75–84. https://doi.org/10.1061/(ASCE)0733-9496(2000)126:2(75)
Loehman E, Orlando J, Tschirhart J, Whinston A (1979) Cost allocation for a regional wastewater treatment system. Water Resour Res 15:193–202. https://doi.org/10.1029/WR015i002p00193
López-Corona O, Padilla P, Escolero O, Morales-Casique E (2018) Heuristic formulation of a contextual statistic theory for groundwater. Found Sci 23:75–83
Madani K (2011) Hydropower licensing and climate change: Insights from cooperative game theory. Adv Water Resour 34:174–183. https://doi.org/10.1016/j.advwatres.2010.10.003
Madani K (2010) Game theory and water resources. J Hydrol 381:225–238. https://doi.org/10.1016/j.jhydrol.2009.11.045
Martin JO (2004) An introduction to game theory. Oxford University Press, New York
Mehrparvar M, Ahmadi A, Safavi HR (2016) Social resolution of conflicts over water resources allocation in a river basin using cooperative game theory approaches: a case study. Int J River Basin Manag 14:33–45
Menga F (2016) Domestic and international dimensions of transboundary water politics. Water Altern 9:704–723

Mogomotsi GEJ, Mogomotsi PK, Mosepele K (2019) Legal aspects of transboundary water management: an analysis of the intergovernmental institutional arrangements in the Okavango River Basin. Leiden J Int Law 33:391–408. https://doi.org/10.1017/S0922156519000736

Mukherjee P (2017) Role of private investment in groundwater irrigation and emerging changes in seasonal land leasing contract during Boro Rice cultivation in West Bengal Since 1990s: an application of game theory. Indian J Agric Econ 72:262–271

Myerson RB (2013) Game theory. Harvard University Press

Nash J (1953) Two-person cooperative games. Econometrica 21:128. https://doi.org/10.2307/1906951

Nazari S, Ahmadi A (2019) Non-cooperative stability assessments of groundwater resources management based on the tradeoff between the economy and the environment. J Hydrol 578:X. https://doi.org/10.1016/j.jhydrol.2019.124075

Nazari S, Ahmadi A, Rad SK, Ebrahimi B (2020) Application of non-cooperative dynamic game theory for groundwater conflict resolution. J Environ Manag 270:110889

Oftadeh E, Shourian M, Saghafian B (2017) An ultimatum game theory based approach for basin scale water allocation conflict resolution. Water Resour Manag 31:4293–4308

Omidvar M, Honar1 T, Nikoo MR, Sepaskhah AR (2016) Developing a fuzzy crop pattern and water allocation optimization model based on cooperative game theory: a case study. Ordibehesht Canal at the Doroodzan Irrigation Network, Northwest of Fars Province in Iran. J Water Soil Sci 20:1–13. https://doi.org/10.18869/acadpub.jstnar.20.76.1

Owen G (n.d.) Game theory

Peter JB, Stoner AW, C.M . Y. (1994) The Ethical Import of Patient Selection. AJOB Neurosci 4:42–43. https://doi.org/10.1016/0022

Rahmoun T, Rahmoun M (2019) A step towards building environmental protection strategy: game theory framework for Al-Abrash River Basin management. Int J Environ Sci, Dev, p 10

Ristić B, Madani K (2019) A game theory warning to blind drivers playing chicken with public goods. Water Resour Res 55:2000–2013. https://doi.org/10.1029/2018WR023575

Schlager E, Resource WB (n.d.) Heterogeneity and common pool resource management. Des Inst Environ Resour Manag Cheltenham

Schmeidler D (1969) The nucleolus of a characteristic function game. SIAM J Appl Math 17:1163–1170

Sechi GM, Zucca R, Zuddas P (2013) Water costs allocation in complex systems using a cooperative game theory approach. Water Resour Manag 27:1781–1796. https://doi.org/10.1007/s11269-012-0171-5

Seyam I, Savenije HHG, Aerts J, Schepel M (2000) Algorithms for water resources distribution in international river basins. In: Physics and chemistry of the Earth, Part B: hydrology, oceans and atmosphere. Pergamon, pp 309–314. https://doi.org/10.1016/S1464-1909(00)00020-4

Shapley LS (1971) Cores of convex games. Int J Game Theory 1:11–26. https://doi.org/10.1007/BF01753431

Shapley, L.S., 1952. A Value for n-Person Games:

Shapley LS, Shubik M (1954) A method for evaluating the distribution of power in a committee system. Am Polit Sci Rev 48:787–792. https://doi.org/10.2307/1951053

Shi GM, Wang JN, Zhang B, Zhang Z, Zhang YL (2016) Pollution control costs of a transboundary river basin: Empirical tests of the fairness and stability of cost allocation mechanisms using game theory. J Environ Manag 177:145–152. https://doi.org/10.1016/j.jenvman.2016.04.015

Shubik M (1982) Game theory in the social sciences: concepts and solutions

Shubik M (1982a) Game theory in the social sciences : concepts and solutions. MIT Press

Shubik M (1982b) Game theory in the social sciences : concepts and solutions. MIT Press

Snyder LJ (2017) Stanford Encyclopedia of Philosophy Stanford Encyclopedia of Philosophy, pp 1–22. https://doi.org/10.1111/1467-9973.00225

Stahl S (1999) A gentle introduction to game theory. American Mathematical Society

Straffin J, Philip D (1993) Game theory and strategy

Straffin PD, Heaney JP (1981) Game theory and the tennessee valley authority. Int J Game Theory 10:35–43. https://doi.org/10.1007/BF01770069

Tayia A (2019) Transboundary water conflict resolution mechanisms: substitutes or complements. Water 11:1337

Tian R, Wu J (2019) Groundwater quality appraisal by improved set pair analysis with game theory weightage and health risk estimation of contaminants for Xuecha drinking water source in a loess area in Northwest China. Hum Ecol Risk Assess An Int J 25:132–157

Tisdell JG, Harrison SR (1992) Estimating an optimal distribution of water entitlements. Water Resour Res 28:3111–3117. https://doi.org/10.1029/92WR01795

Van Der Zaag P, Seyam IM, Savenije HHG (2002) Towards measurable criteria for the equitable sharing of international water resources. Water Policy 4:19–32. https://doi.org/10.1016/S1366-7017(02)00003-X

Varouchakis EA, Apostolakis A, Siaka M, Vasilopoulos K, Tasiopoulos A (2018) Alternatives for domestic water tariff policy in the municipality of Chania, Greece, toward water saving using game theory. Water Policy 20:175–188. https://doi.org/10.2166/wp.2017.182

Von Neumann J, Morgenstern O (1953) Theory of games and economic behavior. Princeton University Press

Vrieze OJ (1995) The environmental game. Control Game Theor Model, Environ

Wang LZ (2003) Water resources allocation: a cooperative game theoretic approach. J Environ Inf 2:11–22. https://doi.org/10.3808/jei.200300019

Wei S (2008) on the use of game theoretic Models for water resources management. Brandenburg University of Technology, Cottbus

William R, Garg J, Stillwell AS (2017) A game theory analysis of green infrastructure stormwater management policies. Water Resour Res 53:8003–8019

Williams MA (1988) An empirical test of cooperative game solution concepts. Behav Sci 33:224–237. https://doi.org/10.1002/bs.3830330306

Xiao Y, Hipel KW, Fang L (2016) Incorporating water demand management into a cooperative water allocation framework. Water Resour Manag 30:2997–3012

Xing X, Yuan R (2017) Game theory research on agricultural irrigation water compensation. Water Sav, Irrig, p 18

Xu J, Hou S, Yao L, Li C (2017) Integrated waste load allocation for river water pollution control under uncertainty: a case study of Tuojiang River, China. Environ Sci Pollut Res 24:17741–17759

Yong W, Hu X, Zhang L (2017) Research on rural water pollution control based on game theory. DEStech Trans Soc Sci Educ Hum Sci

Young H, Okada N, Hashimoto T (1980) Cost allocation in water resources development—a case study of Sweden, RR, pp 80–32

Yu Y, Zhao J, Li D, Wang Z (2019) Effects of hydrologic conditions and reservoir operation on transboundary cooperation in the Lancang-Mekong River Basin. J Water Resour Plan Manag 145:04019020. https://doi.org/10.1061/(ASCE)WR.1943-5452.0001075

Yuan L, He W, Degefu DM, Liao Z, Wu X (2017) Water allocation model in the lancing-mekong river basin based on bankruptcy theory and bargaining game. In: World environmental and water resources congress 2017: International perspectives, history and heritage, emerging technologies, and student papers - selected papers from the world environmental and water resources congress 2017. American Society of Civil Engineers (ASCE), pp 78–92. https://doi.org/10.1061/9780784480595.008

Zanjanian H, Abdolabadi H, Niksokhan MH, Sarang A (2018) Influential third party on water right conflict: a game theory approach to achieve the desired equilibrium (case study: Ilam dam, Iran). J Environ Manage 214:283–294

Zarei A, Mousavi SF, Eshaghi Gordji M, Karami H (2019) Optimal reservoir operation using bat and particle swarm algorithm and game theory based on optimal water allocation among consumers. Water Resour Manag 33:3071–3093. https://doi.org/10.1007/s11269-019-02286-9

Zeng Y, Li J, Cai Y, Tan Q, Dai C (2019) A hybrid game theory and mathematical programming model for solving trans-boundary water conflicts. J Hydrol 570:666–681. https://doi.org/10.1016/j.jhydrol.2018.12.053
Zhang Z, He W, An M, Degefu DM, Juqin S, Yuan L (2019) Multi-game theory analysis of cooperation stability of trans-boundary water pollution governance. Nat Environ Pollut Technol 18:471–478
Zomorodian M, Lai SH, Homayounfar M, Ibrahim S, Pender G (2017) Development and application of coupled system dynamics and game theory: a dynamic water conflict resolution method. PLoS ONE 12:e0188489

Chapter 10
Conflict Resolution: The Gist of the Matter in Water Resources Planning and Management

Maedeh Enayati, Omid Bozorg-Haddad, and Mohsen Tahmasebi Nasab

Abstract Conflict may emerge in different aspects of life when different objectives, solutions, stakeholders, and beneficiaries are obliged to interact within a constricted environment to represent and, in turn, preserve their interests. In essence, when stakeholders could not agree on a subject, conflict may arise. This is mainly attributed to the fact that different stakeholders have different demands, requirements, and priorities. The interest of one stakeholder, often and in most real-world problems, is not in line or even in contrast to the interest of other stakeholders. Thus, when personal interests are considered, emerging conflicts are inevitable. In light of water resources planning and management, this could be a very crucial matter since conflict of interest is a general and critical part of decision-making. This chapter aims to shed light on this matter from the water resources perspective. Particularly, it seeks to tackle some mind-provocative questions such as: What is water conflict? How do such conflicts emerge? And how can these conflicts be resolved? Furthermore, real-world examples are provided throughout the chapter to understand the complex nature of conflicts in the context of water resources management.

Keywords Conflict resolution · Water resources management · Stakeholders

M. Enayati · O. Bozorg-Haddad (✉)
Department of Irrigation and Reclamation Engineering, Faculty of Agricultural Engineering and Technology, College of Agriculture and Natural Resources, University of Tehran, 31587-77871 Karaj, Iran
e-mail: OBHaddad@ut.ac.ir

M. Enayati
e-mail: Maedeh.Enayati@ut.ac.ir

M. T. Nasab
Department of Civil and Environmental Engineering, Bucknell University, One Dent Dr, Lewisburg, PA 17837, USA
e-mail: mohsen.nasab@bucknell.edu

O. Bozorg-Haddad (ed.), *Essential Tools for Water Resources Analysis, Planning, and Management*, Springer Water, https://doi.org/10.1007/978-981-33-4295-8_10

10.1 Introduction

Conflict is an inseparable part of human life and social interactions. In such interactions, it is opposing viewpoints that mainly leads to conflict. These conflicts may be rooted in opposition of interests or perhaps believes of the stakeholders (Aubert 1963). Often enough, however, negotiations can help the involved parties to reach an agreement concerning the conflicting situation. However, it should be noted that reaching an agreement does not necessarily resolve the conflict, and by no means can it guarantee to solve the problem at hand. In fact, the first step to truly address, and in turn, resolve a conflicting situation is to understand the nature of the problem, acknowledge the reasons behind the formation of such conflicts, then, and only then, one can provide possible remedies that can be potentially effective, logical, and fair.

As stated, the *opposition of interests* has been cited as one of the main reasons for conflict situations. The opposition of interests refers to those cases in which securing one's interest would impose some sort of loss to other parties. These situations, commonly, arose when limited, non-renewable resources are involved. For instance, this could be the case where multiple stakeholders are fighting for their fair share of water in a basin that high demands have caused *water shortages* crisis. In this hypothetical case, though not that far-fetched-from-reality case, the unbalanced equilibrium between available resources and demands can be seen as the main reason for the conflict between the stakeholders (Madani 2010; Bozorg-Haddad et al. 2018a).

Generally speaking, the water resources stored in a basin are used to meet the many needs of different sectors of that region, which may include municipal, industrial, agricultural, environmental, and, even, recreational demands (Bozorg-Haddad et al. 2018a). The effective and timely response to these demands can have direct and indirect impacts on the regions' socio-economic, political, and environmental status. For instance, meeting the agricultural, municipal, and hydropower demands can have direct economic effects on a community, whereas meeting the recreational and natural ecosystem demands would affect the social and environmental sectors, respectively (Lund and Palmer 1997). Note that meeting each of these demands mentioned above may only be in line with one of the social, economic, or environmental goals of a community. Thus, utilizing the available water resources to focus on a singular sector can, inevitably, lead to a severe conflicting situation (Lund and Palmer 1997), for in such a scenario, the other sectors' interests are being neglected. From a water resources planning and management perspective, however, in most practical, real-world cases when it comes to meeting the demands of different sectors, there is often a sense of priority which in essence rank the order of which these needs are to be addressed (Bozorg-Haddad et al. 2018a). Consequently, and somewhat inevitably, conflicts of interest is a common enough phenomenon in such cases. Thus resolving these conflicts of interests in a timely and effective manner is quite a crucial task, for failing to do so can amplify the problem on a regional, national, or even international scale. This matter elucidates the importance of applying conflict resolution techniques in water resources management. In fact, over the last five decades, a clear boost has been reported for the application of these methods for water resources planning and management purposes (Madani 2010; Dinar and Hogarth 2015). These methods have been employed to cover a vast and diverse set of topics, including but not limited to water resources management (Parrachino et al. 2006; Carraro et al. 2007), operation of water resources systems (Lund and Palmer 1997), optimal design of water distribution networks (Fallah-Mehdipour et al. 2011), and allocation of water rights (Wang et al. 2003).

10.2 Definition and Terms

In their highly revered book titled "*theory of games and economic behavior,*" published in (1944), von Neumann and Morgenstern introduced and lay the theoretical foundation of *game theory*. The gist of this theory was to provide a mathematical framework to capture and model the strategic interaction of rational decision-makers. According to this concept, each interaction, here referred to as a *game*, has the following three main characteristics (Colman 2013):

I. At least two stakeholders, here referred to as *players*, which interact with one another in the game.
II. Each *player* has at least two actions, here referred to as *strategies,* to interact and participate in the game.
III. The players have well-defined preferences among the possible outcomes so that numerical *pay-offs* reflecting these preferences can be assigned to all players for all outcomes.

Thus, in the game theory terminology, a *game* can indeed refer to any socio-economic interaction. In fact, to certain degrees, most economical, political, and social conflicts can be recaptured through these three properties.

Generally, there are three main classes of games that are games of skill, games of chance, and games of strategy (Colman 2013). In the game of skills, the outcome of the game solely revolves around the actions of one key player. In such situations, the game's result is determined and controlled by the key player's skillset, and as such, chance does not affect the outcome of the game. On the contrary, there are situations in which chance can have significant impacts on the emerging outcomes. These games are referred to as games of chance. Here, both the random nature of the surrounding environment and players' skill sets can affect the outcome of the games. Finally, in the game of strategies, each player is partially responsible for the outcome of the game. Here, based on the skillset of oneself and other involved parties, each player opts for the optimal strategies that ensure one's interests in the game. From this point onward, the term "game" mutually refers to this type of situation.

The strategic games can also break down into two main subclasses, namely, *cooperative* and *non-cooperative* games (Rapoport 2012). In cooperative games, as the name suggests, players tend to choose their strategies by cooperating with other involved parties. These situations occur when the players would gain more by working with, rather than against, other players. Thus, a group of two or perhaps more players would form an *alliance*, here referred to as *coalition* during the game. In non-cooperative games, however, each player assumes their strategies somewhat independently and without building any collaboration with other involved parties. In most cases, the nature of such games would naturally dictate that individual players would play against one another, and thus no alliance can be formed.

In real-world situations, players can interact with one another and make connections. These interactions can, in turn, lead to negotiations between the involved parties. Through these negotiations, if the players could form an alliance to pursue and assume a set of inline strategies, cooperative games would be developed. Otherwise, the interaction between involved parties can be captured through non-cooperative games. When it comes to non-cooperative games, the focus would be on the strategic interactions of the players. In cooperative games, however, the emphasis is on the potential gains that could be achieved through cooperation.

Given the nature of water resources, the situation could be redefined in a way the involved parties could share a common goal. As such, stakeholders' benefits or losses would be intertwined, and thus the involved parties are more inclined to cooperate by forming alliances. Hence, cooperative games are commonly used for real-world water resources problems such as water allocation (Wang et al. 2003; Mahjouri and Ardestani 2010), water pricing (Sechi et al. 2013), groundwater management (Esteban and Dinar 2013), and water quality control (Mahjouri and Ardestani 2011).

Based on the game theory's principles, different conflict resolution methods have been proposed by scholars and researchers over the course of the last decades. Among these methods, three methods clearly stand out in terms of popularity and applicability for most real-world problems, which are Nash solution, Shapley Value, and Nucleolus. Nash solution borrowed his name from a famous American mathematician and Nobel Prize-winner, John Forbes Nash Jr. (1928-2015), who proposed and theorized this method in (1953). A year later in 1954, the Shapley Value method was introduced, which was named in honor of Lloyd Stowell Shapley (1923-2016), another American mathematician and Nobel Prize-winning economist. Lastly, David Schmeidler, an Israeli mathematician and economic theorist, introduced a new concept called Nucleolus in 1969.

10.3 Basic Principles and Logics

When it comes to conflict resolution, two main points need to be addressed before bothering with any other computational effort toward solving the problem. First, it is necessary to identify the type of conflict so that the problem could be adequately and accurately captured via the right methods. Next, one needs to identify the potential causations of the conflict so that the feasible solutions could be assessed accordingly.

As for conflicts in water-related projects, four basic types of situations could arise, namely, no conflict, superficial conflict, latent conflict, and open conflict (Bijani and Hayati 2011). In superficial conflicts, with the courtesy of common understanding and

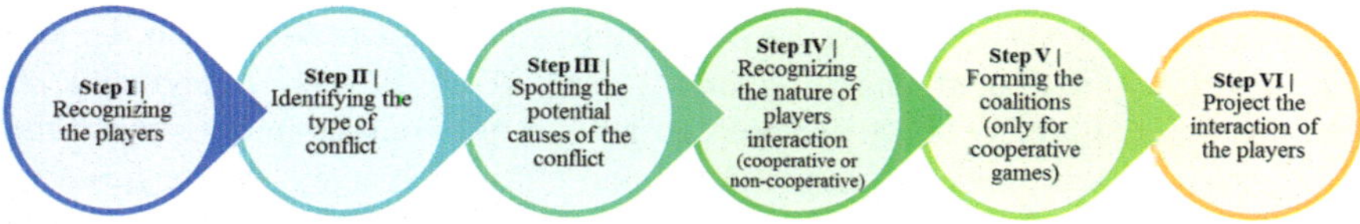

Fig. 10.1 The major steps to conflict resolution

compromise, the problem could potentially be resolved through a series of negations in a rather short time period. In a latent conflict situation, although the problems are potentially there, they have not surfaced yet. The situation could be triggered, in which case the problem would emerge either in the form of superficial or open conflicts. Lastly, open conflict is the most instance form of problem that is observed in water-related projects. These conflicts could be formed and intensified over a long period of time. In such cases, it is necessary to, first, identify the potential causes of these conflicts.

As a strategic commodity, water resources status is always tied to the security of a community (Zolghadr-Asli et al. 2017a). More importantly, water availably can naturally fluctuate through time. In other words, natural-driven phenomena such as prolonged droughts can have a dramatic adverse impact on water resources. While, ostensibly, such circumstances could indeed help create conflicting situations, it should be noted that, in most cases, water shortages are a natural phenomenon, and as such, cannot be the sole cause for disturbing the water security of a given community. In other words, water shortages on its own cannot be the main cause of water-related conflicts; But rather, it is the unfair allocation of these resources, which itself could be rooted in mismanagement, can potentially ignite a water conflict situation (Dabelko and Aaron 2004).

Any conflict resolution is composed of six key steps, which are depicted in Fig. 10.1. As illustrated, steps one to four help form the main structure of the conflict resolution problem at hand. The fifth step is only necessary if the players' interactions indicate that cooperative gameplay is at hand. Lastly, when the limited resources are to be allocated in the format of a cooperative game, two criteria should be taken into account: *individual* and *group intelligence*. The former indicates that each player's gain through the formed alliance must at least be equal to what that player would have achieved should the player decided not to participate and form a union. The latter criterion states similar conditions, but this time for the whole group of participants of the coalitions.

10.4 Importance and Necessity

The total quantity of water in the world is immense, but most are stored either in the form of saltwater (97.5%) or locked in ice caps (1.75%). All that is economically available for human consumption is estimated to be only 0.007% of the total water resources (Wolf 2007). More to the point, freshwater resources are also unevenly and irregularly distributed, and as a result, some parts of the world, inevitably, face extreme water shortages. Furthermore, the growth of human populations and their living standard, excessive water pollution, and climate changes impose new challenges to water resources planning and management (Zolghadr-Asli et al. 2017a). This notion invokes that water-related decisions are of multi-dimensional nature that involves different stakeholders with often enough conflicting interests.

Conflict over water resources may arise in a regional or perhaps national scale when limited water is to be allocated to various stakeholders with opposing interests. In the case of transboundary resources, where two or more countries share a source, these conflicts may escalate to an international scale. Thus water resources can be a source of conflict or even wars between regions or nations (Starr 1991; Zolghadr-Asli et al. 2017b). According to the United Nations (UN), more than 1,831 water conflicts have been reported in the last five decades, 21 of which had led to some sort of military action (Bijani and Hayati 2011). In 1975, for instance, Iraq's government allegedly claimed that Syria is actively trying to tamper with their transboundary rivers and stopping the rivers from running pass Iraqi borders. While both nations' military had lined up in the shared border in response, Saudi Arabia's intervention helped deescalate the situation (Dinar 2002).

Every decision made concerning water resources can potentially have an impact on the stakeholders, and in turn, can leave some involved parties dissatisfied. When the stakeholders are not content with the quality, quantity, the temporal or spatial distribution of water, this could create a potentially conflicting environment (Bozorg-Haddad et al. 2020). As a strategic commodity, while water can be the mean for a community's flourish and prosperity (Zolghadr-Asli et al. 2017a), in practice, opposing views on how to allocate water resources on a national or international scale, mismanagement, one-sided and short-sited decisions, unsustainable plans for developments, and uneven distribution of power can often put water resources at the center of conflicts. Thus, the challenges imposed by conflicting situation has always been felt and present when it comes to water resources planning and management. Therefore, applying conflict resolution methods in the past years in the literature has gained momentum, for these methods can, to some degree, provide a platform for sustainable and integrated management of natural resources while minimizing the risk for any conflicting situation among the stakeholders with opposing viewpoints.

10.5 Methodology

As stated earlier, different methods were formed on the premise of game theory. It was also established that game theory has two main classes: cooperative and non-cooperative games. Due to the nature of most commonly observed water-related projects (e.g., irrigation, wastewater treatment, etc.), the cost functions exhibit strong convexity. As such, there is always an incentive for joining the grand coalition, for it can help reduce the cost imposed on those stakeholders that joined the alliance (Dinar and Hogarth 2015). As a result, cooperative-based games gained more attention when it comes to water resources planning and management. Numerous cooperative methods have been proposed and developed over the years to cope with conflicting situations that can arise in decision-making regarding many disciplines, including but not limited to water resources management. The core idea behind this branch of methods is to maximize the benefits of the members of the coalition. In other words, the main incentive that encourages the stakeholders to join the colation is that they could gain more by participating in an alliance than they could gain individually. Although it seems a sound and solid ideology in the paper, applying this principle could be easier said than done in practice (Curiel 2013). The main challenge in a practical point-of-view is to find a way to allocate the benefits resulted from joining the coalition among the stakeholders in a manner that all involved parties would be satisfied (Bozorg-Haddad et al. 2018a); otherwise, this could be seen as a serious disincentive for stakeholders to create an alliance (Dinar and Hogarth 2015). That being said, cooperative methods are still considered as some of the most popular and practical ways to address real-world conflicts in water resources problems (Madani 2010).

Based on the above-mentioned same principle, many cooperative methods have been proposed over the years, such as the kernel (Davis and Maschler 1965), the generalized Shapley value (Loehman and Whinston 1976), and many others; although some methods such as Nash solution, Shapley value, and Nucleolus gained more popularity for water resources planning and management purposes (e.g., Kucukmehmetoglu 2002; Dinar and Nigatu 2013; Safari et al. 2014).

Nash solution is perhaps one of the most well-known and revered examples of such methods, which is a method for fair and efficient sharing of the obtained benefits under cooperation (Nash 1953). The more generalized form of this method is the Nash-Harsanyi solution (Harsanyi 1959), which can handle an n-person bargaining game. Both Shapley value (Shapley 1953) and Nucleolus (Schmeidler 1969) are other known alternative cooperative methods, each of which provides a unique perspective on how to share the obtained benefits fairly and efficiently. See Madani and Dinar (2012), Dinar and Hogarth, (2015), and Curiel (2013) for more detailed information regarding these methods.

The other alternative method is *bankruptcy methods,* which are also examples of cooperative games theory solutions (Young 1994; Sheikhmohammady and Madani 2008; Madani et al. 2014). These are methods that are used for situations in which a group of players faces a total debt that exceeds the total amount of available resources to be credited among the players. This branch contains a class of rules, which some have been more common to cope with water-related situations, which are as proportional (P), adjusted proportional (AP), constrained equal award (CEA), constrained equal loss (CEL), Talmud (Tal), and Piniles (Pin) rules. It is worth noting that while bankruptcy methods are technically classified as cooperative games theory solutions, they slightly differ from the common representations of this class of game theory solutions. In most cooperative methods, the core idea is to share the gained benefits among the stakeholders, while in bankruptcy methods, the gist is to allocate the amount of deficit among the involved parties (Madani et al. 2014). For more information on these branch of methods, see Madani and Zarezadeh (2012).

10.6 Practical Examples

A common challenge in water-system management is to allocate water resources under scarcity conditions. Water allocation under such circumstances could be represented as a game in which bankruptcy prevailed. In other words, that the total demands for water exceed the amount of available water resources within the region, and as such, stakeholders' expectations surpass the available assets that can be divided among them (Sechi and Zucca 2015). The foundations for bankruptcy models have started with the works of O'Neill (1982) and Aumann and Maschler (1985). Some of these bankruptcy methods have been cited in the previous section of this chapter. The CEA rule, for instance, aims to satisfy the lower claims to the extent possible in order to minimize the number of unsatisfied creditors. According to the CEA rule, the initial allocation to all stakeholders is equal to the lowest claim, provided that the sum of initial distributions does not exceed the demand. The fully satisfied creditor is then excluded, and the process continues with the remaining creditors after updating their unsatisfied claims as well as the residual resource value. This process is repeated up until the point that allocating an amount equal to the lowest demand to all remaining creditors is not feasible, in which the remaining resource is distributed equally among all the remaining stakeholders (Madani et al. 2014; Degefu and He 2016).

As stated earlier on, bankruptcy methods are quite applicable when it comes to water resources planning and management. Consider the case of the Qezelozan-Sefid Rud River Basin, Iran, for instance, which is a transboundary river basin shared by eight provinces. As a water-bankrupt basin, the riparian areas of this region face water shortages, which in turn have triggered new conflicts between the parties above. Zarezadeh et al. (2012) resorted to bankruptcy methods to resolve the problems mentioned above by extracting new allocation policies. They explored through their study, how bankruptcy methods can successfully help navigate through the turmoil caused during conflicting situations where the amount of available water is not sufficient enough to meet all the stakeholders' demands.

Another notable water-challenged region is the case of the Urmia Lake Basin. The basin is located in north-west Iran, which has been named after a saline lake housed in the basin. The main river that charges the lakes is the Zarrineh-Rud River. However, during the past two decades, Urmia Lake has been depleting at an alarming rate (Alizadeh-Choobari et al. 2016). Some have linked this shrinking pattern to the construction of the Zarrineh-Rud Dam, located in the upstream of the lake (AghaKouchak et al. 2015). As such, Bozorg-Haddad et al. (2018b) used a game theory-based framework to address the conflicting interests of different stakeholders involved in this matter. They restated the basin's water resources through bankruptcy rules. Using this representation, they were able to make a balance between many stakeholders (i.e., urban-industrial, agricultural, and environmental sector), in a way that not only the risk of any conflict between the involved parties would be minimized, but also, the lakes' reviving could be potentially achievable in the long run.

10.7 Summary

In socio-economic encounters, the difference in believes, values, and interests make the conflicting situation a rather common sight. Conflicting situations may arise in different shape, form, or scale, but they are mainly driven by the opposition of interests in sharing limited or non-renewable resources. Although water resources are not non-renewable per se, in many cases, the availability of these natural resources is limited. Thus, different stakeholders, with often opposing viewpoints and interests, are to interact and share these resources, which sometimes leads to conflicting situations. In this chapter, we reviewed some of the real-life water-related conflicts around the world.

Acknowledgements The authors thank Iran's National Science Foundation (INSF) for the support of this research.

References

AghaKouchak A, Norouzi H, Madani K, Mirchi A, Azarderakhsh M, Nazemi A, Nasrollahi N, Farahmand A, Mehran A, Hasanzadeh E (2015) Aral Sea syndrome desiccates Lake Urmia: Call for action. J Great Lakes Res 41(1):307–311

Alizadeh-Choobari O, Ahmadi-Givi F, Mirzaei N, Owlad E (2016) Climate change and anthropogenic impacts on the rapid shrinkage of Lake Urmia. Int J Climatol 36(13):4276–4286

Aubert V (1963) Competition and dissensus: two types of conflict and of conflict resolution. J Conflict Resolut 7(1):26–42

Aumann R, Maschler M (1985) Game theoretic analysis of a bankruptcy problem from the Talmud. J Econ Theory 36(2):195–213

Bijani M, Hayati D (2011) Water conflict in agricultural system in Iran: a human ecological analysis. J Ecol Environ Sci 2(2):27–40

Bozorg-Haddad O, Abutalebi M, Chu X, Loáiciga HA (2020) Assessment of potential of intraregional conflicts by developing a transferability index for inter-basin water transfers, and their impacts on the water resources. Environ Monit Assess 192(1):40

Bozorg-Haddad O, Athari E, Fallah-Mehdipour E, Bahrami M, Loáiciga HA (2018a) Allocation of reservoir releases under drought conditions: a conflict-resolution approach. Proc Inst Civil Eng-Water Manag 172(5):218–228

Bozorg-Haddad O, Athari E, Fallah-Mehdipour E, Loáiciga HA (2018b) Real-time water allocation policies calculated with bankruptcy games and genetic programing. Water Sci Technol: Water Supply 18(2):430–449

Carraro C, Marchiori C, Sgobbi A (2007) Negotiating on water: Insights from non-cooperative bargaining theory. Environment and Development Economics 12(2):329–349

Colman AM (2013) Game theory and its applications: in the social and biological sciences. Psychology Press, Padstow, UK
Curiel I (2013) Cooperative game theory and applications: cooperative games arising from combinatorial optimization problems. Springer Science & Business Media, New York, NY
Dabelko D, Aaron T (2004) Water, conflict, and cooperation. Environ Change Secur Project Rep 10:60–66
Davis M, Maschler M (1965) The kernel of a cooperative game. Naval Res Logist Q 12(3):223–259
Degefu DM, He W (2016) Allocating water under bankruptcy scenario. Water Resour Manag 30(11):3949–3964
Dinar S (2002) Water, security, conflict, and cooperation. Sais Rev 22(2):229–253
Dinar A, Hogarth M (2015) Game theory and water resources: Critical review of its contributions, progress and remaining challenges. Found Trends® Microecon 11(1–2):1–139
Dinar A, Nigatu GS (2013) Distributional considerations of international water resources under externality: the case of Ethiopia, Sudan and Egypt on the Blue Nile. Water Resour Econ 2:1–16
Esteban E, Dinar A (2013) Cooperative management of groundwater resources in the presence of environmental externalities. Environ Resour Econ 54(3):443–469
Fallah-Mehdipour E, Bozorg-Haddad O, Beygi S, Mariño MA (2011) Effect of utility function curvature of Young's bargaining method on the design of WDNs. Water Resour Manag 25(9):2197–2218
Harsanyi JC (1959) A bargaining model for the cooperative n-person game. Ph.D. dissertation, Princeton University, Princeton, NJ
Kucukmehmetoglu M (2002) Water resources allocation and conflicts: The case of the Euphrates and the Tigris. Ph.D. dissertation, Department of Higher Education, The Ohio State University, Columbus, OH
Loehman ET, Whinston AB (1976) A generalized cost allocation scheme. Theory Meas Econ Extern. https://doi.org/10.1016/b978-0-12-450450-9.50013-0
Lund JR, Palmer RN (1997) Water resource system modeling for conflict resolution. Water Resour Update 3(108):70–82
Madani K (2010) Game theory and water resources. J Hydrol 381(3–4):225–238
Madani K, Dinar A (2012) Cooperative institutions for sustainable common pool resource management: application to groundwater. Water Resour Res 48(9):W09553
Madani K, Zarezadeh M (2012) Bankruptcy methods for resolving water resources conflicts. In: World Environmental and water resources congress 2012, Albuquerque, New Mexico, United States, 20–24 May 2012
Madani K, Zarezadeh M, Morid S (2014) A new framework for resolving conflicts over transboundary rivers using bankruptcy methods. Hydrol Earth Syst Sci 18(8):3055
Mahjouri N, Ardestani M (2010) A game theoretic approach for interbasin water resources allocation considering the water quality issues. Environ Monit Assess 167(1–4):527–544
Mahjouri N, Ardestani M (2011) Application of cooperative and non-cooperative games in large-scale water quantity and quality management: A case study. Environ Monit Assess 172(1–4):157–169
Morgenstern O, Von Neumann J (1944) Theory of games and economic behavior. Princeton Uuniversity Press, Princeton, NJ
Nash J (1953) Two-person cooperative games. Econometrica 21(1):128–140
O'Neill B (1982) A problem of rights arbitration from the Talmud. Math Soc Sci 2(4):345–371
Parrachino I, Dinar A, Patrone F (2006) Cooperative game theory and its application to natural, environmental, and water resource issues: 3. application to Water Resources. World Bank Policy Research Working Paper, No. 4074, Washington, DC
Rapoport A (ed) (2012) Game theory as a theory of conflict resolution, vol 2. Springer Science and Business Media, Dordrecht, Holland
Safari N, Zarghami M, Szidarovszky F (2014) Nash bargaining and leader–follower models in water allocation: Application to the Zarrinehrud River basin, Iran. Appl Math Model 38(7–8):1959–1968

Schmeidler D (1969) The nucleolus of a characteristic function game. SIAM J Appl Math 17(6):1163–1170
Sechi GM, Zucca R (2015) Water resource allocation in critical scarcity conditions: a bankruptcy game approach. Water Resour Manag 29(2):541–555
Sechi GM, Zucca R, Zuddas P (2013) Water costs allocation in complex systems using a cooperative game theory approach. Water Resour Manag 27(6):1781–1796
Shapley LS (1953) A value of n-person games. Ann Math Stud Contrib Theory Games:307–318
Sheikhmohammady M, Madani K (2008) Sharing a multi-national resource through bankruptcy procedures. In: Proceeding of the 2008 World environmental and water resources congress, Honolulu, Hawaii, United States, 12–16 May 2008
Starr JR (1991) Water wars. Foreign Policy 82:17–36
Wang LZ, Fang L, Hipel KW (2003) Water resources allocation: a cooperative game theoretic approach. J Environ Inf 2(2):11–22
Wolf AT (2007) Shared waters: conflict and cooperation. Annu Rev Environ Resour 32:241–269
Young HP (1994) Cost allocation. In: Handbook of game theory with economic applications. Elsevier, Amsterdam, Netherlands
Zarezadeh M, Madani K, Morid S (2012) Resolving transboundary water conflicts: Lessons learned from the Qezelozan-Sefidrood river bankruptcy problem. In: World environmental and water resources congress 2012, Albuquerque, New Mexico, United States, 20–24 May 2012
Zolghadr-Asli B, Bozorg-Haddad O, Chu X (2017a) Strategic importance and safety of water resources. J Irrig Drainage Eng 143(7):02517001
Zolghadr-Asli B, Bozorg-Haddad O, Chu X (2017b) Brief chronicle of water wars: search for global peace. J Irrig Drainage Eng 143(7):02517002

Chapter 11
Multi-objective Optimization Approaches for Design, Planning, and Management of Water Resource Systems

Ahmad Ferdowsi, Vijay P. Singh, Mohammad Ehteram, and Seyedali Mirjalili

Abstract Problems of water resource systems entail many inherent objectives from cost-related to environmental issues, which are usually in conflict and should be addressed simultaneously in decision-making. Multi-objective optimization algorithms aim to simultaneously satisfy two or more objectives and provide a set of optimal solutions for multi-objective optimization problems. Such algorithms can also improve the accuracy of simulation models, such as artificial neural network and adaptive neuro fuzzy interface system. This chapter first presents essential definitions in multi-objective algorithms and their applications in different areas of water resources systems (i.e., model calibration, water distribution network, reservoir operation and management, water quality, water structure design, and groundwater). It then evaluates performance of a recently proposed and two well-known multi-objective algorithms for solving three problems in the optimal operation of a real multipurpose dam.

Keywords Optimization algorithms · Artificial neural network · Adaptive neuro fuzzy interface system · Water resources systems

---

A. Ferdowsi (✉) · M. Ehteram
Department of Water Engineering and Hydraulic Structures, Faculty of Civil Engineering, Semnan University, Semnan 35131-19111, Iran
e-mail: ahmad.ferdowsi@semnan.ac.ir

M. Ehteram
e-mail: mohammdehteram@semnan.ac.ir

V. P. Singh
Department of Biological and Agricultural Engineering and Zachry Department of Civil Engineering, Texas A&M University, College Station, TX 77843, USA
e-mail: vsingh@tamu.edu

S. Mirjalili
Centre for Artificial Intelligence Research and Optimisation, Torrens University Australia, 90 Bowen Terrace, Fortitude Valley, QLD 4006, Australia
e-mail: ali.mirjalili@laureate.edu.au

O. Bozorg-Haddad (ed.), *Essential Tools for Water Resources Analysis, Planning, and Management*, Springer Water, https://doi.org/10.1007/978-981-33-4295-8_11

11.1 Introduction

Let us begin this study with the exact words of Leonhard Euler (Swiss mathematician in the eighteenth century) "For since the fabric of the universe is most perfect and the work of a wisest Creator, nothing at all takes place in the universe in which some rule of the maximum or minimum does not appear" (Kline 1982), which refer to the current concept of optimization. In other words, what is the best (i.e., maximum or minimum)?

In water resource systems, different demands are simultaneously considered in decisions and designs, including construction and maintenance costs, wildlife and ecological issues, flood damage, irrigation benefits, construction time, power generation, navigation, present and future demands, pollution, materials, and recreational use. Nowadays, with population growth and climate change, water resources management has become an emerging challenge for decision-makers. Climate change alone has a significant effect on a large portion of the available water resources (Mallakpour et al. 2019). Competition for limited water resources is another challenge (Halbe et al. 2018). Thus, optimal design of water resources systems is essential for design, planning, and management. Optimization of water utilization strategies for multipurpose systems is a major consideration in water resources management. On one hand, the reservoir operator should determine how much water is to be stored for the present and future uses and how much should be released for other issues, such as hydropower generation, and ecological and irrigation needs (Yaseen et al. 2019).

Many reservoirs are still operated based on experience and statistical rules fixed at the construction time (Yazdi and Moridi 2018). Thus, it is necessary to employ techniques to design, planning, and manage water and other resources in a way that meet societal and ecological demands for the present time and future. In recent years, optimization using intelligent techniques, in particular metaheuristics have been applied to a wide range of challenging problems (Mohammad-Azari et al. 2020). They are able to cope with the most common difficulties of an optimization problem, including but not limited to, large number of decision variable, highly constrained search space, multiple conflicting objectives, dynamically changes search space, and locally optimal solutions. In this work, we leverage on the most recent multi-objective metaheuristics to find optimal solutions in the area of water resource management.

The basic concepts of these techniques are discussed and three problems are solved using a recently proposed algorithm, namely multi-objective sine cosine algorithm (MOSCA). The problems relate to the optimal operation of multipurpose Mahabad dam (Iran). In addition, the multi-objective particle swarm optimization (MOPSO) algorithm and the multi-objective genetic algorithm (MOGA) are used to investigate the MOSCA performance. The rest of the chapter is organized as follows:

In Sect. 11.2, applications in water resource systems are discussed thoroughly. Section 11.3 provides preliminaries and essential definitions in multi-objective optimization. The experimental results on the case study are discussed in Sect. 11.4. Finally, Sect. 11.5 concludes the work and suggests future directions.

11.2 Applications in Water Resource Systems

Water resources planning and management are often multi-objective (Reed et al. 2013) and multi-objective approaches have therefore been employed from calibration of models to engineering design and operation (Zheng et al. 2016). The goal of multi-objective optimization is to approximate the set of Pareto optimal solutions that represent the best trade-offs between conflicting objectives (Nicklow et al. 2009). In this section, applications of multi-objective approaches for design, planning, and management of water resources systems are categorized. Table 11.1 shows some of the studies from the literature, with their multi-objective algorithms.

11.2.1 Model Calibration

Single and multi-objective algorithms are widely used in calibrating models. In water resources management, the algorithms are mostly used in calibration of rainfall-runoff models, but they have also been used in the calibration of other models, including water quality simulation, hydraulic, and hydrodynamic models. Rainfall-runoff models play an important role in water resources management, especially in risk assessment (Adeyeri et al. 2020) and are essential tools for assessing runoff changes in the catchment area for the evaluation of impacts of climate change (Najafi et al. 2011) and predicting river flow is an important issue (Bomhof et al. 2019). To that end, objective functions can be defined as differences between observed and simulated streamflow and different indices have been used to measure these differences, such as Nash-Sutcliffe, relative bias, correlation coefficient etc. (Ferdowsi 2019). Calibration of rainfall-runoff models (i.e., lumped, distributed, semi-distributed) including SWAT (Bekele and Nicklow 2007; Confesor and Whittaker 2007) and MIKE-SHE (Khu et al. 2008), has been done with different algorithms, basin areas, and parameters i.e., soil moisture capacity, reservoir release coefficient, surface permeability, melt factor etc. Table 11.1 lists some model calibration problems, which have been solved by multi-objective methods.

11.2.2 Water Distribution Networks

Water distribution networks (WDNs) are intrinsic parts of modern cities. Because WDNs are used to transfer water to households, farmlands, industries etc., they, which should be designed for current and future water demands, with suitable pipe diameter and required pressure (Monsef et al. 2019). In addition, they should be able to serve in extraordinary situations, including urban fire (Quintiliani et al. 2019). WDNs problems are complex (involving pipes, pumps, valves, tanks, etc.), include many constraints (pressure, velocity, etc.), and have a large search space (Wu et al. 2013). In

Table 11.1 Review of applications of multi-objective approaches in water resource problems

Problem	Study	Algorithm
Calibration of models	Kapelan et al. (2003)	MOGA
	Kapelan et al. (2005a)	MOGA
	Bekele and Nicklow (2007)	NSGA-II
	Confesor and Whittaker (2007)	NSGA-II
	Khu et al. (2008)	POGA
	Mostafaie et al. (2018)	NSGA-II, MOPSO, PESA-II
	Wang et al. (2018a)	MOPSO
	Zhang et al. (2018)	MOSCEM-UA
	Gutierrez et al. (2019)	NSGA-II, NSGA-III, SPEA-II
	Adeyeri et al. (2020)	MOPSO
Water distribution network	Farmani et al. (2005)	NSGA-II, SPEA2
	Kapelan et al. (2005b)	RNSGAII
	Alfonso et al. (2009)	NSGA-II
	Wu et al. (2011)	MOGA
	Ehsani and Afshar (2010)	NSGA-II
	Fu and Kapelan (2010)	ANN-GA, NSGA-II
	Wu et al. (2013)	MOGA
	Marques et al. (2015)	AMOSA
	Zheng et al. (2016)	NSGA-II, SAMODE, Borg
	Monsef et al. (2019)	MOPSO, NSGA-II, MODE
Reservoir operation and management	Hajiabadi and Zarghami (2014)	NSGA-II
	Ashofteh et al. (2015)	MO-GP
	Bozorg-Haddad et al. (2016)	NSGA-II
	Paseka et al. (2018)	NSGA-II
	Yazdi and Moridi (2018)	NSDE
	KhazaiPoul et al. (2019)	NSDE
	Khorshidi et al. (2019)	NSGA-II
Water quality	Saadatpour and Afshar (2013)	MOPSO
	Aboutalebi et al. (2016)	NSGA-II
	Amirkhani et al. (2016a)	NSGA-II
	Aalami et al. (2018)	MOPSO
	Quintiliani et al. (2019)	AMGA2
Water structures design	Nikoo et al. (2014)	NSGA-II
	Hojjati et al. (2017)	NSGA-II

(continued)

Table 11.1 (continued)

Problem	Study	Algorithm
	Ghorbani Mooselu et al. (2019)	NSGA-II
	Nikoo et al. (2019)	NSGA-II
Groundwater	Reed et al. (2001)	NSGA
	Saafan et al. (2011)	MOGA
	Piscopo et al. (2015)	Borg
	Nouiri et al. (2015)	MOGA
	Mirzaie-Nodoushan et al. (2017)	NSGA-II
	Fatkhutdinov and Stefan (2019)	NSGA-II
	Taravatrooy et al. (2019)	NSGA-II

solving WDNs problems, it is possible to only consider cost as an objective function i.e., treat as a single objective problem and consider pipe diameters as a primary objective function (Moosavian and Lence 2016). However, other problems include several objective functions, such as construction and operation cost, hydraulic or mechanical performance, reliability, water quality (contamination), greenhouse gas emissions etc. so they require multi-objective approaches. In WDNs studies, pipe size (diameter) and material, tank location, etc., can be considered as decision variables. Some of the multi-objective studies are shown in Table 11.1.

11.2.3 Reservoir Operation and Management

Population growth and climate change are stressors in water resources management (Yang et al. 2017). Optimal operation and management is a broad topic, with a wide range of objectives to be considered. In addition, various decision functions and constraints affect reservoir operation and management. On the other hand, due to great uncertainty of climate variability, optimal reservoir operation faces difficulties (Khorshidi et al. 2019). However, as multi-objective methods simultaneously include many objectives, decision variables, and constraints, are effective and are efficient tools in solving problems with a large search space, they can be used to solve reservoir problems. Flood impacts (on downstream areas and upstream of reservoir), water demands (urban, agriculture, environment, etc.), power generation, navigation rate (based on volume of discharge), and sediment flushing, are related to optimal release policy, which can be considered as objective functions. Decision-makers and dam operators can manage reservoirs with scenarios that are provided with multi-objective algorithms. Examples of such problems are listed in Table 11.1.

11.2.4 Water Quality

Floods, human activities, climate change, and flora and fauna may influence water quality of reservoirs, rivers, urban distribution systems, irrigation networks, and basin surface. Government bodies can benefit from multi-objective algorithms in different water-quality problems. One of the applications of multi-objective approaches is determining a trade-off between temperature conditions and reservoir water-quality reduction. Total dissolved solids (TDS), which may enter reservoirs with flood flow, can change reservoir quality. For example, minimization of TDS release from reservoir and minimization of differences between temperature of reservoir's inflow and outflow were considered as objective functions by Amirkhani et al. (2016a). Other models, such as water quality simulation model of CE-QUAL-W2, are used to simulate reservoir conditions and to enhance model reliability. Other objective functions can be employed in these problems to determine cost, reliability, vulnerability, and risk in reservoirs, rivers, and water distribution networks. Table 11.1 lists some of the works.

11.2.5 Water Structure Design

Single objective algorithms have been widely used in optimization of water structures design, as for instance, stepped spillway (Bozorg-Haddad et al. 2010), labyrinth spillway (Ferdowsi et al. 2019, 2020), open channel (Orouji et al. 2016), earth dam (Rezaeeian et al. 2019) and so on. Construction cost (i.e., material volume) and hydraulic performance are two important objective functions in these problems. Considering the advantages of multi-objective algorithms, optimal design of structures can be developed i.e., all concerns, such as construction cost and time, design discharge, energy dissipation, scour depth, cavitation number, safety etc. can be considered simultaneously. In addition, optimum design of water structures and devices, which work together in water networks, can be defined as a multi-objective programming model. In a multi-objective study, Nikoo et al. (2019) considered the hydraulic performance (i.e., minimization of seepage, uplift force, and vertical exit gradient) and project cost of cutoff walls and aprons under a diversion dam as objective functions. The computational fluid dynamics models can be also employed to simulate flow in these problems. Table 11.1 lists examples of such studies.

11.2.6 Groundwater

Groundwater is one of the foremost means of water supply in some regions of the world, providing one-third of the total world's freshwater (Moreaux and Reynaud 2006). Increasing water demands, affected by population growth, climate change, and governments' policies, have caused drawdown of watertable in aquifers. Groundwater contamination can jeopardize public health, which raises public concern. Meyer et al. (1994) used metaheuristic methods for the optimization of objectives for groundwater monitoring. Others have employed multi-objective techniques in optimizing groundwater remediation design (Wang and Zheng 1998), determining well locations (Park and Aral 2004), proposing a cost effective long-term groundwater monitoring model (Reed et al. 2001), determining a dynamic connection between precipitation and water-table depth (Giustolisi et al. 2008), optimizing placement of pumping facilities (Siegfried et al. 2009), and optimizing pumping rates and operation cost (Saafan et al. 2011). In Table 11.1, some such problems are listed.

11.2.7 Other Applications

As multi-objective optimization approaches have unique advantages, these have been implemented in design, planning, and management of water resources systems, such as in treated wastewater allocation (Tayebikhorami et al. 2019), stormwater harvesting systems (Blinco et al. 2018), design and rehabilitation of urban drainage systems (Vojinovic et al. 2014; Ngamalieu-Nengoue et al. 2019), locations of green roofs and permeable pavements in urban catchments (Giacomoni and Joseph 2017), hydropower generation (Bozorg-Haddad et al. 2017), operation of gated spillways (Amirkhani et al. 2016b), and sewer network design (Altarabsheh et al. 2018).

11.3 Basic Concepts

In a multi-objective optimization problem, there is more than one objective, which is often also called criterion or performance. Such problems are formulated as follows:

$$\begin{aligned} \text{Minimize or Maximize } g(\vec{x}) = |g_1(\vec{x}), g_2(\vec{x}), \ldots, g_n(\vec{x})|,\ \vec{x} = [x_1, x_2, \ldots, x_k] \in S \\ \text{Subjected to } f_i(\vec{x}) \leq 0, i = 1, 2, \ldots, I \\ h_j(\vec{x}) = 0,\ j = 1, 2, \ldots, J \\ (\vec{x})^l \leq \vec{x} \leq (\vec{x})^u \end{aligned} \tag{11.1}$$

where S = the set of solutions, $g_1(\vec{x})$, $g_2(\vec{x})$, ..., and $g_n(\vec{x})$= the objective functions, $f_i(\vec{x})$= the inequality constraints, $h_j(\vec{x})$= the equality constraints,$(\vec{x})^l$= the lower bounds of decision variables, and $(\vec{x})^u$= the upper bounds of decision variables.

As shown in Eq. (11.1), in multi-objective problems there are n functions, which should be optimized simultaneously. These n functions may have one of these three conditions: (1) all n functions need minimizing; (2) all n functions need maximizing; and (3) some of them need to be minimized and others maximized. The multi-objective approach is derived from single objective problems where there is one

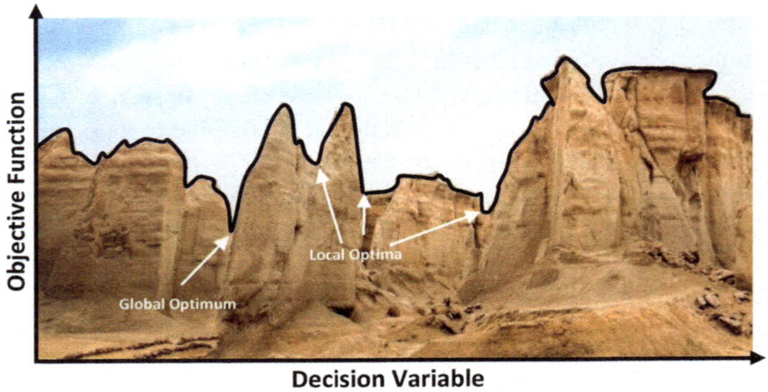

Fig. 11.1 Illustrations of global optimum and local optima (picture shows a portion of the Stars Valley in Qeshm, Iran)

function to be optimized. If the problem goal is minimizing (i.e., in problems which their objective functions are cost, time, environmental damage, error evaluation etc.), the best solution is called the global optimum—for example, the deepest point in a valley (Fig. 11.1). However, there may exist one or some solutions, which have less value than others, which are known as local optima (Fig. 11.1).

In multi-objective problems, instead of a unique solution, there is a set of optimal solutions. In this case, for a problem with two fitness functions (two-objective problem), one may need the least value, while the second one need the highest. In other words, in most cases, objectives are in conflict with each other. For example, optimal solution in the problem of reservoir water release for downstream environmental needs is probably in conflict with drinking water supply. Hence, the multi-objective problems need a decision-maker to choose the most suitable answer among the set of solutions (Coello et al. 2007). Using the Pareto Optimality Theory (Ehrgott 2005), the set of solutions can be found. The following terms are essential parts of an optimization problem.

Decision variables: The numerical quantities, which are calculated during optimization, are called decision (or design) variables. These variables are limited between lower and upper bounds.

Constraints: Finding optimum solutions is not allowed at any cost. In other words, there are some intrinsic restrictions in almost all optimization problems, which eliminate unsatisfactory solutions from the search space. These restrictions are known as constraints, which divide solutions into feasible and infeasible ones (Zhou et al. 2011). Time restrictions and maximum error value are such examples.

Search Space: The search space (decision space) includes all feasible or allowable solutions. By limiting this space in real engineering problems, it is expected to find an optimal solution in the least time. For instance, a reservoir operation problem with 52 decision variables (i.e., 52 weeks) and 5 different releases has a search space size of 5^{52} i.e., 2.22E+36 (Maier et al. 2019).

In multi-objective algorithms, the following terms also should be defined:

Dominance: X_2 is dominated by X_1, if

- Solution X_1 is no worse than X_2 in all objective functions.
- Solution X_1 is strictly better than X_2 in at least one objective.

Non-dominated Solution: A non-dominated solution is the one that generates a suitable compromise between all objectives without decaying any of them. In other words, these are solutions which are worth considering and evaluating them as a final solution of the problem.

Pareto Optimal Set: The Pareto Optimal Set is regarded as the set of entire feasible decision space.

Pareto Optimal Front: Instead of a single optimum, there is a set of trade-off solutions, generally regarded as Pareto Optimal Front. In other words, that is a collection of non-dominated solutions. (True) Pareto Optimal Front can have different conditions, based on its shape and continuity i.e., it may be continuous, discontinuous, or concave.

Figure 11.2 shows a multi-objective problem that has two objectives and both of them should be minimized. The ideal point is a hypothetical solution, which has zero value for both objectives—never reachable in real engineering problems. Thus, distances between this point and solutions are important. The aforementioned concepts (i.e., non-dominated and dominated solutions and Pareto optimal front) are illustrated in Fig. 11.2.

The final problem solution is chosen from non-dominated (Pareto) solutions. Also, the quality of solutions produced by different multi-objective algorithms is an important issue in the literature. Various metrics have been proposed to evaluate Pareto

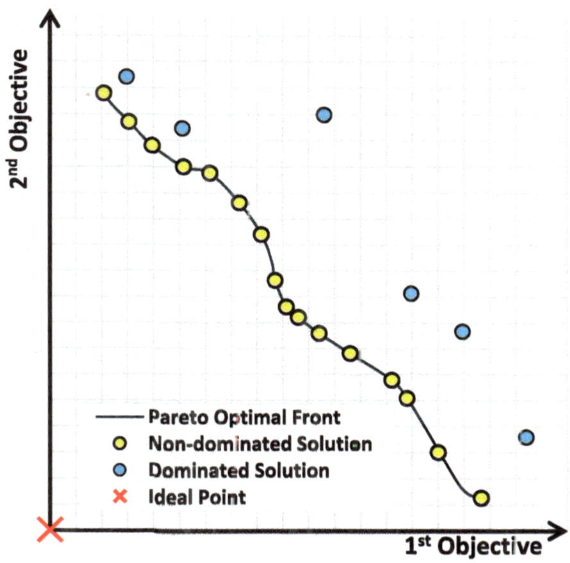

Fig. 11.2 Sample solutions of a two-objective problem, which both of objectives need minimizing

solutions. Some of these methods are presented in Section 4 (problem definitions section).

11.4 Practical Examples in Reservoir Optimal Operation

Three real engineering problems are solved for Mahabad dam using m MOSCA, MOPSO, and MOGA. The Sine Cosine Algorithm (SCA) was proposed by Mirjalili et al. (2016), which has been used successfully in a number of engineering problems, such as cancer classification (Majhi 2018) and predicting wind speed (Wang et al. 2018b).

The Mahabad dam is a multipurpose dam, which is located in the northwest part of Iran. It is used to supply irrigation demands and control seasonal floods. Table 11.2 shows the average inflow to reservoir, drinking and agricultural demands, net evaporation, and maximum release of the reservoir. The months of summer (May, June, and July) see the most drinking and agricultural demands. The average annual rainfall and temperature are 471 mm and 12.1 °C, respectively. The reservoir should supply drinking and agricultural demands for summer months while the inflow to the reservoir is not high for summer months. On the other hand, the most net evaporation occurs in summer months. Thus, the optimal operation of Mahabad reservoir is a real challenge for decision-makers in the Mahabad River basin.

Table 11.2 Hydrological parameters at Mahabad dam site 1990–2007

Month	Average inflow (MCM)	Drinking agricultural demand (MCM)	Net evaporation (mm)	Maximum release of the reservoir (MCM)
September	1.28	20.65	121.2	51.84
October	6.12	8.14	40.69	51.84
November	10.05	1.51	–	51.84
December	17.12	1.41	–	51.84
January	21.12	1.38	–	51.84
February	54	1.3	51.78	51.84
March	96.12	5.87	154.23	53.57
April	53.78	25.02	269.43	53.57
May	9.78	32.06	319.21	53.57
June	2.47	28.76	312.23	53.57
July	1.12	30.22	311.4	53.57
August	0.8	26.78	241.23	53.57

MCM: Million Cubic Meter

Source Jahad-e-Keshavazi Organization's guideline and Water and company, West Azarbaijan Province, Mahabad Town, Iran

11.4.1 Problem Definitions

First Example

In the first example, the inflow to the Mahabad dam was simulated using Artificial Neural Network (ANN) and multi-objective optimization algorithms. ANN models are widely used for predicting hydrological variables, such as rainfall (Yaseen et al. 2018), drought (Khan et al. 2018), and streamflow (Meng et al. 2019). An ANN model includes a number of neurons arranged into three basic layers. An ANN model has an input, middle (hidden layer), and the output layer. The input variables are received by the input layer. The transfer function is a fundamental unit of the ANN to compute output for a given input.

The following equation shows the relationship between x (inputs) and y (outputs):

$$y_m = f\left(B_m + \sum_{i=1}^{n} W_{im}x_i\right) \tag{11.2}$$

where y is the output, x_i represents the input, B_m indicates the bias, W_{im} shows the weight connection and $n=$ the number of input variables. It is evident that bias and weight parameters are the important parameters for ANN models. Although the common training algorithms, such as backpropagation algorithm, are widely used to fine-tune the ANN parameters (weight connection and bias parameter), they do not have the fast convergence rate. Thus, it is necessary to use a robust optimization algorithm for training ANN algorithms. In this chapter, multi-objective algorithms are used to train the ANN models.

Equation 11.3 shows inputs for predicting reservoir inflow.

$$I_t = (I_t, I_{t-1}, I_{t-2}, \ldots, I_{t-12}) \tag{11.3}$$

where $I_t$is the inflow to the reservoir in the current month, I_{t-1}indicates the inflow in one previous month, I_{t-2}shows the inflow in two previous months, and I_{t-12}represents the inflow in the twelve previous months. One of the challenges in preparing simulation models is the determination of their architecture. In this study, multi-objective optimization algorithms were used to fine-tune the weight and bias parameters. In this regard, the structure of an agent's position in hybrid ANN-multi-objective optimization algorithms includes three parts. The first part includes the value of weight parameters. The second part includes the value of bias parameters and the third includes the combination of inputs. In fact, each agent shows the ANN parameters and the input combinations.

Two objective functions were used. The first one was used to select the best ANN parameters. The second one was used to select the best-input combination. In other words, the first objective function is minimizing root mean square error (RMSE). The second objective function is minimizing the mean absolute error (MAE).

Second Example

In the second example, multi-objective functions were used to optimize the Mahabad reservoir operation using two objective functions. The first objective function was the reliability index. The reliability index is the probability that the water meets irrigation demands during the period of operation. The second objective function was the resiliency index. Resiliency index is the probability that a system recovers from a failure period:

$$D_i^t = \begin{bmatrix} X^i_{T\arg et,t} - X^i_{\sup plied,t} \leftarrow if\left(X^i_{T\arg et,t} > X^i_{\sup plied,t}\right) \\ 0 \leftarrow \left(if\left(X^i_{t\arg et}\right) = X^i_{\sup plied,t}\right) \end{bmatrix} \tag{11.4}$$

$$\text{maxmize}(z_1)\text{Rel}^i = \frac{No.of\left(times\left(D_i^t\right) = 0\right)}{n} \tag{11.5}$$

$$\text{Maximize}(z_2)(\text{Resilience}) = \frac{No.\,(of)\,times\,\left(D_i^t\right) = 0\,follows\,(D_i) > 0}{No.\,(of)\,times\,\left(D_i^t\right) > 0\,(occured)} \tag{11.6}$$

where Rel^i is the reliability index, $X^i_{T\arg et,t}$ shows the water demand for the ith water user at time t, $X^i_{\sup plied,t}$ represents the water supplied for the ith user at time t, $No.(of)times\left(D_i^t\right) > 0(occured)$ indicates all failure periods, n: the number of months and $No.(of)times\left(D_i^t\right) = 0follows(D_i)$ is a successful period recovering from a failure period (Loucks 1997; McMahon et al. 2006; Sandoval-Solis et al. 2011).

The continuity equation is as follows:

$$S_{t+1} = S_t + Q_t - Loss_t - R_t - SP_t \tag{11.7}$$

where S_{t+1} is the reservoir storage at time t+1 (MCM), Q_t shows the inflow at time t, $Loss_t$ shows the loss volume of water (MCM), R_t indicates the released water. The following constraints are used for the reservoir operation problem:

$$\begin{gathered} 0 \le R_t \le D_{et} \\ S_{\text{min}} \le S_t \le S_{\text{max}} \end{gathered} \tag{11.8}$$

where S_{min} shows the minimum reservoir storage, S_{max} is the maximum reservoir storage, and D_{et} is the irrigation demand. If the constraints were not satisfied, the following penalty function was used:

$$P_{1,t} = \begin{bmatrix} 0 \leftarrow if(S_{t+1}) > S_{\min} \\ \frac{(S_{\min} - S_{t+1})^2}{S_{\min}} \leftarrow otherwise \end{bmatrix} \tag{11.9}$$

$$P_{2,t} = \begin{bmatrix} 0 \leftarrow if(S_{t+1}) < S_{\max} \\ \frac{(S_{t+1} - S_{\max})^2}{S_{\max}} \leftarrow otherwise \end{bmatrix} \tag{11.10}$$

$$P_{3,t} = \begin{bmatrix} 0 \leftarrow if(R_t < D_t) \\ \frac{(R_t - D_t)^2}{D_{\max}} \end{bmatrix} \tag{11.11}$$

where $P_{1,t}$, $P_{2,t}$, $P_{3,t}$ are the penalty functions and $D_{\max}$ shows the maximum irrigation demand during the total operational periods. The following indices were used to evaluate the multi-objective algorithms.

- Root mean square error (RMSE):

$$RMSE = \sqrt{\frac{\sum_{t=1}^{T} (D_t - R_t)^2}{T}} \tag{11.12}$$

- MAE

$$MAE = \frac{\sum_{t=1}^{T} |D_t - R_t|}{T} \tag{11.13}$$

- Nash-Sutcliffe efficiency

$$NSE = 1 - \frac{\sum_{t=1}^{T} (D_t - R_t)^2}{\sum_{t=1}^{T} (D_t - \bar{D}_t)^2} \tag{11.14}$$

where T= the number of operational periods, and $\bar{D}_t$ is the average irrigation demand.

Third Example

To show the importance of the multi-objective model, operation policies for different inflow scenarios into the Mahabad dam reservoir were considered. Thus, the third example was solved based on three inflow scenarios:

1. Average monthly inflow −0.5 standard deviation
2. Average monthly inflow

3. Average monthly inflows +0.5 standard deviation.

11.4.2 Selection of the Best Solution in Pareto Front (PF)

After extracting the PF based on multi-objective algorithms, the most appropriate solution was identified via multi-criteria decision. In this study, compromise programming (CP) was used to select the optimal solution. The following equation was used to find the optimal solution among other solutions:

$$L_p(POS_i) = \left[\sum_{j=}^{n}\left(\frac{x_j^+ - x_{ij}}{x_j^+ - x_j^-}\right)\right] \quad (11.15)$$

where $L_p(POS_i)$is the distance of Pareto optimal solution to the ideal solution or ideal point (illustrated in Fig. 11.2),x_{ij}shows the preference of alternative I with respect to criterion j, p indicates the control parameter for CP $x_j^+$is the largest value for maximizing criteria or the lowest value for minimizing criteria, and x_j^- = the lowest value for maximizing criteria or the largest value for maximizing criteria. The following indices were used to verify the generated Pareto solutions using the optimization algorithms:

$$IGD = \sqrt{\frac{\sum_{i=1}^{nt} (d_i')^2}{n}} \quad (11.16)$$

$$SP = \sqrt{\frac{1}{n-1}\sum_{i=1}^{n} (\bar{d} - d_i)^2} \quad (11.17)$$

$$MS = \sqrt{\sum_{i=1}^{o} \max(d(a_i, b_i))} \quad (11.18)$$

where $IGD$is the inverse general distance, nt indicates the number of true Pareto optimal solutions, d_i' represents the Euclidean distance between the ith true Pareto optimal solution and the nearest Pareto optimal solution obtained in the reference net, SP shows the spacing spread measured, MS = the maximum spacing measured, d = a function to compute the Euclidean distance, a_i= the maximum value in the ith objective, b_i= the minimum value in the ith objective, and o= the number of objectives. The lower value for SP and IGD is suitable. The higher value for MS is ideal.

11.4.3 Results and Discussion

Selection of Random Parameters in Multi-objective Optimization Algorithms

The performance of multi-objective optimization algorithms depends on the value of random parameters. Thus, it was necessary to select the appropriate values of random parameters in the optimization algorithms (Ehteram et al. 2019). The Taguchi model is one of the robust models for designing parameters of experimental and computational methods (Alizadeh and Omrani 2019). The orthogonal arrays were used in the Taguchi model to reduce the number of experiments or runs of parameters that affected the outputs. Parameter setting was necessary to run evolutionary algorithms. The total number of experiments for finding the value of random parameters of optimization algorithms was computed as follows:

$$total(number)of(experiments) = (number\ of\ levels)^{number\ of\ \text{factors}} * (number\ of\ \text{iterations}) \tag{11.19}$$

when the values of a parameter are changed, the other parameters are fixed. Table 11.3 shows the algorithm parameters and their levels for the first example.

Table 11.3 Algorithm parameters and their levels for the first example for a = MOGGA, b = MOSPSO and c = MOSSCA

(a)				
Parameter	Level 1	Level 2	Level 3	Level 4
Population size	100	200	300	400
Rate of crossover	0.30	0.50	0.7	0.90
Rate of mutation	0.10	0.30	0.50	0.70
Iteration	50	100	150	200
(b)				
Parameter	Level 1	Level 2	Level 3	Level 4
Population size	100	200	300	400
Acceleration coefficient	0.2	0.4	0.6	0.80
Inertia weight	0.10	0.30	0.50	0.70
Iteration	50	100	150	200
(c)				
Parameter	Level 1	Level 2	Level 3	Level 4
Population size	50	100	150	200
r_1	0.2	0.4	0.6	0.80
r_3	0.10	0.30	0.50	0.70
Iteration	50	100	150	200

The signal to noise (S/N) ratio was used to show the impact of each parameter on the performance of optimization algorithms:

$$\frac{S}{N}ratio = -10\log(objective\ function)^2 \quad (11.20)$$

The aim was to maximize the S/N ratio. Table 11.4 shows the computed S/N ratio for each level of parameters. The most value of S/N ratio indicates the best value of parameters.

Results for the First Example

Figure 11.3 shows the Pareto front for MOSCA. The CP method was used to select the best solution. Table 11.5 compares the performance of MOSCA, MOPSO, and MOGA. It is worth mentioning that the outputs were compared, based on model performance in 200 runs.

Inspecting the results in Table 11.5, it is observed that MOSCA was able to converge with good distribution of non-dominated solutions for the first example. As far as the IGD and SP were concerned, MOSCA performed better, since MOSCA had the least values for IGD and SP compared to MOPSO and MOGA. Table 11.5 also indicates MS for 200 simulation runs for multi-objective optimization algorithms. It is observed that MOSCA had the highest value for MS. Generally, results indicated

Table 11.4 Computed S/N ratio for each level of parameters for a = MOGA, b = MOPSO and c = MO SCA (bold values indicate the best value of parameters)

(a)				
Parameter	Level 1	Level 2	Level 3	Level 4
Population size	S/N:1.2	S/N:1.8	S/N:1.4	**S/N:1.1**
Rate of crossover	S/N:1.3	**S/N:1.5**	S/N:1.2	S/N:1.0
Rate of mutation	S/N:1.3	**S/N:1.6**	S/N:1.2	S/N:1.0
Iteration	S/N:1.4	S/N:1.3	**S/N:1.6**	S/N:1.2
(b)				
Parameter	Level 1	Level 2	Level 3	Level 4
Population size	S/N:1.2	**S/N:1.5**	S/N:1.1	S/N:1.3
Acceleration coefficient	S/N:1.4	S/N:1.7	**S/N:1.9**	S/N:1.2
Inertia weight	S/N:1.2	S/N:1.3	S/N:1.4	**S/N:1.7**
Iteration	S/N:1.2	**S/N:1.5**	S/N:1.3	S/N:1.1
(c)				
Parameter	Level 1	Level 2	Level 3	Level 4
Population size	S/N:1.4	**S/N:1.7**	S/N:1.2	S/N:1.5
r_1	S/N:1.3	S/N:1.6	**S/N:1.8**	S/N:1.5
r_3	S/N:1.4	**S/N:1.6**	S/N:1.8	S/N:1.4
Iteration	S/N:1.4	S/N:1.7	**S/N:1.8**	S/N:1.9

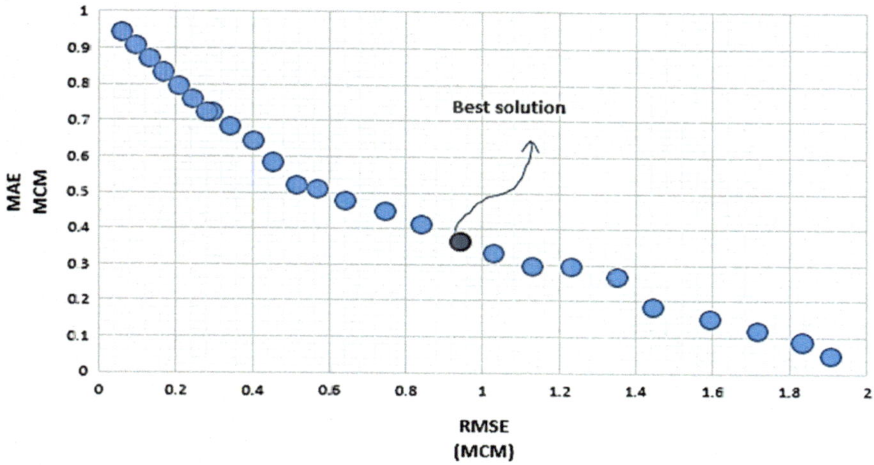

Fig. 11.3 Pareto front for MOSCA

Table 11.5 Investigation of multi-objective algorithms for the Pareto optimal fronts obtained

Algorithm	IGD	SP	MS
MOSCA	0.12	0.011	0.72
MOPSO	0.23	0.015	0.64
MOGA	0.29	0.019	0.58

that MOSCA outperformed MOPSO and MOGA. Table 11.6 shows the results of hybrid ANN models for predicting inflow, and indicated that ANN-MOSCA provided a more reasonable estimation for inflow to the reservoir as compared to MOPSO and MOGA. It is evident that the RMSE and MAE values for ANN-MOSCA were low, while NSE for this model was high. Results also indicated that ANN-MOPSO performed better than MOGA. Figure 11.4 shows the scatterplots for the hybrid ANN models. Based on the plots for ANN-MOSCA, it can be seen that ANN-MOSCA was

Table 11.6 Results of hybrid ANN models for predicting inflow

Training period			
Model	RMSE (MCM)	MAE	NSE
ANN-MOSCA	0.90	0.33	0.93
ANN-MOPSO	1.12	0.47	0.92
ANN-MOGA	1.19	0.65	0.91
Testing period			
ANN-MOSCA	1.24	0.50	0.90
ANN-MOPSO	1.34	0.71	0.89
ANN-MOGA	1.45	0.89	0.87

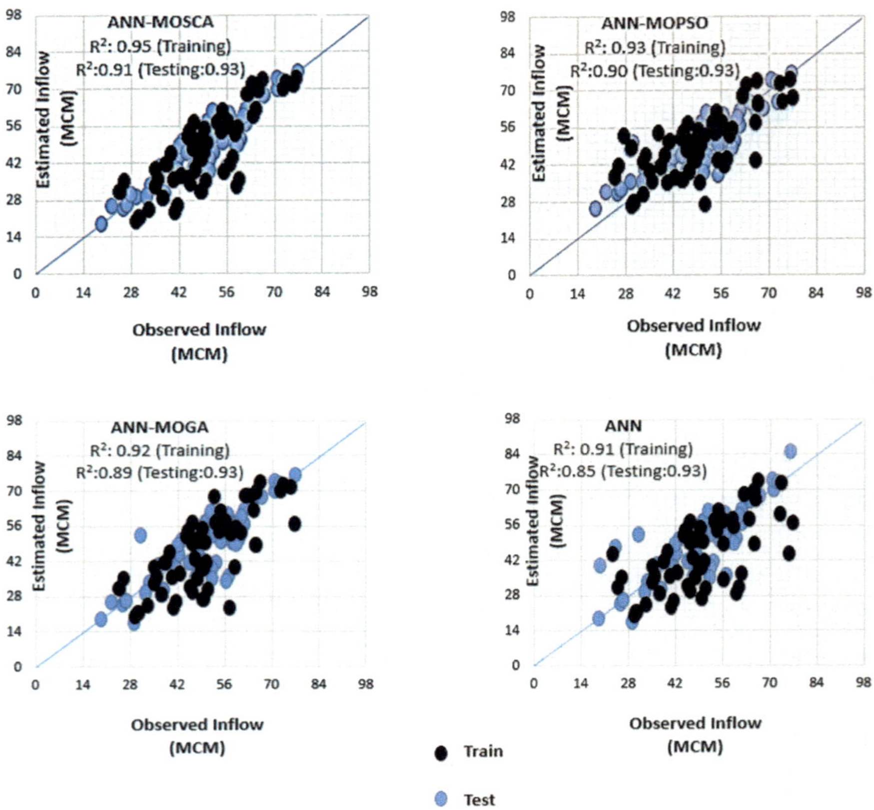

Fig. 11.4 Scatter plots for the models for the first example

more accurate than ANN-MOGA and ANN-MOPSO. Figure 11.4 and Table 11.5 showed that the hybrid ANN models had better performance than the standalone ANN model. The standalone ANN model used the back propagation algorithm for training level.

Results for the Second Example

Figure 11.5 shows the Pareto optimal front for the second example. The best solution was selected using the CP method. As observed in Table 11.7, three indexes were used for comparing multi-objective algorithms. According to Table 11.7, IGD and SP of MOSCA had the least values. It was concluded from the results in Table 11.7 that MOGA had worse performance than MOSCA and MOPSO. In the case of the coefficient of variation, it was clear that MOSCA had the lowest value among other multi-objective algorithms. Among multi-objective algorithms, MOGA had the highest coefficient of variation.

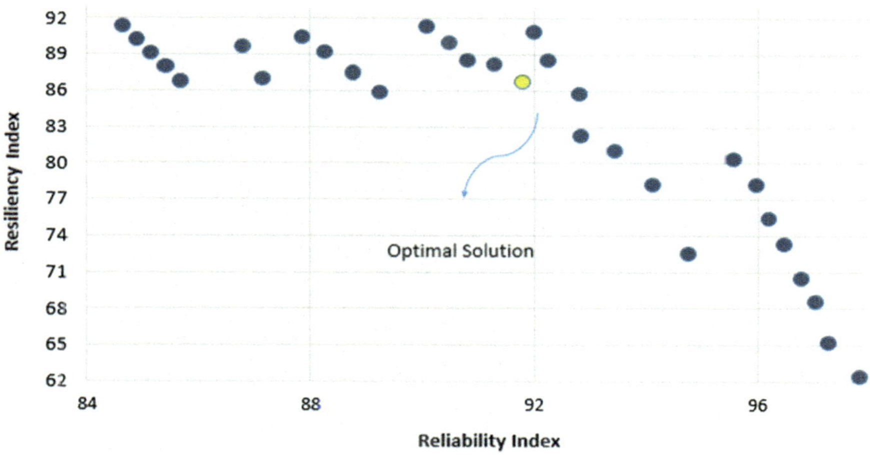

Fig. 11.5 Pareto front for the second problem

Table 11.7 Analysis of provided Pareto fronts for the second example

Algorithm	Parameter	IGD	SP	MS
MOSCA	Worst	0.25	0.019	0.54
	best	0.11	0.010	0.67
	Average	0.18	0.012	0.59
	Coefficient of variation	0.02	0.01	0.02
MOPSO	Worst	0.29	0.025	0.45
	best	0.18	0.014	0.57
	Average	0.20	0.019	0.50
	Coefficient of variation	0.05	0.03	0.03
MOGA	Worst	0.38	0.029	0.41
	best	0.20	0.017	0.52
	Average	0.29	0.022	0.48
	Coefficient of variation	0.09	0.05	0.04

Figure 11.6 compares the performances of multi-objective algorithms based on RMSE, MAE, and NSE indices. MOSCA had the best performance, as verified by the values of indices (i.e., RMSE= 2.12 MCM, MAE=1.1 MCM, and NSE= 0.93). Results indicated that MOGA had the highest value of RMSE and MAE among other multi-objective algorithms.

Figure 11.7 shows the performance criteria for different water demands. For example, for a 50% decrease in water demand, results of reliability indices of MOSCA, MOPSO, and MOGA were 81, 78, and 70%, respectively. In the case of reliability criterion, when the water demand of reservoir system increased from

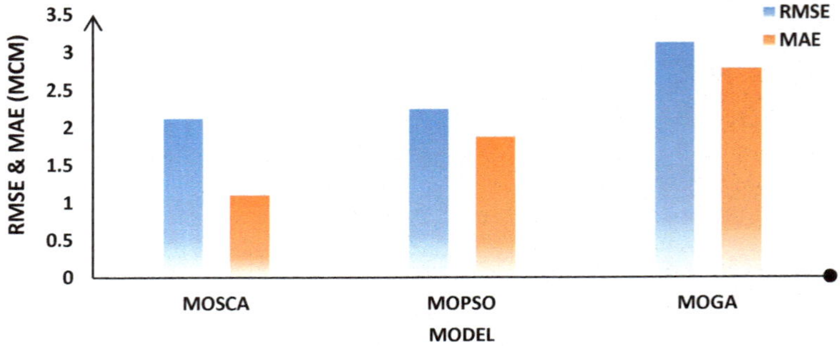

(A)

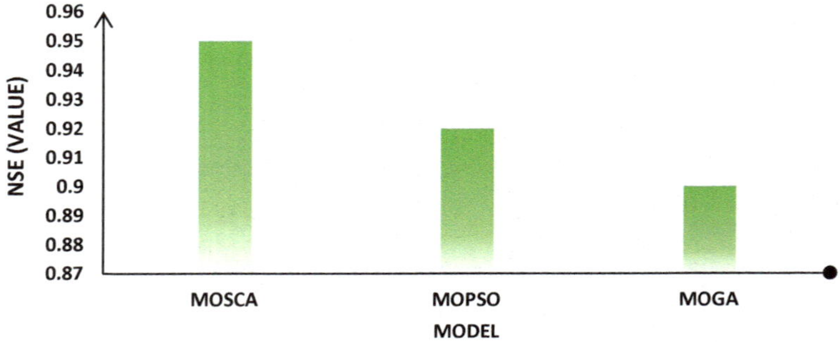

(B)

Fig. 11.6 **a** Comparison of multi-objective algorithms based on RMSE and MAE indices and **b** Comparison of multi-objective algorithms based on NSE index

20 to 100%, the reliability index of MOSCA reduced from 93 to 72%. Overall, the reliability index of MOGA decreased 23%, when the water demand increased from 20 to 100%.

Figure 11.8 shows the variation of resiliency indices for different water demands. For instance, for a 50% reduction in water demand, the resilience indices of MOSCA, MOPSO, and MOGA were 69, 65, and 63%. When the water demand was 100%, the resilience indices were 72, 70, and 59% for MOSCA, MOPSO, and MOGA, respectively. The high value of resilience index indicated that MOSCA was able to follow successfully a failure period in comparison to multi-objective MOPSO and multi-objective MOGA.

Results for the Third Example

The third example considered different inflow scenarios for the second example. The second example used the reliability index and resilience index as objective functions.

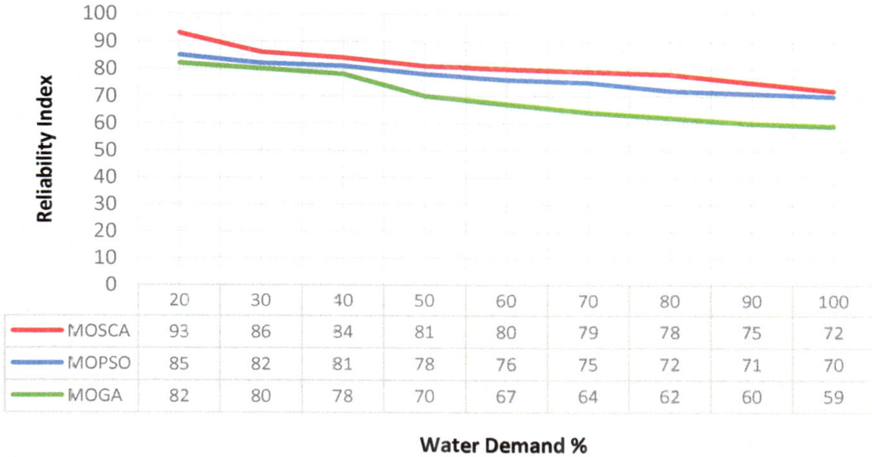

	20	30	40	50	60	70	80	90	100
MOSCA	93	86	84	81	80	79	78	75	72
MOPSO	85	82	81	78	76	75	72	71	70
MOGA	82	80	78	70	67	64	62	60	59

Fig. 11.7 Variation of reliability index for different water demands

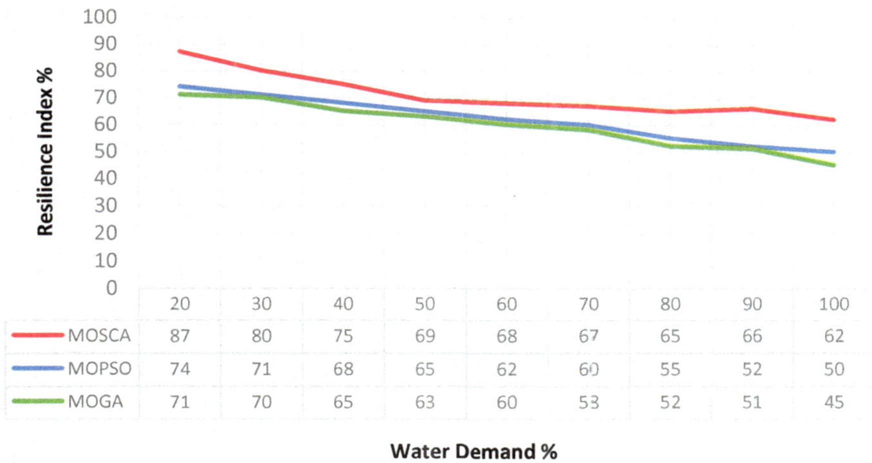

	20	30	40	50	60	70	80	90	100
MOSCA	87	80	75	69	68	67	65	66	62
MOPSO	74	71	68	65	62	60	55	52	50
MOGA	71	70	65	63	60	53	52	51	45

Fig. 11.8 Variation of resilience index for different water demands

Figure 11.9 shows the Pareto front for the third example. The optimal solutions were selected by the CP method.

As observed in Table 11.8, three indices were used for comparing multi-objective algorithms. Results indicated that the best Pareto optimal front was obtained by MOSCA for the second and third scenarios, where the best values of IGD, SP, and MS were 0.019, 0.14, and 0.6, respectively, for MOSCA (second scenario), and were 0.020, 0.16, and 0.62, respectively, for MOSCA (third scenario).

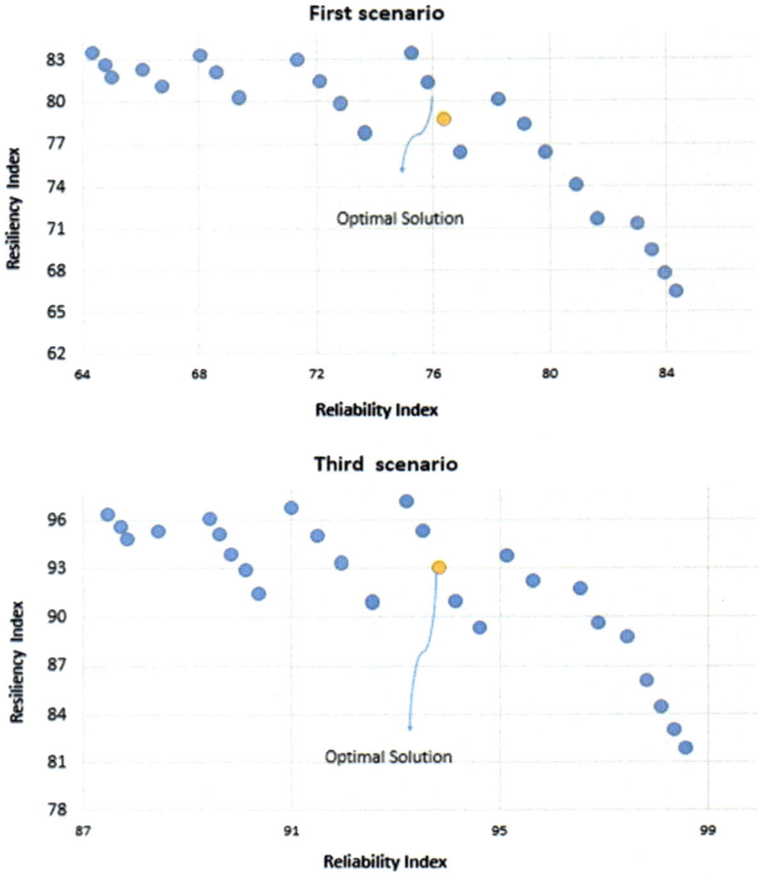

Fig. 11.9 The Pareto front for the third example

Figure 11.10 compares the performances of different models for different inflow scenarios. Results of the second scenario (SS) were presented by the second example (normal inflow scenario), and indicated that the multi-objective optimization algorithms that employed the first scenario had the highest RMSE and MAE. This indicated that the first scenario (FS) had a significant effect on the operation of the reservoir. As observed in Fig 11.10, the third scenario (TS) provided better conditions.

Thus, decision-makers should consider the effect of uncertainty of different parameters on the reservoir operation. The climate change condition, such as wet periods and dry periods, affect stream flow and inflow to the reservoir. Table 11.9 shows the variation of reliability index and resilience index versus different water demands. For example, for a 50% decrease in water demand, results of reliability index of MOSCA were 81, 79, and 82% for the second scenario, first scenario, and third scenario respectively. When the water demand was 100%, the resilience index

Table 11.8 The Pareto front for the first and third scenarios

First scenario				
Algorithm	Parameter	IGD	SP	MS
MOSCA	Worst	0.27	0.021	0.52
	best	0.12	0.011	0.69
	Average	0.19	0.014	0.60
	Coefficient of variation	0.04	0.020	0.05
MOPSO	Worst	0.32	0.027	0.47
	best	0.19	0.017	0.58
	Average	0.20	0.020	0.52
	Coefficient of variation	0.04	0.04	0.04
MOGA	Worst	0.39	0.030	0.42
	best	0.22	0.018	0.53
	Average	0.30	0.023	0.49
	Coefficient of variation	0.08	0.07	0.05
	best	0.23	0.019	0.54
	Average	0.32	0.024	0.50
	Coefficient of variation	0.09	0.08	0.06
Third				
Algorithm	Parameter	IGD	SP	MS
MOSCA	Worst	0.25	0.022	0.54
	best	0.14	0.012	0.70
	Average	0.20	0.016	0.62
	Coefficient of variation	0.05	0.040	0.07
MOPSO	Worst	0.33	0.028	0.48
	best	0.17	0.015	0.59
	Average	0.21	0.019	0.53
	Coefficient of variation	0.05	0.04	0.05
MOGA	Worst	0.40	0.032	0.45
	best	0.23	0.019	0.54
	Average	0.32	0.024	0.50
	Coefficient of variation	0.09	0.08	0.06

of MOSCA was 62, 61, and 63% for the second scenario, first scenario and third scenario, respectively. In general, results indicated that the multi-objective optimization algorithms had better performance in the third scenario than in the first and second scenarios.

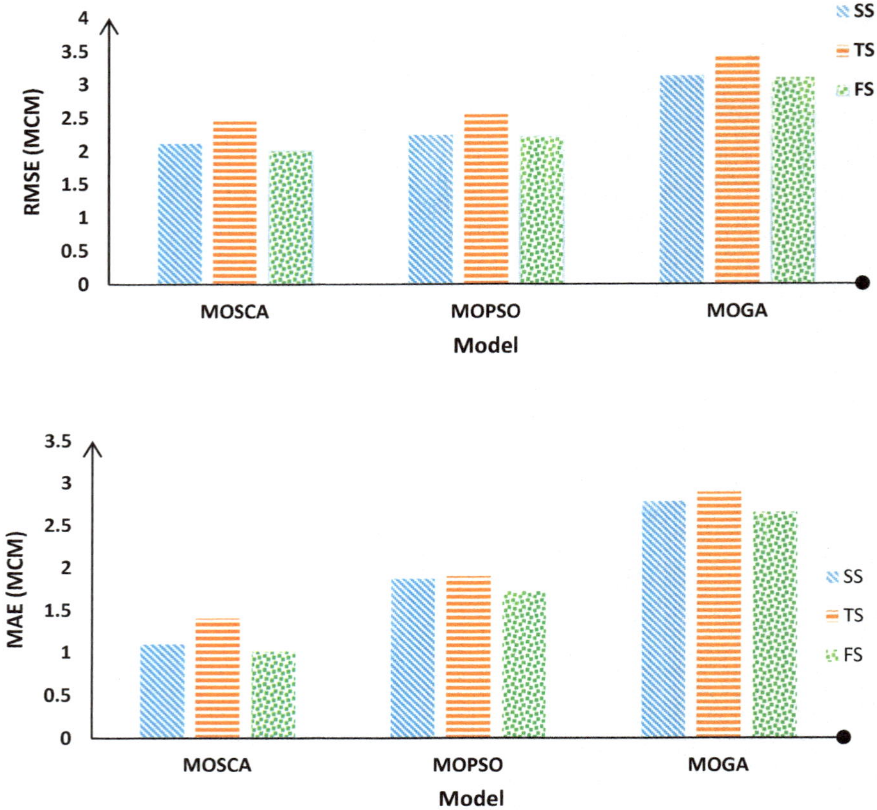

Fig. 11.10 Comparison of different scenarios for the third example

11.5 Summary

Problems in design, planning, and management of water resources systems can be solved using multi-objective optimization approaches. In this chapter, three multi-objective optimization algorithms were used to solve three complex problems. Firstly, the MOSCA, MOPSO, and MOGA were used to train the ANN models for modelling inflow. Secondly, the multi-objective optimization algorithms were used to increase the resiliency index and volumetric reliability index for optimal operation of a reservoir. Thirdly, the multi-objective optimization algorithms were used to solve the second problem using different inflow scenarios. It was observed that the RMSE and MAE values for ANN-MOSCA were low while NSE for this model was high. Results also indicated that ANN-MOPSO performed better than MOGA. It was concluded from the results in the second example that MOGA had worse performance than MOSCA and MOPSO. In the case of the coefficient of variation, it was

Table 11.9 Results of indices for different water demands (Third example)

Water demand	20%	30%	40%	50%	60%	70%	80%	90%	100%
First scenario (Reliability index)									
MOSCA	92	85	83	79	78	77	76	65	61
MOPSO	82	81	80	75	74	73	71	51	50
MOGA	81	77	77	68	65	62	60	50	49
Third scenario (Reliability index)									
MOSCA	94	87	85	82	82	80	79	77	74
MOPSO	87	84	82	79	77	76	74	73	72
MOGA	83	83	79	71	69	66	64	62	61
First scenario (Resilience index)									
MOSCA	86	78	74	67	67	66	64	62	61
MOPSO	72	70	67	63	61	59	54	50	49
MOGA	70	68	64	61	59	57	51	49	44
Third scenario (Resiliency index)									
MOSCA	88	81	77	72	69	68	67	66	63
MOPSO	76	73	71	67	63	62	59	54	52
MOGA	72	71	66	65	64	60	59	53	49

clear that MOSCA had the lowest value among other multi-objective algorithms for the second example. For the third example, for a 50% decrease in water demand, results of reliability index of MOSCA were 81, 79, and 82% for the second scenario, first scenario, and third scenario, respectively. When the water demand was 100%, the resilience index of MOSCA were 62, 61, and 63% for the second scenario, first scenario, and third scenario, respectively. Generally, the results indicated that the new multi-objective optimization algorithm (i.e., SCA) had superior ability for solving complex and nonlinear problems. However, the multi-objective algorithms such as SCA require the accurate design of random parameters. Additionally, the multi criteria decisions should be used to find the best solutions among the other solutions.

Acknowledgements The authors thank Iran's National Science Foundation (INSF) for the support of this research.

References

Aalami MT, Abbasi H, Nourani V (2018) Sustainable management of reservoir water quality and quantity through reservoir operational strategy and watershed control strategies. Int J Environ Res 12(6):773–788

Aboutalebi M, Bozorg-Haddad O, Loáiciga HA (2016) Multiobjective design of water-quality monitoring networks in river-reservoir systems. J Environ Eng 143(1):04016070

Adeyeri OE, Laux P, Arnault J, Lawin AE, Kunstmann H (2020) Conceptual hydrological model calibration using multi-objective optimization techniques over the transboundary Komadugu-Yobe basin, Lake Chad Area, West Africa. J Hydrol: Region Stud 27:100655

Alfonso L, Jonoski A, Solomatine D (2009) Multiobjective optimization of operational responses for contaminant flushing in water distribution networks. J Water Resour Plann and Manag 136(1):48–58

Alizadeh A, Omrani H (2019) An integrated multi response Taguchi-neural network-robust data envelopment analysis model for CO2 laser cutting. Measurement 131:69–78

Altarabsheh A, Ventresca M, Kandil A (2018) Evolutionary algorithm for selecting wastewater system configuration. J Comput Civil Eng 32(6):04018048

Amirkhani M, Bozorg-Haddad O, Fallah-Mehdipour E, Loáiciga HA (2016a) Multiobjective reservoir operation for water quality optimization. J Irrig Drain Eng 142(12):04016065

Amirkhani M, Bozorg-Haddad O, Azarnivand A, Loáiciga HA (2016b) Multiobjective optimal operation of gated spillways. J Irrig Drain Eng 143(2):04016078

Ashofteh PS, Bozorg-Haddad O, Loáiciga HA (2015) Evaluation of climatic-change impacts on multiobjective reservoir operation with multiobjective genetic programming. J Water Resour Plann Manag 141(11):04015030

Bekele EG, Nicklow JW (2007) Multi-objective automatic calibration of SWAT using NSGA-II. J Hydrol 341(3–4):165–176

Blinco LJ, Simpson AR, Lambert MF, Marchi A (2018) Optimization of pumping costs and harvested volume for a stormwater harvesting system. J Water Resour Plann Manag 144(8):05018011

Bomhof J, Tolson BA, Kouwen N (2019) Comparing single and multi-objective hydrologic model calibration considering reservoir inflow and streamflow observations. Can Water Resour J/Revue canadienne des ressources hydriques:1–18

Bozorg-Haddad O, Azarnivand A, Hosseini-Moghari SM, Loáiciga HA (2016) Development of a comparative multiple criteria framework for ranking Pareto optimal solutions of a multiobjective reservoir operation problem. J Irrig Drain Eng 142(7):04016019

Bozorg-Haddad O, Garousi-Nejad I, Loáiciga HA (2017) Extended multi-objective firefly algorithm for hydropower energy generation. J Hydroinf 19(5):734–751

Bozorg-Haddad O, Mirmomeni M, Mariño MA (2010) Optimal design of stepped spillways using the HBMO algorithm. Civil Eng Environ Syst 27(1):81–94

Coello CAC, Lamont GB, Van Veldhuizen DA (2007) Evolutionary algorithms for solving multi-objective problems. Springer, New York

Confesor RB, Whittaker GW (2007) Automatic calibration of hydrologic models with multi-objective evolutionary algorithm and Pareto optimization. J Am Water Resour Assoc 43(4):981–989

Ehrgott M (2005) Multicriteria optimization, 2nd edn. Springer, Berlin

Ehsani N, Afshar A (2010) Optimization of contaminant sensor placement in water distribution networks: multi-objective approach. Water Distrib Syst Anal:338–346

Ehteram M, Singh VP, Ferdowsi A, Mousavi SF, Farzin S, Karami H, Mohd NS, Afan HA, Lai SH, Kisi O, Malek MA, Ahmed AN, El-Shafie A (2019) An improved model based on the support vector machine and cuckoo algorithm for simulating reference evapotranspiration. PLoS ONE 14(5):e0217499

Farmani R, Savic DA, Walters GA (2005) Evolutionary multi-objective optimization in water distribution network design. Eng Optim 37(2):167–183

Fatkhutdinov A, Stefan C (2019) Multi-Objective optimization of managed aquifer recharge. Groundwater 57(2):238–244

Ferdowsi A (2019) Optimization of labyrinth spillway geometry using metaheuristic algorithms under climate change. M.Sc. thesis, Semnan University, Semnan, Iran

Ferdowsi A, Farzin S, Mousavi SF, Karami H (2019) Hybrid bat and particle swarm algorithm for optimization of labyrinth spillway based on half and quarter round crest shapes. Flow Meas Instrum 66:209–217

Ferdowsi A, Mousavi SF, Farzin S, Karami H (2020) Optimization of dam's spillway design under climate change conditions. J Hydroinf (in Press)

Fu G, Kapelan Z (2010) Embedding neural networks in multiobjective genetic algorithms for water distribution system design. Water Distrib Syst Anal:888–898

Ghorbani Mooselu M, Nikoo MR, Rayani NB, Izady A (2019). Fuzzy multi-objective simulation-optimization of stepped spillways considering flood uncertainty. Water Resour Manag:1–15

Giacomoni MH, Joseph J (2017) Multi-objective evolutionary optimization and Monte Carlo simulation for placement of low impact development in the catchment scale. J Water Resour Plann Manag 143(9):04017053

Giustolisi O, Doglioni A, Savic DA, Di Pierro F (2008). An evolutionary multi-objective strategy for the effective management of groundwater resources. Water Resour Res 44(1)

Gutierrez JCT, Adamatti DS, Bravo JM (2019) A new stopping criterion for multi-objective evolutionary algorithms: application in the calibration of a hydrologic model. Comput Geosci 23(6):1219–1235

Hadka D, Reed PM (2013) Borg: an auto-adaptive many-objective framework. Evol Comput 21(2):231–259

Hajiabadi R, Zarghami M (2014) Multi-objective reservoir operation with sediment flushing; case study of Sefidrud reservoir. Water Resour Manag 28(15):5357–5376

Hojjati SH, Hojjati SH, Salehi Neyshabouri SAA (2017) The objective design of triangular bucket for dam's spillway using non-dominated sorting genetic algorithm II: NSGA-II. Scientia Iranica 24(1):19–27

Halbe J, Pahl-Wostl C, Adamowski J (2018) A methodological framework to support the initiation, design and institutionalization of participatory modeling processes in water resources management. J Hydrol 556:701–716

Kapelan ZS, Savic DA, Walters GA (2003) Multiobjective sampling design for water distribution model calibration. J Water Resour Plann Manag 129(6):466–479

Kapelan ZS, Savic DA, Walters GA (2005a) Optimal sampling design methodologies for water distribution model calibration. J Hydraulic Eng 131(3):190–200

Kapelan ZS, Savic DA, Walters GA (2005b) Multiobjective design of water distribution systems under uncertainty. Water Resour Res 41(11)

Khan M, Muhammad N, El-Shafie A (2018) Wavelet-ANN versus ANN-based model for hydrometeorological drought forecasting. Water 10(8):998

KhazaiPoul A, Moridi A, Yazdi J (2019) Multi-objective optimization for interactive reservoir-irrigation planning considering environmental issues by using parallel processes technique. Water Resour Manag:1–15

Khorshidi MS, Nikoo MR, Sadegh M, Nematollahi B (2019) A multi-objective risk-based game theoretic approach to reservoir operation policy in potential future drought condition. Water Resour Manag 33(6):1999–2014

Khu ST, Madsen H, Di Pierro F (2008) Incorporating multiple observations for distributed hydrologic model calibration: an approach using a multi-objective evolutionary algorithm and clustering. Adv Water Resour 31(10):1387–1398

Kline M (1982) Mathematics: the loss of certainty, vol 686. Oxford University Press, USA

Maier HR, Razavi S, Kapelan Z, Matott LS, Kasprzyk J, Tolson BA (2019) Introductory overview: optimization using evolutionary algorithms and other metaheuristics. Environ Model Softw 114:195–213

Loucks DP (1997) Quantifying trends in system sustainability. Hydrol Sci J 42(4):513–530

Majhi SK (2018) An efficient feed foreword network model with sine cosine algorithm for breast cancer classification. Int J Syst Dyn Appl 7(2):1–14

Mallakpour I, AghaKouchak A, Sadegh M (2019) Climate-induced changes in the risk of hydrological failure of major dams in California. Geophys Res Lett 46(4):2130–2139

Marques J, Cunha M, Savić DA (2015) Multi-objective optimization of water distribution systems based on a real options approach. Environ Model Softw 63:1–13

McMahon TA, Adeloye AJ, Sen-Lin Z (2006) Understanding performance measures of reservoirs. J Hydrol (Amsterdam) 324:359–382

Meng E, Huang S, Huang Q, Fang W, Wu L, Wang L (2019) A robust method for non-stationary streamflow prediction based on improved EMD-SVM model. J Hydrol 568:462–478

Meyer PD, Valocchi AJ, Eheart JW (1994) Monitoring network design to provide initial detection of groundwater contamination. Water Resour Res 30(9):2647–2659

Mirjalili S (2016) SCA: a sine cosine algorithm for solving optimization problems. Knowl-Based Syst 96:120–133

Mirzaie-Nodoushan F, Bozorg-Haddad O, Loáiciga HA (2017) Optimal design of groundwater-level monitoring networks. J Hydroinf 19(6):920–929

Mohammad-Azari S, Bozorg-Haddad O, Loáiciga HA (2020) State-of-art of genetic programming applications in water-resources systems analysis. Environ Monit Assess 192(73)

Monsef H, Naghashzadegan M, Jamali A, Farmani R (2019) Comparison of evolutionary multi objective optimization algorithms in optimum design of water distribution network. Ain Shams Eng J 10(1):103–111

Moosavian N, Lence BJ (2016) Nondominated sorting differential evolution algorithms for multiobjective optimization of water distribution systems. J Water Resour Plann Manag 143(4):04016082

Moreaux M, Reynaud A (2006) Urban freshwater needs and spatial cost externalities for coastal aquifers: a theoretical approach. Region Sci Urban Econ 36(2):163–186

Mostafaie A, Forootan E, Safari A, Schumacher M (2018) Comparing multi-objective optimization techniques to calibrate a conceptual hydrological model using in situ runoff and daily GRACE data. Comput Geosci 22(3):789–814

Najafi MR, Moradkhani H, Jung I (2011) Assessing the uncertainties of hydrologic model selection in climate change impact studies. Hydrol Process 25(18):2814–2826

Ngamalieu-Nengoue UA, Martínez-Solano FJ, Iglesias-Rey PL, Mora-Meliá D (2019) Multi-objective optimization for urban drainage or sewer networks rehabilitation through pipes substitution and storage tanks installation. Water 11(5):935

Nicklow J, Reed P, Savic D, Dessalegne T, Harrell L, Chan-Hilton A, Zechman E (2009) State of the art for genetic algorithms and beyond in water resources planning and management. J Water Resour Plann Manag 136(4):412–432

Nikoo MR, Gavahi K, Khoramshokooh N (2019) A multi-objective simulation–optimization approach for Design of Cutoff Walls and Apron of diversion dams. Iranian J Sci Technol Trans Civil Eng 43(2):241–252

Nikoo MR, Varjavand I, Kerachian R, Pirooz MD, Karimi A (2014) Multi-objective optimum design of double-layer perforated-wall breakwaters: Application of NSGA-II and bargaining models. Appl Ocean Res 47:47–52

Nouiri I, Yitayew M, Maßmann J, Tarhouni J (2015) Multi-objective optimization tool for integrated groundwater management. Water Resour Manag 29(14):5353–5375

Orouji H, Mahmoudi N, Pazoki M (2016) Shuffled frog-leaping algorithm for optimal design of open channels. J Irrig Drain Eng 142(10):06016008

Park CH, Aral MM (2004) Multi-objective optimization of pumping rates and well placement in coastal aquifers. J Hydrol 290:80–99

Paseka S, Kapelan Z, Marton D (2018) Multi-objective optimization of resilient design of the multipurpose reservoir in conditions of uncertain climate change. Water 10(9):1110

Piscopo AN, Kasprzyk JR, Neupauer RM (2015) An iterative approach to multi-objective engineering design: Optimization of engineered injection and extraction for enhanced groundwater remediation. Environ Model Softw 69:253–261

Quintiliani C, Marquez-Calvo O, Alfonso L, Di Cristo C, Leopardi A, Solomatine DP, De Marinis G (2019) Multiobjective valve management optimization formulations for water quality enhancement in water distribution networks. J Water Resour Plann Manag 145(12):04019061

Reed PM, Hadka D, Herman JD, Kasprzyk JR, Kollat JB (2013) Evolutionary multiobjective optimization in water resources: the past, present, and future. Adv Water Resour 51:438–456
Reed P, Minsker BS, Goldberg DE (2001) A multi-objective approach to cost effective long-term groundwater monitoring using an elitist nondominated sorted genetic algorithm with historical data. J Hydroinf 3(2):71–89
Rezaeeian A, Davoodi M, Jafari MK (2019) Determination of optimum cross-section of earth dams using ant colony optimization algorithm. Scientia Iranica 26(3):1104–1121
Saadatpour M, Afshar A (2013) Multi objective simulation-optimization approach in pollution spill response management model in reservoirs. Water Resour Manag 27(6):1851–1865
Saafan TA, Moharram SH, Gad MI, KhalafAllah S (2011) A multi-objective optimization approach to groundwater management using genetic algorithm. Int J Water Resour Environ Eng 3:139–149
Sandoval-Solis S, McKinney DC, Loucks DP (2011) Sustainability index for water resources planning and management. J Water Resour Plann Manag 137(5):381–390
Siegfried T, Bleuler S, Laumanns M, Zitzler E, Kinzelbach W (2009) Multi-objective groundwater management using evolutionary algorithms. IEEE Trans Evol Comput 13:229–242
Taravatrooy N, Nikoo MR, Adamowski JF, Khoramshokooh N (2019) Fuzzy-based conflict resolution management of groundwater in-situ bioremediation under hydrogeological uncertainty. J Hydrol 571:376–389
Tayebikhorami S, Nikoo MR, Sadegh M (2019) A fuzzy multi-objective optimization approach for treated wastewater allocation. Environ Monit Assess 191(7):468
Vojinovic Z, Sahlu S, Torres AS, Seyoum SD, Anvarifar F, Matungulu H, Kapelan Z (2014) Multi-objective rehabilitation of urban drainage systems under uncertainties. J Hydroinf 16(5):1044–1061
Wang Y, Hua Z, Wang L (2018a) Parameter estimation of water quality models using an improved multi-objective particle swarm optimization. Water 10(1):32
Wang J, Yang W, Du P, Niu T (2018b) A novel hybrid forecasting system of wind speed based on a newly developed multi-objective sine cosine algorithm. Energy Convers Manag 163:134–150
Wang M, Zheng C (1998) Ground water management optimization using genetic algorithms and simulated annealing: formulation and comparison. J Am Water Resour Assoc 34(3):519–530
Wu W, Maier HR, Simpson AR (2013) Multiobjective optimization of water distribution systems accounting for economic cost, hydraulic reliability, and greenhouse gas emissions. Water Resour Res 49(3):1211–1225
Wu W, Simpson AR, Maier HR, Marchi A (2011) Incorporation of variable-speed pumping in multiobjective genetic algorithm optimization of the design of water transmission systems. J Water Resour Plann Manag 138(5):543–552
Yang G, Guo S, Liu P, Li L, Xu C (2017) Multiobjective reservoir operating rules based on cascade reservoir input variable selection method. Water Resour Res 53(4):3446–3463
Yaseen ZM, Ehteram M, Hossain M, Fai CM, Binti Koting S, Mohd NS, Ahmed AN (2019) A novel hybrid evolutionary data-intelligence algorithm for irrigation and power production management: application to multi-purpose reservoir systems. Sustainability 11(7):1953
Yaseen ZM, Ghareb MI, Ebtehaj I, Bonakdari H, Siddique R, Heddam S, Deo R (2018) Rainfall pattern forecasting using novel hybrid intelligent model based ANFIS-FFA. Water Resour Manag 32(1):105–122
Yazdi J, Moridi A (2018) Multi-objective differential evolution for design of cascade hydropower reservoir systems. Water Resour Manag 32(14):4779–4791
Zhang R, Liu J, Gao H, Mao G (2018) Can multi-objective calibration of streamflow guarantee better hydrological model accuracy? J Hydroinf 20(3):687–698
Zheng F, Zecchin AC, Maier HR, Simpson AR (2016) Comparison of the searching behavior of NSGA-II, SAMODE, and Borg MOEAs applied to water distribution system design problems. J Water Resour Plann Manag 142(7):04016017
Zhou A, Qu BY, Li H, Zhao SZ, Suganthan PN, Zhang Q (2011) Multiobjective evolutionary algorithms: A survey of the state of the art. Swarm Evol Comput 1(1):32–49

Chapter 12
Multi-attribute Decision-Making: A View of the World of Decision-Making

Babak Zolghadr-Asli, Omid Bozorg-Haddad, and Nora van Cauwenbergh

Abstract Water resources planning and management is marked by competing interests and different values of multiple stakeholders. Thus, when it comes to making any decision with regard to water resources, different - and often conflicting - interests have to be taken into consideration. Multi-attribute decision-making (MADM) is an umbrella term to describe mathematical frameworks that enable decision-makers to quantify the desirability of alternative options with respect to a set of evaluation attributes. As these evaluation attributes represent the multitude of competing interests, MADM are especially helpful for water resources planning and management. This chapter aims to shed light on the logic behind MADM methods, different schools of thought in MADM, and the characteristics that made these methods crucial for water resources planning and management.

Keywords Water resources planning and management · Multi-attribute decision-making · Multi-attribute analysis

12.1 Introduction

Decision-making is an inseparable part of *water resources planning and management*. Arguably, the way in which decision-making is done, will determine the success and prosperity of a project (e.g., Godinez-Madrigal et al. 2018). Considering the long-life expectancy of water resources infrastructure, their economic impacts, and so many other reasons of that nature, the decisions that help shape these projects can

B. Zolghadr-Asli · O. Bozorg-Haddad (✉)
Department of Irrigation and Reclamation Engineering, Faculty of Agricultural Engineering and Technology, College of Agriculture and Natural Resources, University of Tehran, Karaj, Tehran, Iran
e-mail: OBHaddad@ut.ac.ir

B. Zolghadr-Asli
e-mail: ZolghadrBabak@ut.ac.ir

N. van Cauwenbergh
Water Resources Planning, Department of Land and Water Management, IHE Delft, 2601 DA Delft, The Netherlands
e-mail: n.vancauwenbergh@un-ihe.org

O. Bozorg-Haddad (ed.), *Essential Tools for Water Resources Analysis, Planning, and Management*, Springer Water, https://doi.org/10.1007/978-981-33-4295-8_12

have profound impacts on society. While the right decision may set the community to flourish, in contrast, a hasty decision could indeed jeopardize the efforts and financial resources that went into supporting these projects. Thus, in terms of sound decision-making, what it comes downs to, is to find an answer to the question, "*what is a good project?*"

Before exploring the notion of *a good* choice, one must first define decision-making. Accordingly, the *Oxford dictionary* defines this term as "*the process of deciding about something important, especially in a group of people or in an organization.*" From a psychological point-of-view, decision-making is regarded as the cognitive process resulting in the selection of a belief or a course of action among several alternative possibilities (Buchanan and O'Connell 2006). Notably, the end-product of the decision-making process is a final choice, which may or may not prompt action. Thus, the process of decision-making is merely concerned with the selection of the alternative options (e.g., in a water resources development plan), rather than the procedural requirements of implementing the selected set of alternatives.

Strictly speaking, decision-making is founded on a four-stage analytical process (Vroom and Jago 1974; Weber and Coskunoglu 1990), which is depicted in Fig. 12.1. The first stage of the decision-making process is the *descriptive analysis*. Descriptive analysis, also known as the positive analytics, is mainly concerned with describing the observed behavior of the stakeholders through studying their previous choices. The term stakeholder here refers to any involved parties whose interests are at stake via the decision-making process (e.g., Reed et al. 2009). The analysis attempts to understand and describe these choices through behavioral and social psychology, neurology, or even data sciences (e.g., data mining) to identify the set of motivations

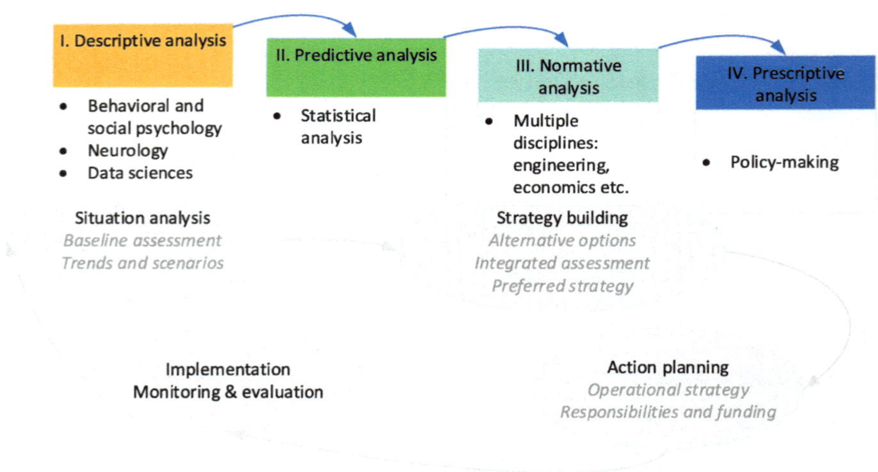

Fig. 12.1 Four stage analytical process of decision-making and its relation with steps in water resources planning

that can help describe the course of actions made by the stakeholders as individuals or a group of decision-makers (Peng et al. 2008; Santos and Rosati 2015). It is worth noting that data sciences are mainly concerned with identifying the *correlation* rather than the *causation*; however, they can provide practical hints on how to unravel the mysteries of decision-making. In water resources planning, this stage is referred to as situation analysis (Loucks and van Beek 2017), mobilizing a range of methods to include cause-effect relations and (future) trends in different issues.

Through the *predictive analysis,* the analyst attempts to quantify the likelihood in which an outcome would occur given a specific circumstance. To do so, the statistical analyst would explore the data sets to determine the probability of an outcome or the likelihood of a situation's occurrence (Bell et al. 1988; Kleindorfer et al. 1993). Note that the application of predictive analysis is limited to the decision-making under uncertainty; subsequently, not all decision-makings would consist of predictive analysis.

The third stage of the decision-making process is the *normative analysis*. The term normative generally refers to connecting an item to an evaluative standard through studying the item's behavior (Kahneman and Tversky 1984; Tversky and Kahneman 1986). Hence, normative analysis is mainly concerned with techniques through which the decision-makers can evaluate the feasible alternatives in a mathematical sense (Kleindorfer et al. 1993). In other words, it provides a framework to see whether alternatives score high or low on a given criteria. One of the main assumptions in the conventional normative analysis is *rationalism*, which, loosely speaking, describes a situation in which decision-makers actively attempt to secure their interests and goals, which have been identified and described through previous stages of decision-making (Kørnøv and Thissen 2000). This would mean one selects the alternative with the highest score. In practice, however, stakeholders do not often show perfect rationality, which can impose challenges for decision-makers (Nehring 2000). In such cases, either rational choice or the normative analysis is not well or not uniformly understood; hence, those involved in the decision-making process value things differently from what has been described mathematically in previous stages. For instance, when designing a rehabilitation plan for a groundwater system, for various reasons including limited financial resources or available time, the decision-makers may not be able to have access to all the datasets that describe the interest of stakeholders (e.g., the farmers that harness water from the aquifer) (Sharifi and Rodriguez 2002). In such cases, the stakeholders' behaviors are to be represented using bounded rationality, a particular class in which perfect rationality could not be held. In addition, uncertainty related to different knowledge frames and ambiguity can complicate the translation of predictions into normative values (e.g., Brugnach et al. 2008; Godinez-Madrigal et al. 2018).

The final stage of the decision-making process is the *prescriptive analysis*, through which the decision-makers would determine which alternatives to consider as the most desirable solution to the problem at hand (Saad 2001). The analysts, herein, combine the pieces of information gathered through studying the behavioral patterns of the stakeholders (descriptive analysis) and the randomness embedded in the decision-making process (predictive analysis), which are expressed in mathematical-oriented frameworks (normative analysis), and obtain the best course of actions for the decision-makers (prescriptive analysis). Furthermore, decision-makers can explore the possible options on how to take advantage of the future opportunity or cope with the future risks, and, eventually, evaluate the implication of each feasible decision options based on the unique characters of the decision-making problem (Tzeng and Huang 2011). In water resources planning, these decision making stages are followed by the operationalizing of the decision in action plans (where governance mode and funding/financing is agreed upon) and the implementation and monitoring and evaluation of these plans (see Fig. 12.1).

12.2 Definition and Terms

As stated, the main principle of decision-making is identifying the *best choice* (Sahaf 2009). While that may seem a straightforward principle, it poses a challenging dilemma for decision-makers. Indeed, the concept of a "good choice" may vary from one decision-maker to another, due to their different preferences, experience, values, shared knowledge, and backgrounds (Sobel and Clark 2014). In other words, a universal and unanimous agreement between different decision-makers on the very notion of *goodness* does often not exist. Furthermore, the selection procedure between the alternatives may differ from one decision-maker to another, even when facing the same decision-making problem. Nevertheless, the core idea of sound decision-making is to opt for a set of solutions that would outperform the other alternatives, based on a set of evaluation criteria defined either explicitly or implicitly by the decision-makers (Zanakis et al. 1998). This is the general idea behind the *multi-criteria decision-making* (MCDM). MCDM is a sub-discipline of operations research and can be further divided into MODM and MADM methods, the latter being the focus of this chapter. In practice, almost everyone faces an MCDM problem on a daily basis, to which most would aggregate their preferences through an intuition-oriented weighting mechanism. When complexity is higher, and multiple functions and stakeholders are involved, however, implementing a systematic MCDM approach is vital to make an informed and logical decision.

In technical terms, MCDM refers to any procedure in which the decision-maker explicitly evaluates a set of alternatives with regard to multiple and usually competing criteria (Alinezhad and Khalili 2019; Bozorg-Haddad et al. 2020). Through MCDM methods, the decision-makers is able to restructure and redefine the decision-making problem in a mathematical form. Although developing and implementing MCDM methods are not a novel modern idea per se, there have been undeniable advances in this field since the blooming era of computational intelligence (CI) during the early 1960s and 1970s. Computational developments have led to novel and often hybrid MCDM methods, which particularly facilitate and improve the decision-making process. Ever since MCDM has been an active area of research, and played a crucial role in an array of disciplines, ranging from politics and business to the environment and energy.

To facilitate studying the systematic approaches to tackle problems in which the decision-makers attempt to consider multiple evaluation criteria simultaneously, Hwang and Yoon (1981) proposed clustering MCDM methods into two main categories, namely, *multi-objective decision-making* (MODM), and *multi-attribute decision-making* (MADM). Fundamentally, this classification is based on whether the solutions and the evaluation criteria are defined either *explicitly* or *implicitly* (Mendoza and Martins 2006; Velasquez and Hester 2013). MODM methods are used where decision-makers are searching for a set of optimal solutions that would not only satisfy the constraints surrounding the given problem, but optimize the decision-makers' pre-defined objectives as well (Hwang and Yoon 1981). In essence, MODM is suitable for the *design/operation* problems, in which the decision-makers aim to achieve the optimal or aspired goals by considering the various interactions within the given constraints. For example, a groundwater pumping regime can be optimized within constraints of minimum levels and guarantee of supply (Van Cauwenbergh et al. 2008). The decision space of MODM problems can be described as a multi-dimensional *Cartesian space*, with each objectives as an axiom, and a set of constraints that separate the feasible and infeasible solutions. Even though MODM is conventionally defined with a continuous decision space, problems with discrete decision spaces can also be solved via MODM methods, as long as the feasible alternatives are mathematically described as an infinite set of entities (Hwang and Yoon 1981). In practice, the main challenges imposed on MODM methods are the trade-off and the scale problems. Application of MODM methods in water resources planning and management are found in the development of operation policy for reservoirs (e.g., Aboutalebi et al. 2015), water distribution systems (e.g., Farmani et al. 2006), and water quality monitoring network (e.g., Aboutalebi et al. 2016) amongst others. For more information on these challenges and how they can be addressed, the readers can refer to Alinezhad and Khalili (2019) and Bozorg-Haddad et al. (2020). MADM methods, on the other hand, are used for situations in which the decision-makers are to evaluate a finite number of pre-defined alternatives that are explicitly known to decision-makers. In that case, MADM helps decision-makers to systematically assess each given alternative via a discrete preference rating mechanism, defined for each evaluating attribute (Hwang and Yoon 1981). In general, MADM methods are suited for the *evaluation/choosing*. In the context of water resources planning and management, decision-makers are often presented with a set of discretely defined alternative options, each with their own unique merits and drawbacks. In those cases, MADM methods can aid the decision-maker in the process of comparison and choice. Since the early 1970s, applications of the MADM techniques in this field is on the rise (Hajkowicz and Collins 2007). Examples of such applications include demand management (e.g., Banihabib and Shabestari 2017), groundwater management (e.g., Barzegar et al. 2019), irrigation and drainage network (e.g., Cetinkaya and Gunacti 2018), supply management (e.g., Goodwin et al. 2019), wastewater management (e.g., Nawaz and Ali 2018), assessment of water distribution networks (e.g., Islam et al. 2013), flood management (e.g., Lee et al. 2014), and wastewater treatment (e.g.,

Table 12.1 the main characteristics of MODM and MADM methods (Tzeng and Huang 2011)

	MODM	MADM
Criteria are defined by	Objectives	Attributes
Objectives are defined	Explicitly	Implicitly
Attributes are defined	Implicitly	Explicitly
Constraints are defined	Explicitly	Implicitly
Alternatives are defined	Implicitly	Explicitly
Number of alternatives are	Infinite	Finite
Relative to	Design/operation	Evaluation/choice

Kim et al. 2013).and water governance (e.g., Vila et al. 2018). Table 12.1 summarizes the main characteristics of MODM and MADM methods.

In the following sections, this chapter will take a deep dive into the world of MADM to shed light on why this method can be useful for water resources planning and management. Before that, however, it is crucial to have a better understanding of MADM methods' basic principles and the logic behind these methods.

12.3 Basic Principles and Logics: Different Viewpoints on MADM

The basic paradigm of MADM is to evaluate a set of feasible attributes with respect to a set of, often, competing and even conflicting attributes. From the introduction of the utility theory in the early eighteenth century (Bernoulli 1738) to the era of French school of thought in decision-making in the 1960s to 1980s [i.e., elimination et Choix traduisant la realité (ELECTRE) (Benayoun et al. 1966; Roy and Bertier 1971; Roy 1978; Roy and Hugonnard 1982); and preference ranking organization methods for enrichment evaluation (PROMETHEE) (Brans 1982; Vincke and Brans 1985; Brans et al. 1986; Mladineo et al. 1987)] to the development of the state-of-the-art methods such as the best-worst method (BWM) (Rezaei et al. 2015), MADM has managed to root in, practically, most operational and scientific fields including water resources planning and management. A chronicle overview of some of the most fundamental and influential MADM methods is presented in Fig. 12.2.

As can be seen in Fig. 12.2, numerous and seemingly vastly different MADM methods exist and recent developments have led to novel and often hybrid MADM methods. What all methods have in common is that they establish mechanisms to evaluate preferences, assign a mathematical nature to attributes and prescribe how to handle different viewpoints of participating decision-makers. The particular expression of these principles could be used to cluster similar MADM methods. Accordingly, different branches of MADM methods are identified based on (1) the decision-makers' prioritizing system, (2) the mathematical nature of attributes' values, and (3)

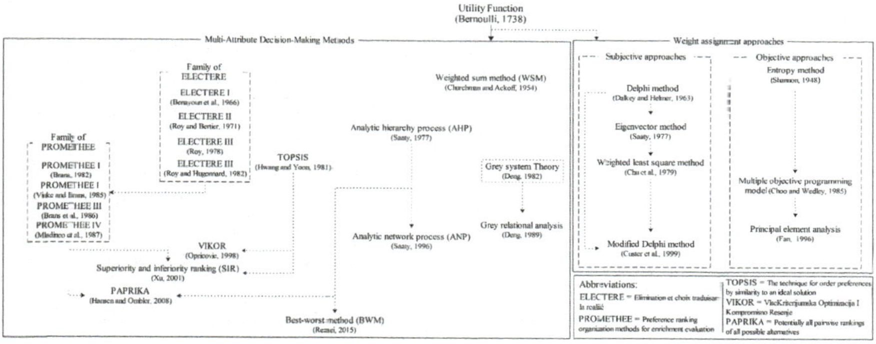

Fig. 12.2 A choronological overview on some of the most influential MADM methods

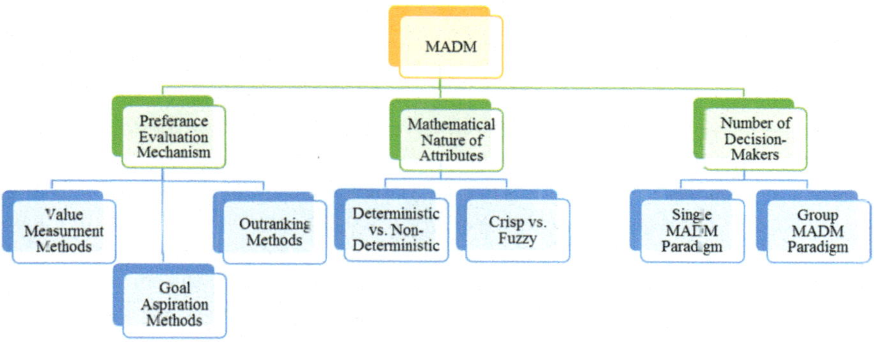

Fig. 12.3 Classification/clustering of MADM methods

the number of decision-makers. Figure 12.3 illustrates these different ways to cluster MADM methods. The following sections explore these concepts and demonstrates how they can be used to cluster MADM methods.

12.3.1 Clustering by Preference Evaluation Mechanism

Every MADM method requires a preference evaluation mechanism, the main functionality of which would be to reflect the stakeholders' ideology. These mechanisms perform as a sort of scale that enables the decision-makers to evaluate the preference for each alternative. The preference evaluation mechanism can be defined either *explicitly or implicitly*. In the explicit valuation, preference values are computed through a set of mathematical functions, whereas the *implicit valuation*, is a reflection of the decision-makers' experience, expertise, perception, and instinct.

Based on the notion of preference evaluation mechanism, a traditional classification would divide MADM methods into multi-attribute utility theory (MAUT) and outranking methods (Mendoza and Martins 2006). Founded on the premise of the Bernoulli's utility theory, MAUT methods obtain the decision-makers' preferences and represent them as a hierarchical structure by employing a proper utility function. When using this method, the alternative with the highest utility value emerges as the solution. Although MAUT-based methods benefit from a solid axiomatic background, their main criticism is the unrealistic assumption of preferential independencies (Tzeng and Huang 2011). In MADM terminology, preferential independence refers to a situation in which priorities of the alternatives based on a given criterion are not influenced by the other criteria. Although this is a sound concept on paper, it might not be an accurate description of real-world MADM problems in practice. Alternatively, instead of evaluating the alternatives via complex utility functions, outranking-based methods provide a paradigm to compare the preference relations among alternatives to identify the best alternative. Although outranking methods were proposed to overcome the practical difficulties experienced with the utility function, they are criticized for lacking an axiomatic foundation and solid mathematical structural (Tzeng and Huang 2011; Bozorg-Haddad et al. 2020).

Although the previous classification serves to categorize conventional MADM methods, it may face difficulties categorizing some of the hybrid and novel MADM methods, which do not fit to either the categories' descriptions. Eventually, a more sophisticated classification system was proposed, in which MADM methods are categorized into three classes, namely, *value measurement*, *goal aspirations*, and *outranking* methods (Belton and Stewart 2002). In value measurement methods, numerical scales are implemented to represent the degree to which a feasible alternative may be preferred over another. The scores obtained for each alternative are estimated with respect to each individual evaluation attribute and are then aggregated to estimate the overall preference of the decision-maker. In goal aspirations methods, however, the decision-maker defines a satisfactory level of achievement for each criterion. The alternative closest to these defined reference points is then identified as the solution to the MADM problem. Lastly, the outranking MADM methods evaluate the alternatives' relative performances against one another using a comparison-oriented framework.

12.3.2 Clustering by the Attributes' Mathematical Nature

From a mathematical perspective, evaluation attributes can be either *deterministic* or *non-deterministic* and *fuzzy* or *crisp*. In a deterministic point-of-view, the decision-makers are certain about every aspect of the decision-making problem. In a non-deterministic viewpoint, however, some aspects of the decision-making paradigm have a stochastic component (Pearl 1996). On the other hand, when the decision-makers' preferences cannot be expressed with a single numeric value (i.e., crisp viewing paradigm), a fuzzy set can be employed to express these preferences (Bellman and Zadeh 1970).

It should be noted however, that the fuzzy-based and the probability-based uncertainty follow two distinctive logics and are not interchangeable per se. When a decision-maker faces a process which outcome cannot be expressed with absolute certainty, probability-uncertainty logic is conventionally employed. However, when the decision-makers are not certain on how to express their preferences, fuzzy logic is the most conventional perspective to express uncertainty. To recap, while the probability-uncertainty logic deals with the likelihood in which an event could take place, the fuzzy logic addressee the degree to which the decision-makers are certain about their priorities (Bozorg-Haddad et al. 2020).

12.3.3 Clustering by the Number of Decision-Makers

Based on the number of involved individuals, the MADM frameworks could be clustered into two major categories, namely single and group decision-making paradigms (Black 1948). In single MADM paradigms, the opinion of that single individual would form the preference evaluation mechanism of the decision-making process. In group decision-making paradigms, at least two decision-makers with different viewpoints contribute and influence the decision-making process (Kiesler and Sproull 1992). Though group decision-making paradigms are, in the core, founded based on single decision-making methods, they would require an additional synthesizing strategy through which individual decision-makers' opinions would be aggregated and integrated to form the final result. Who belongs to the decision-making group itself has changed, given the consensus in (international) policies and donor regulations that inclusion of stakeholders in water related decisions is a prerequisite for their success. Decision-making processes have therefore increasingly shifted from purely top-down or single decision-making to bottom-up and group decision-making using participatory and collaborative approaches. Messner (2006) gives an overview of experiences with participatory MADM methods.

12.4 Importance and Necessity: Water-Food-Energy Security Nexus

Water resources problems are complicated since they are multi-dimensional by nature. As a result, in most practical problems, multiple evaluation criteria are considered to reflect these points. Furthermore, commonly there is more than one stakeholder involved in any decision-making required to address these issues, which results in different preference values to reflect the difference of interest among the involved parties. Decision-making under such characteristics, thus, benefits from MADM-based frameworks. In this section we further explore the importance of MADM methods and their application in water resources planning and management. We do so within the scope of water security and the (decisions on) water distribution between key societal sectors in relation to core elements of the water-food-energy security nexus: water, energy, food and the environment.

Whereas water is a renewable natural resource that is continuously re-distributed throughout the planet through the *hydrological cycle*, its uneven distribution in both time and space, creates issues with continuous water availability. These issues are at the core of many challenges of water resources planning and management (Zolghadr-Asli et al. 2017). Population growth and changing living standards as well as growing environmental concerns and the projected changes in the climate add complexity to this already complicated problem (Bidoglio et al. 2019; Zolghadr-Asli et al. 2018; Zolghadr-Asli et al. 2020a, b). Given the crucial role of water in most, if not all strategic sectors in modern society, a sound decision in water resources requires addressing these dynamic and complex interactions between different sectors and multiple dimensions.

The definition of water-energy-food-environment (WEFE) security nexus is broad because the term itself covers a broad realm. Perhaps, as a result, there is no consensus on the WEFE security nexus definition (Keskinen et al. 2016; Smajgl et al. 2016; Endo et al. 2017). The basic paradigm of the WEFF security nexus is portrayed in Fig. 12.4. In general, the nexus notion merely states that the security of water, energy, food and the environment are codependent in a way that one cannot truly address one of the securities without discussing the others (Zhang et al. 2018, 2019). The attention to sectoral interdependencies and coordination in water is not new; it is at the heart of familiar concepts such as integrated water resources management (IWRM). However, whereas IWRM is mainly aimed at the process, the nexus point-of-view is goal-oriented and explicitly focuses on the trade-offs In short, the concept of WEFE security nexus indicates that sustainable water management must be integrated with

Fig. 12.4 Water-energy-food-environment (WEFE) security nexus paradigm

a cross-sector coordinating platform, in which often conflicting attributes reflecting different stakeholders' interests are to be considered simultaneously. As the priorities of one sector are not necessarily in line with others, decision-making with regard to water resources is quite challenging. It is in this context that MADM provides a structured and formal platform in which crucial and challenging decisions can be made. Through MADM-based frameworks, the decision-makers can consider various, often conflicting attributes, and the emerged solutions from these methods can potentially promote the WEFE security nexus.

12.5 Methodology: Steps to Conduct MADM

Two schools of thought exist in MADM methods, namely, *compensatory* and *non-compensatory* (Jeffreys 2004). When the MADM method permits compensation of an alternative's poor performance with respect to a specific attribute by high performance in some other attribute, this method is considered as compensatory. In non-compensatory methods, however, the significant poor performance of an alternative with regard to an attribute cannot be compensated, even when the alternative performs high with respect to all other attributes (Banihabib et al. 2017). Both methods have their own merits and drawbacks and the selection of the method in a specific case will depend on the decision-making context. The nature of some cases, for instance, does not permits applying the compensatory methods, as key stakeholders may not be interested in or willing to accept compensation of an alternative's low performances for a given criterion or attribute.

Figure 12.5 shows the basic steps to a MADM method. Decision-makers first have to capture the premise of the decision-making problem in a mathematical context, which is referred to in technical terms as the *decision matrix*. Here, the decision-makers define a set of feasible alternatives as well as a set of evaluation attributes to assess them. In a second step, the decision-makers set their priorities regarding the importance of the attributes through a weight assignment mechanism (*see* Fig. 12.5). In step 3, the performance of each individual alternative is evaluated with regard to each of the defined attributes. Finally, the decision-makers are to aggregate these pieces of information to measure the overall performance of the alternatives, and in

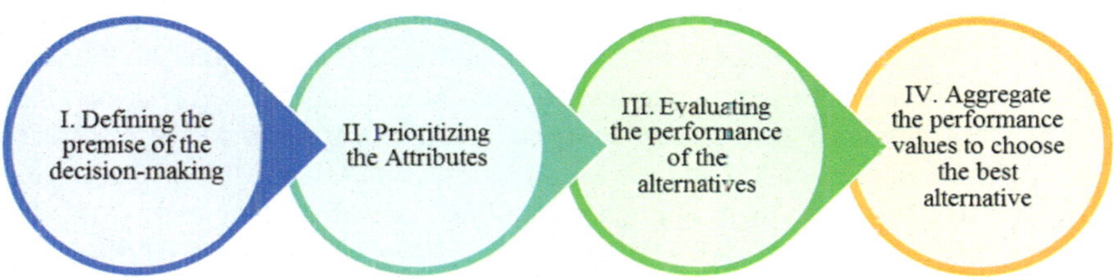

Fig. 12.5 The basic steps to a MADM method

turn, select the solution in step 4 (Bozorg-Haddad et al. 2020). Each MADM method proposes different approaches to address these steps, with potential differences in the resulting ranking of alternatives (e.g., Van Cauwenbergh 2008). To avoid that the result is purely defined by the method chosen and underlying assumptions are not understood, Myšiak (2006) recommends the use of different methods as well as a sensitivity analysis to validate the MADM results. For more information on different MADM methods, the readers can refer to Kallis et al. (2006), Tzeng and Huang (2011), and Bozorg-Haddad et al. (2020).

12.6 Practical Examples

As stated, in this day-and-age, decision-makers have numerous methods and techniques at their disposal to deal with MADM problems ranging from simple, easy-to-use approaches to complex, yet structured methods. In this section, we illustrate the range of available methods with two examples in the context of water resources and planning and management.

12.6.1 Water Management Plans in Jacuípe River basin, Brazil Using AHP

The analytic hierarchy process (AHP) is a *subjective, compensatory* MADM method that was first theorized by Saaty (1977) and further developed in late 1970s (Saaty 1980). Two mechanisms enable the AHP to cope with MADM problems in a structured and pragmatic manner; the *hierarchy structure* and the *pairwise comparisons*. The hierarchy structure of AHP enables the decision-maker to recompile the MADM problem in smaller and independent units, each of which has a structure of a single attribute decision-making problem. Using a series of pairwise comparisons, AHP then taps the decision-makers' cognitive understanding, expertise, and experiences to gather subjective-based assessments to solve the decision-making problems depicted in these units. Next, AHP employs a bottom-up approach to aggregate these assessments and identify the most suitable alternative.

In light of water resources planning and management, AHP is one of the most popular MADM methods (Hajkowicz and Collins 2007). Srdjevic and Medeiros (2008), for instance, created a three-level hierarchy structure to select the best long-term water management plan for the Jacuípe River basin, eastern Brazil. The decision-matrix was composed of 3 management alternatives and 24 attributes that covered a broad range of topics including economic, environmental, political, social, and technical attributes. Using the AHP method, Srdjevic and Medeiros (2008) were able to conduct a thorough subjective assessment to find the most desirable solution.

There is a critical point that needs to be stated at this stage regarding the fundamental perspective toward decision-making. Evidence suggests that from time to time, these methods are being treated as merely number-crunching frameworks; this is especially troubling when subjective methods, such as AHP, are being used (Hajkowicz and Collins 2007). Relying solely on a subjective weight assignment mechanism can act like a double-sided blade. While they can incorporate years of experience and various viewpoints into the decision-making process, they are one misstep away from leading to confusing and misleading results. So one should these methods should be implemented with cautious, and a checking mechanism should be in place to make sure that decision-makers' subjective assessment is consistent and in line with reality.

12.6.2 Water Development Plans in Iran with TOPSIS

The technique for order preferences by similarity to an ideal solution (TOPSIS) is an *objective, compensatory* MADM method that is rooted in the basic principles of reference-dependence theory. Although TOPSIS method was first promoted by Hwang and Yoon (1981), the method primarily flourished through the works of Yoon (1987), and Hwang et al. (1993). Chiefly, the TOPSIS method attempts to identify the alternative, which has the shortest distance from the ideal solution and the farthest distance from the inferior solution, a characteristic that makes it a suitable risk-avoidance MADM method (Opricovic and Tzeng 2004; Jahan et al. 2012; Aghajani-Mir et al. 2016).

Due to its objective viewpoint and efficient computational algorithm, TOPSIS has gained popularity to address MADM problems in various research fields, including water resources management (Bozorg-Haddad et al. 2020). Afshar et al. (2011), for instance, applied this method to analyze and, in turn, evaluate the potential of the feasible water development plans for the Karun River basin, Iran. The decision-matrix of this particular problem consisted of 7 feasible alternatives that depicted different development plans which were to be evaluated against 30 evaluation attributes that cover environmental, socio-economic, and technical topics. Notably, this study adopted a fuzzy-based perspective to address the uncertainties surrounding the decision-making process. Through this framework, Afshar et al. (2011) were able to select the best water development plan for the basin mentioned above. Furthermore, they concluded that this method is suitable to be employed for larger-scale problems.

It should be noted that uncertainty plays a crucial role in the field of decision-making. In water resources planning and management, hydrological, economic and other models can be used to assess the impact of infrastructural and other measures under uncertainty of climate change, population growth, new technologies, etc. In addition to uncertain variability in the events, uncertainty can manifest itself as incomplete knowledge and ambiguity (Dourojeanni 2019). Whereas addressing these uncertainties is quite vital to make sound and pragmatic decisions, evidence, however, suggests that this notion is commonly being ignored either wholly or partially in most water resources related researches (Hajkowicz and Collins 2007), with most of the focus going to understand and manage variability. To deal with different types of uncertainty such as deep uncertainty, incomplete knowledge and ambiguity in the decision-making, frameworks as adaptive management (e.g., Dewulf et al. 2008), *dynamic adaptive policy pathways* (Haasnoot et al. 2013) and (participatory) adaptive planning approaches (Zandvoort et al. 2018; Van Cauwenbergh et al. 2020) can be used.

12.7 Summary

Decision-making is something we all face on a daily basis. When dealing with complex and multi-dimensional problems, decision-makers often rely on structured and formal processes to support the identification of the most viable solutions. In such cases, the decision-makers would be presented with a set of feasible alternatives, which are to be evaluated using a set of, often, conflicting attributes. MADM provides a proper mathematical-based paradigm to find the solution to such a question. From the introduction of the utility theory in the early eighteenth century to the hybrid state-of-the-art MADM methods, numerous techniques have been proposed to ensure a sound and smooth decision-making process.

The nexus role of water further complicates the already challenging process of decision-making in water management and planning. MADM enables the decision-makers to not only address the technical aspect of decision-making, but also, the environmental, socio-economic, and even legal attributes that are involved in water-related problems. Resultantly, the MADM-oriented studies in the field of water resources planning and management have dramatically grown in number in recent years. Two main areas for improvement are these frameworks viewpoint on uncertainties associated with water resources problems and a better underrating toward the subjective mechanism embedded in some of the MADM methods to avoid misleading results in the ranking of alternatives.

Acknowledgements The authors thank Iran's National Science Foundation (INSF) for the support of this research.

References

Aboutalebi M, Bozorg-Haddad O, Loáiciga HA (2015) Optimal monthly reservoir operation rules for hydropower generation derived with SVR-NSGAII. J Water Resour Plann Manag 141(11):04015029

Aboutalebi M, Bozorg-Haddad O, Loáiciga HA (2016) Multiobjective design of water-quality monitoring networks in river-reservoir systems. J Environ Eng 143(1):04016070

Afshar A, Mariño MA, Saadatpour M, Afshar A (2011) Fuzzy TOPSIS multi-criteria decision analysis applied to Karun reservoirs system. Water Resour Manag 25(2):545–563

Aghajani-Mir Joshi R, Banwet DK, Shankar R (2011) A Delphi-AHP-TOPSIS based benchmarking framework for performance improvement of a cold chain. Expert Syst Appl 38(8):10170–10182

Alinezhad A, Khalili J (2019) New methods and applications in multiple attribute decision making (MADM). Springer International Publishing, Switzerland

Banihabib ME, Shabestari MH (2017) Fuzzy hybrid MCDM model for ranking the agricultural water demand management strategies in arid areas. Water Resour Manag 31(1):495–513

Banihabib ME, Hashemi-Madani FS, Forghani A (2017) Comparison of compensatory and non-compensatory multi criteria decision making models in water resources strategic management. Water Resour Manag 31(12):3745–3759

Barzegar R, Moghaddam AA, Adamowski J, Nazemi AH (2019) Delimitation of groundwater zones under contamination risk using a bagged ensemble of optimized DRASTIC frameworks. Environ Sci Pollut Res 26(8):8325–8339

Bell DE, Raiffa H, Tversky A (1988) Decision making: descriptive, normative, and prescriptive interactions. Cambridge University Press, Cambridge, UK

Bellman RE, Zadeh LA (1970) Decision-making in a fuzzy environment. Manag Sci 17(4):141–164

Belton V, Stewart T (2002) Multiple criteria decision analysis: an integrated approach. Kluwer Academic Publishers, Massachusetts, BST

Benayoun R, Roy B, Sussman B (1966) ELECTRE: Une méthode pour guider le choix en présence de points de vue multiples. Note de travail 49, SEMA-METRA International, Direction Scientifique, Paris, France

Bernoulli D (1738) Specimen theoriae novae de mensura sortis. Comentarii Academiae Scientiarum Imperiales Petropolitanae 5:175–192

Bidoglio G, Vanham D, Bouraoui F, Barchiesi S (2019) The water-energy-food-ecosystems (WEFE) nexus. In: Encyclopedia of ecology. Elsevier. doi:10.1016/B978-0-12-409548-9.11036-X

Black D (1948) On the rationale of group decision-making. J Polit Econ 56(1):23–34

Bozorg-Haddad O, Zolghadr-Asli B, Loáiciga HA (2020) A handbook on multi-attribute decision-making methods. Wiley. ISBN: 1119563496

Brans JP (1982) L'ingénierie de la decision. Elaboration d'instruments d'aide a la décision: Methode PROMETHEE. L'aide a la Decision: Nature, Instruments et Perspectives D'avenir, Presses de Universite Laval, Québec, Canada

Brans JP, Vincke P, Mareschal B (1986) How to select and how to rank projects: the PROMETHEE method. Eur J Oper Res 24(2):228–238

Brugnach M, Dewulf A, Pahl-Wostl C, Taillieu T (2008) Toward a relational concept of uncertainty: about knowing too little, knowing too differently, and accepting not to know. Ecol Soc 13(2):30

Buchanan L, O'Connell A (2006) A brief history of decision making. Harvard Bus Rev 84(1):32

Cetinkaya CP, Gunacti MC (2018) Multi-criteria analysis of water allocation scenarios in a water scarce basin. Water Resour Manage 32(8):2867–2884

Dewulf A, Francois G, Brugnach M, Isendahl N, Taillieu T, Pahl-Wostl C, Moellenkamp S (2008). The Role of uncertainty, ambiguity and framing in transition to adaptive management. About knowing to little, accepting not to know and knowing too differently. Deliverable of the EU 6th FP NeWater project. www.newater.info

Dourojeanni P (2019) Understanding the role of uncertainty in the mainstreaming of Nature-Based Solutions for adaptation to climate change. Master's thesis, IHE Delft

Endo A, Tsurita I, Burnett K, Orencio PM (2017) A review of the current state of research on the water, energy, and food nexus. J Hydrol: Reg Stud 11:20–30

Farmani R, Walters G, Savic D (2006) Evolutionary multi-objective optimization of the design and operation of water distribution network: total cost vs. reliability vs. water quality. J Hydroinf 8(3):165–179

Godinez-Madrigal J, Zaag PVD, Van Cauwenbergh N (2018) A half-baked solution: drivers of water crises in Mexico. Proc Int Assoc Hydrol Sci 376:57–62

Goodwin D, Raffin M, Jeffrey P, Smith HM (2019) Stakeholder evaluations of risk interventions for non-potable recycled water schemes: a case study. Sci Total Environ 674:439–450

Haasnoot M, Kwakkel JH, Walker WE, ter Maat J (2013) Dynamic adaptive policy pathways: a method for crafting robust decisions for a deeply uncertain world. Glob Environ Change 23(2):485–498

Hajkowicz S, Collins K (2007) A review of multiple criteria analysis for water resource planning and management. Water Resour Manag 21(9):1553–1566

Hwang CL, Lai YJ, Liu TY (1993) A new approach for multiple objective decision making. Comput Oper Res 20(8):889–899

Hwang CL, Yoon K (1981) Methods for multiple attribute decision making. In: Multiple attribute decision making: lecture notes in economics and mathematical systems. Springer Publication Company, Heidelberg, Germany

Islam MS, Sadiq R, Rodriguez MJ, Najjaran H, Francisque A, Hoorfar M (2013) Evaluating water quality failure potential in water distribution systems: A fuzzy-TOPSIS-OWA-based methodology. Water Resour Manag 27(7):2195–2216

Jahan A, Mustapha F, Sapuan SM, Ismail MY, Bahraminasab M (2012) A framework for weighting of criteria in ranking stage of material selection process. Int J Adv Manuf Technol 58(1):411–420

Jeffreys I (2004) The use of compensatory and non-compensatory multi-criteria analysis for small-scale forestry. Small-scale Forest Econ Manag Policy 3(1):99–117

Kahneman D, Tversky A (1984) Choices, values, and frames. Am Psychol 39(4):341

Kallis G, Videira N, Antunes P, Pereira ÂG, Spash CL, Coccossis H, Quintana SC, del Moral L, Hatzilacou D, Lobo G, Mexa A, Paneque P, Mateos BP, Santos R (2006) Participatory methods for water resources planning. Environm Plann C: Gov Policy 24(2):215–234

Keskinen M, Guillaume J, Kattelus M, Porkka M, Räsänen T, Varis O (2016) The water-energy-food nexus and the transboundary context: insights from large Asian rivers. Water 8(5):193

Kiesler S, Sproull L (1992) Group decision making and communication technology. Organ Behav Hum Decis Process 52(1):96–123

Kim Y, Chung ES, Jun SM, Kim SU (2013) Prioritizing the best sites for treated wastewater instream use in an urban watershed using fuzzy TOPSIS. Resour Conserv Recycl 73:23–32

Kleindorfer PR, Kunreuther H, Schoemaker PJ (1993) Decision sciences: an integrative perspective. Cambridge University Press, Cambridge, UK

Kørnøv L, Thissen WA (2000) Rationality in decision-and policy-making: implications for strategic environmental assessment. Impact Assess Project Appraisal 18(3):191–200

Lee G, Jun KS, Chung ES (2014) Robust spatial flood vulnerability assessment for Han River using fuzzy TOPSIS with α-cut level set. Expert Syst Appl 41(2):644–654

Loucks DP, van Beek E (2017) Water resource systems planning and management: an introduction to methods, models, and applications. Springer International Publishing, Cham, Switzerland

Mendoza GA, Martins H (2006) Multi-criteria decision analysis in natural resource management: a critical review of methods and new modelling paradigms. For Ecol Manag 230(1–3):1–22

Messner F (2006) Applying participatory multicriteria methods to river basin management: Improving the implementation of the water framework directive. Environ Plann C: Gov Policy 24:159–167

Mladineo N, Margeta J, Brans JP, Mareschal B (1987) Multicriteria ranking of alternative locations for small scale hydro plants. Eur J Oper Res 31(2):215–222

Myšiak J (2006) Consistency of the results of different MCA methods: a critical review. Environ Plann C: Gov Policy 24(2):257–277

Nawaz S, Ali Y (2018) Factors affecting the performance of water treatment plants in Pakistan. Water Conserv Sci Eng 3(3):191–203

Nehring K (2000) A theory of rational choice under ignorance. Theor Decis 48(3):205–240

Opricovic S, Tzeng GH (2004) Compromise solution by MCDM methods: a comparative analysis of VIKOR and TOPSIS. Eur J Oper Res 156(2):445–455

Pearl J (1996) Decision making under uncertainty. ACM Comput Surv 28(1):89–92

Peng Y, Kou G, Shi Y, Chen Z (2008) A descriptive framework for the field of data mining and knowledge discovery. Int J Inf Technol Decis Making 7(04):639–682
Reed MS, Graves A, Dandy N, Posthumus H, Hubacek K, Morris J, Prell C, Quinn CH, Stringer LC (2009) Who's in and why? A typology of stakeholder analysis methods for natural resource management. J Environ Manag 90(5):1933–1949
Rezaei J, Wang J, Tavasszy L (2015) Linking supplier development to supplier segmentation using best-worst method. Expert Syst Appl 42(23):9152–9164
Roy B (1978) ELECTRE III: Une algorithme de classement fondé sur une représentation floue des préférences en présence de critères multiples. Cahiers du Centre d'Etudes de Recherche Opérationnelle 20(1):3–24
Roy B, Bertier P (1971) La méthode ELECTRE II: Une methode de classement en presence de critteres multiples. Direction Scientifique, Note de Travail No. 142, Paris
Roy B, Hugonnard JC (1982) Ranking of suburban line extension projects on the Paris metro system by a multicriteria method. Transp Res Part A: Gen 16(4):301–312
Saad GH (2001) Strategic performance evaluation: descriptive and prescriptive analysis. Ind Manag Data Syst 101(8):390–399
Saaty TL (1977) A scaling method for priorities in hierarchical structures. J Math Psychol 15(3):234–281
Saaty TL (1980) The analytic hierarchy process. McGraw-Hill Education, New York, NY
Santos LR, Rosati AG (2015) The evolutionary roots of human decision making. Ann Rev Psychol 66:321–347
Sahaf MA (2009) Management accounting: principles and practice. Vikas Publishing House, New Delhi, India
Sharifi MA, Rodriguez E (2002) Design and development of a planning support system for policy formulation in water resources rehabilitation: The case of Alcazar De San Juan District in Aquifer 23, La Mancha, Spain. J Hydroinf 4(3):157–175
Smajgl A, Ward J, Pluschke L (2016) The water-food-energy nexus—Realising a new paradigm. J Hydrol 533:533–540
Sobel RS, Clark JR (2014) Preference formation, choice sets, and the creative destruction of preferences. J Ayn Rand Stud 14(1):55–74
Srdjevic B, Medeiros YDP (2008) Fuzzy AHP assessment of water management plans. Water Resour Manag 22(7):877–894
Tversky A, Kahneman D (1986) Rational choice and the framing of decisions. J Bus:S251–S278
Tzeng GH, Huang JJ (2011) Multiple attribute decision making: methods and applications. CRC Press, Boca Raton, FL
Van Cauwenbergh N, Pinte D, Tilmant A, Frances I, Pulido-Bosch A, Vanclooster M (2008) Multi-objective, multiple participant decision support for water management in the Andarax catchment, Almeria. Environ Geol 54(3):479–489
Van Cauwenbergh N (2008) Expert and local knowledge for decision support in natural resource management : analysis of capture and use. Ph.D. thesis, Universite catholique de Louvain (UCL)
Van Cauwenbergh N, Dourojeanni P, Mayor B, Altamirano M, Dartee K, Basco-Carrera L, Piton G, Tacnet JM, Manez M, Lopez-Gunn E (2020) Guidelines for the definition of implementation and investment plans for adaptation. EU Horizon 2020 NAIAD Project, Grant Agreement, N°730497
Velasquez M, Hester PT (2013) An analysis of multi-criteria decision making methods. Int J Oper Res 10(2):56–66
Vila M, Afsordegan A, Agell N, Sánchez M, Costa G (2018) Influential factors in water planning for sustainable tourism destinations. J Sustain Tour 26(7):1241–1256
Vincke JP, Brans P (1985) A preference ranking organization method" The PROMETHEE method for MCDM. Manag Sci 31(6):647–656
Vroom VH, Jago AG (1974) Decision making as a social process: normative and descriptive models of leader behavior. Decis Sci 5(4):743–769
Weber EU, Coskunoglu O (1990) Descriptive and prescriptive models of decision-making: implications for the development of decision aids. IEEE Trans Syst Man Cybern 20(2):310–317

Yoon K (1987) A reconciliation among discrete compromise solutions. J Oper Res Soc 38:277–286
Zanakis SH, Solomon A, Wishart N, Dublish S (1998) Multi-attribute decision making: a simulation comparison of select methods. Eur J Oper Res 107(3):507–529
Zandvoort M, van der Vlist MJ, van den Brink A (2018) Handling uncertainty through adaptiveness in planning approaches: comparing adaptive delta management and the water diplomacy framework. J Environ Plann Policy Manag 20(2):183–197
Zhang C, Chen X, Li Y, Ding W, Fu G (2018) Water-energy-food nexus: concepts, questions and methodologies. J Clean Prod 195:625–639
Zhang P, Zhang L, Chang Y, Xu M, Hao Y, Liang S, Liu G, Yang Z, Wang C (2019) Food-energy-water (FEW) nexus for urban sustainability: a comprehensive review. Resour Conserv Recycl 142:215–224
Zolghadr-Asli B, Bozorg-Haddad O, Chu X (2017) Strategic importance and safety of water resources. J Irrig Drainage Eng 143(7):02517001
Zolghadr-Asli B, Bozorg-Haddad O, Chu X (2018) Chapter 1: Introduction. In: Advanced optimization by nature-inspired algorithms. Springer International Publishing AG, Singapore
Zolghadr-Asli B, Bozorg-Haddad O, Chu X (2020a) Hydropower in climate change. In: Encyclopedia of water: science, technology, & society. Wiley. doi:10.1002/9781119300762.wsts0089
Zolghadr-Asli B, Bozorg-Haddad O, Chu X (2020b) Hot hand fallacy: A contagious misinterpretation in water resources sectors. In: Encyclopedia of water: science, technology, & society. Wiley. doi:10.1002/9781119300762.wsts0088

Chapter 13
Risk, Uncertainty, and Reliability in Water Resources Management

Ali Arefinia, Omid Bozorg-Haddad, and Arturo A. Keller

Abstract The real world is full of random phenomena, complicating predictions of future changes in water resources, hence probabilistic distributions can assist in understanding their potential effects. Uncertainties could be unintended consequences of intentional changes in accidental phenomena which are part of decision making and policymaking in water resource management, and their disregard for decreasing the reliability and risk of water resource systems. Uncertainty analysis and sensitivity analysis are two processes that can be used to simulate unexpected changes and improve the usefulness of the results of modeling water resource systems. Also, because water resource systems are always associated with natural phenomena, they can be analyzed by the aid of the uncertainty and sensitivity analysis to examine different effects of the phenomena on these systems. In this chapter, the general components of water resources systems, and the concepts and terminology related to uncertainty in these systems are presented. furthermore, it has been proposed to introduce the uncertainty and necessity of its review in water resources systems, the basic principles of statistical methods and probability, and various methods. The analysis and evaluation of the failure to solve problems in modeling, simulation, and optimization of water resources systems have been addressed. In the end, due to the importance of the performance criteria of water resource systems in evaluating and analyzing these systems, it introduces common performance criteria in their performance evaluation.

A. Arefinia · O. Bozorg-Haddad (✉)
Department of Irrigation and Reclamation Engineering, Faculty of Agricultural Engineering and Technology, College of Agriculture and Natural Resources, University of Tehran, Karaj, Tehran 31587-77871, Iran
e-mail: OBHaddad@ut.ac.ir

A. Arefinia
e-mail: Ali.Arefinia@ut.ac.ir

A. A. Keller
Bren School of Environmental Science and Management, University of California, Santa Barbara, CA, USA
e-mail: keller@bren.ucsb.edu

O. Bozorg-Haddad (ed.), *Essential Tools for Water Resources Analysis, Planning, and Management*, Springer Water, https://doi.org/10.1007/978-981-33-4295-8_13

Keywords Uncertainty · Risk · Reliability · Performance measures · Water resources management

13.1 Introduction

Investigating uncertainty and other probabilistic and stochastic parameters is the first step in risk and reliability studies of water systems. Most mathematical models use deterministic strategies and methods to examine different processes and systems, while all real-world issues are uncertain. The nature of atmospheric processes such as evaporation, rainfall, and temperature may cause uncertainty. Also, the cities' future population, per capita water consumption, irrigation patterns, and water use priorities are not certain (Loucks et al. 1981). Loucks (1997) developed an index for evaluating the sustainability of water resources systems using the three criteria of reversibility, reliability, and system vulnerability. Jariaj and Vedula (2000) investigated the role of input uncertainty in optimal reservoir utilization in a linear programming model. Huang et al. (2009a) examined the uncertainty in the target functions and the constraints associated with the flow into a reservoir, costs, allocation constraints, and access to water resources. Stefano et al. (2004) designed a scenario support decision system that can manage and plan the allocation of water resources under climatic and hydraulic uncertainties. In their study, the fuzzy two-stage randomization method with interval parameters was used for designing the water resources management system of this region under uncertainty conditions. Stefano et al. (2005) examined the management of water resources under uncertainty using a support decision system. Webby et al. (2007) assessed the risk and damage caused by changes in Australia's Burley Lake due to the uncertainty of input and precipitation flows. Marie et al. (2009) assessed the uncertainty of climate change on water resource management and its effects in the Preobanka River Basin. Huang et al. (2009b) researched the application of planning techniques for the design of water resources systems under conditions of uncertainty. The research findings showed that this method could be an effective way to achieve sustainable development. Huang et al. (2008), in a fuzzy programming study, uses multistage intervals to manage water resources under uncertainty conditions. The results showed that their applied technology could be used for other resources and environmental management issues that include policy analysis and system programming under conditions of uncertainty. Zhuang et al. (2009), designed a water supply system in an assumed area under uncertainty conditions. They used optimization methods with conservative controller parameters or strong optimization to apply uncertainty conditions in the model. Yamen et al. (2012), using quality data, examined the reliability of water quality, flexibility, and vulnerability of water resources under uncertainty conditions. Chen et al. (2013), using a robust analysis approach, examined the risk of water allocation under uncertain conditions. Annika et al. (2015) evaluated cost and profit as well as an uncertainty assessment of water loss reduction methods. Their case study was the Guttenberg drinking water distribution system. Safavi et al. (2016) investigated and analyzed the planning and management of water resources under uncertainty conditions in the Zayandehrud river basin. Sarzaeim et al. (2017) estimated the precipitation, temperature, runoff, and environmental requirements and their uncertainty in climate change conditions. Sarzaeim et al. (2018) examined the uncertainty of hydroelectric energy production under various climate change scenarios. Zhuang et al. (2018) studied the effect of climate change on water resources under uncertainty conditions using simulation integration and optimization methods. Zhu et al. (2018) evaluated and analyzed the uncertainty of groundwater. Liu (2019), to optimize the irrigation planning structure, taking into account the effects of water under unpredictable conditions. Ezbakhe and Pérez-Foguet (2019), uncertainty Data were analyzed in water resources decision making. Their case study was in the assessment of water and wastewater supply in Spain.

13.2 Definitions and Terms

In this section, we will introduce important terms in the discussion of uncertainty:

- Management: a purposeful process in which material and human resources are efficiently and effectively implemented in the planning, organization, and integration of resources and tools, and the guidance and control of the system to achieve a specific organizational purpose or goals.
- Analysis: A comprehensive approach to system issues where it first divides the system into smaller components and then each component is examined, and ultimately the final summing up of the entire system takes place.
- Planning: Thinking and organizing various activities, including decision making about the time and place of the activities, visualizing the desirable situation of the system in the future, and creating the appropriate tools and methods for achieving this desirable situation. Program Introduce yourself
- Decision Making: A set of mental processes that can be used to select an option from among different options.
- Prediction: The process of identifying uncertain amounts of phenomena in the future.
- Modeling: Describing and simulating natural phenomena via mathematical relationships.
- Simulation: One of the methods for solving scheduling models in situations where it is not possible to use algebraic analysis methods or real-world testing.
- Optimization: In a variety of disciplines such as management, mathematics, economics, etc., the best choice is chosen from a set of optimized members.

Calibration: Adjustment of coefficient and parameter values in a model to best represent measured system conditions and behavior.

- Validation: A test to determine if the calibration values for the coefficients and parameters of the model are valid, using measured values that were not employed in the calibration process.
- Random variable: An indicator that the amount of its changes over time and place is not predictable.
- Non-deterministic events: The consequences or phenomena whose exact prediction is not possible are risky or uncertain events.
- Uncertainty analysis: A process in which sets of input values or model parameters are randomly selected using their probabilistic distribution functions, and these values are used to obtain distributions related to the output value of the model.
- Sensitivity analysis: A process that explores the change in model output based on the changes in model input values.
- Sensitivity coefficients: One of the sensitivity analyses that are derived from the derivative of a model output variable relative to a variable or input parameter.
- Fuzzy set: A set in which the membership of a number is not known.
- Membership function: A function that indicates the degree of membership, different members to a fuzzy set.
- Performance metrics for the water resources System: Indicators that serve to assess whether a water resources system meets particular goals, in response to management plans and policies.
- Water resources reliability: Degree to which a water resources system will perform at an acceptable level (Hashimoto et al. 1982).
- Reversibility of the water resource system: The probability that a system returns to the desired state after failure (Hashimoto et al. 1982). In the other definition, the maximum consecutive periods that a system had before returned to be considered as reversibility or resilience of water resource systems (Moy et al. 1986).
- Water resource vulnerability: Likelihood of severe system failure after a major change. Loucks et al. (2005) described the vulnerability as the mean of failures, Hashimoto et al. (1982), the mean of the maximum deficiencies over a succession of failures in the system, and Menzoda et al. (1997) the likelihood of a shortage in one or more periods of a certain limit.
- Sustainability Index: Summarizing system performance metrics in a general indicator to facilitate comparison and decision making between different options for managing and planning water resources (Lane et al. 2014).

13.3 Basics and Logic of the Subject

Risk events are phenomena that can not be accurately predicted. They are called risky events in the context of the existence of probabilities for different phenomena. In fact, if these probabilities are not quantitative or the prediction of events is not possible, the phenomena are considered an uncertain event. Uncertainty arises from the lack of proper assumptions, the variability of natural processes, and/or insufficient information. In issues related to water resource management, the uncertainty and risk associated with the systems must be determined for the change in inputs and the values of the parameters.

The effect of input uncertainties on the output uncertainty of a model is one of the most important issues in the discussion of uncertainty in the modeling of water resource systems, especially when the modeling conditions are different from the actual conditions of the system under investigation. There are uncertainties in the various stages of designing, locating and exploiting systems; this uncertainty, often due to our low understanding of the phenomena and the events involved in the problem, includes various types, for example hydraulic, hydrological, hydro-meteorological, geotechnical, structural, environmental, economic and social. In issues related to water resources, more attention is diverted to the hydraulic and hydrological aspects of uncertainty.

Using the uncertainty analysis, the properties of the output parameters of the model can be obtained as a function of the input parameters. Another advantage of uncertainty is that designers, by analyzing uncertainty, obtain the effect of each input parameter on the overall uncertainty of the output of the model, and thus identify the more effective parameters and pay more attention to their better evaluation by reducing output uncertainty. For example, uncertainty analysis can be used to forecast and flood control plans, predicting the water requirement for urban use in designing water distribution networks, predicting river flow in river engineering designs, determining the roughness of water channels in water transfer plans and determining the type of land use in designing irrigation networks. There is also uncertainty in the operation of the systems. The analysis of economic benefits from the utilization of the water resources system and the timing of water harvesting from reservoirs are examples of uncertainties in utilizing water resources systems.

Uncertainty sources are errors and approximations existing in measuring input data, parameter values, model structure, and algorithms for solving models. There are many ways to reduce these errors, but they can never be eliminated; therefore, in issues related to the management of water resource systems, there will be risky and uncertain conditions and they need to be updated with new data and knowledge.

How people deal with uncertain events, their decisions, and their reaction to these events is another important issue in the discussion of uncertainty. Although uncertainty analysis methods do not consider this kind of uncertainty, it is a very effective parameter in uncertainty analysis. As a result, design and utilization policies of water resources systems should be flexible enough to this kind of uncertainty, rather than the system it can perform its task with minimum failure and maximum efficiency.

Investigating uncertainty and other probabilistic and stochastic parameters is the first step in risk and reliability studies in water systems. Uncertainty is the occurrence of quantitative or qualitative changes, which are not controlled by humans. If the necessary decisions are made in designing and operating water supply systems, due to different types of uncertainty, the variables that affect them will increase the reliability of water supply systems. To estimate the quantitative reliability of water systems, probabilistic and statistical models should be used (Mays and Tung 1992); therefore, some basic rules of statistics and probabilities are discussed.

13.3.1 *Categorization of Uncertainty Types in Water Resources*

So far, different categories have been presented for types of uncertainties in water resources, such as:

1. Modeling uncertainty: information uncertainties, algorithmic approximations, errors, and quantitative approximations.
2. Uncertainties in water resources projects: hydrological uncertainties, hydraulic uncertainty, instrumental uncertainties, and economic uncertainties.
3. In a general categorization, uncertainty is divided into three groups of natural variability of phenomena (such as rainfall variability, evapotranspiration, water consumption, etc.), the uncertainty of knowledge (usually due to lack of adequate and complete knowledge of phenomena and models) and the uncertainty of decision making that results from impotence and ignorance of future decisions.

Figure 13.1 shows a schematic of the effect of input data uncertainty on model output. The output of the model varies over time or place due to the variability and uncertainty of each input data. Each of the lines indicates the time or spatial variations in model output due to input changes.

Where x, y, and z ranges represent the maximum range of output variations of the model due to input changes or uncertainty of inputs; therefore, for each input, the effect of the input uncertainty can be determined in the model output.

Figure 13.2 shows the upper and lower case curves showing the effect of the variability and uncertainty of the collection of different inputs on the output results of the model over time and place.

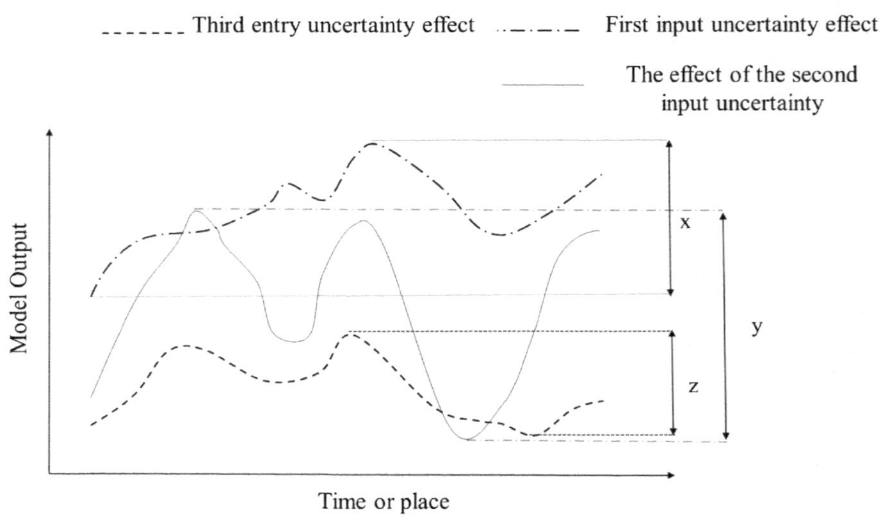

Fig. 13.1 Model output changes due to the uncertainty of different input data

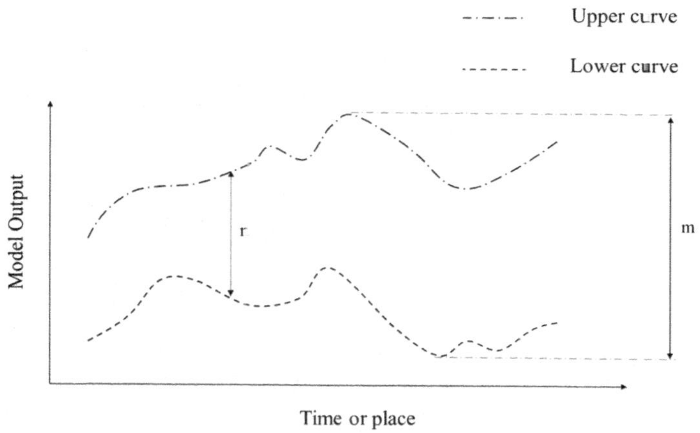

Fig. 13.2 Cover curves Model output changes due to the uncertainty of different input data

Where n = the distance between two curves at any time or location and indicates the domain of the model's variability due to the variability of the input parameters, and m = represents the maximum range of output variability due to the uncertainties of the inputs.

13.3.2 Basic Principles of Probabilities

In the theory of probabilities, the result of an observation is an experiment, and to all possible states of an experiment is a sample space. One of the scenarios in the sample space is the event. Events are examined as a set, so the proper functions are used as community, subscription, and complement. The occurrence of event *A* or occurrence *B* (community *A* and *B*) is shown as if the simultaneous occurrence of events *A* and *B* (subscriptions *A* and *B*) is represented either in a form *(A, B)*. In one set, if two occurrences of *A* and *B* do not have a common denominator, they are called two disjoint or incompatible sets that represent it in the form $(A, B) = \varphi$, and if the occurrence of occurrence *A* does not affect the occurrence of event *B*, these two events are independent. If the occurrence of event A is related to event *B*, then *A* is a conditional event that is represented as $A|B$; in this case, the complement of the event *A* represents *A*'. The probability of a quantitative criterion is the probability of an event occurring. The probability of occurrence of any occurrence, such as event *A*, can be calculated in two ways: (1) dependent or late probabilities based on observations of the occurrence of the incident, and (2) initial or prior probabilities based on experience and judgment. The basic theorems of probabilities are as follows (Mays and Tung 1992):

1. The probability of occurrence of any incident is negligible

2. There is a general theorem that is in shape $P_r(S) = 1$ And S are in that sample space.
3. To evaluate the probability of the association, two events A and B are referred to as Eq. 13.1:

$$P_r(A \cup B) = P_r(A) + P_r(B) - P_r(A \cap B) \tag{13.1}$$

If we generalize it for K events, we will have the following Eq. 13.2:

$$P_r\left(\bigcup_{i=1}^{K} A_i\right) = \sum_{i=1}^{K} P_r(A_i) - \sum_{i\neq}^{K} \sum_{j}^{K} P_r(A_i, A_j) + \sum_{i}^{K} \sum_{j}^{K} \sum_{l}^{K} P_r(A_i, A_j, A_l) - \cdots + (-1)^{K+1} P_r(A_1, A_2, \ldots, A_K) \tag{13.2}$$

If two incidents are incompatible, their likelihood of association is in the form $P_r(A \cup B) = P_r(A) + P_r(B)$ and if we generalize this for K events, then it is an Eq. 13.3:

$$P_r(A_1 \cup A_2 \cup \ldots \cup A_K) = P_r\left(\bigcup_{i=1}^{K} A_i\right) = \sum_{i=1}^{K} P_r(A_i) \tag{13.3}$$

If two events are independent, the probability of their subscription will be as follows, and the probability of the occurrence of the K event of an independent event is given by Eqs. 13.4 and 13.5:

$$P_r(A \cap B) = P_r(A, B) = P_r(A) \times P_r(B) \tag{13.4}$$

$$P_r(A_1 \cap A_2 \cap \ldots \cap A_K) = P_r\left(\bigcup_{i=1}^{K} A_i\right) = \prod_{i=1}^{K} P_r(A_i) \tag{13.5}$$

The probability of a conditional event is called a conditional probability. The conditional probability is calculated as Eq. (13.6):

$$P_r(A|B) = \frac{P_r(A, B)}{P_r(B)} \tag{13.6}$$

That $P_r(A|B)$ the probability of the occurrence of A provided that event B occurs.

Variables are usually three groups: (1) definitive variables that can be measured at any location and time without error; (2) probabilistic variables that in some cases are not known and probabilistic distributions must be used for their definition, and (3) stochastic variables which must be partly obtained by equations and mathematical relations and the other part by probabilities that these variables have uncertainty. To analyze the statistical properties of the performance of water systems, many efficient events can be defined based on the related probabilistic variables. In probability theory, the probability variable is a fundamental concept. The result of a random event determines the value of the probability variable and cannot determine the value of the variable before the results are obtained. For example, determining the amount of rainfall in a given month, or determining the number of times a water level in a reservoir has fallen below a certain level, are items that depend on future events and cannot be quantified before they occur. The former example mentioned above, is a contingent variable, and it can be used to express the values of a continuous probability set, such as real numbers, to the application, but the latter example is a discrete probability variable whose expression is to be from a set of discrepancies of values such as numbers. On a general contract basis, the probabilistic variables are represented in lowercase Latin letters with uppercase letters and special values that they can take (Loucks et al. 1981).

Statistics are used to specify a complete set of all possible values for the probability variable of the population. A subset of the population is a sample. In general, for describing the statistical properties of probabilistic variables, they use indices that include the following three groups (Mays and Tung 1992):

- Indicators that indicate the center's inclination.
- Indicators that indicate the dispersion of the center value.
- Indices that indicate the amount of distribution skewness.

Modeling of some of the probable or probabilistic phenomena is one of the tasks commonly used in water resources planning. To do this, a probability distribution function must be fitted to a set of values from the observed variable. These fitted distributions are used to obtain the possible values of the probabilistic variable. In probability distributions of the vertical axis, the probability of occurrence and the horizontal axis of the probability number is shown.

Various probabilistic distributions for the analysis of water systems are used. Probabilistic distributions, according to the type of probability variables can be divided into two groups of discrete distributions and continuous distributions. Binomial distribution and Poisson distribution are discrete distributions commonly used in water resource system reliability assay. Normal distributions, normal log, gamma, vibrato, and exponential are commonly used continuous distributions.

13.3.3 Uncertainty Analysis Methods

The quantities used in the analysis and design of the water systems are dependent on a large number of variables, some of which lead to uncertainty. If we are to examine uncertainty in real issues, then we must use the methods to quantify the effect of that uncertainty so that it can be used in relations. The following is a discussion of uncertainty testing methods.

13.3.3.1 Sensitivity Analysis Method

One of the easiest approaches for evaluating the effect of uncertainty on the performance of the system is the sensitivity analysis method. In this method, by changing the values of the parameters or non-deterministic variables (individually or together), the results of the changes in the model output can be quantified. The sensitivity analysis method is a method by which one can identify the parameters or variables whose model output is particularly sensitive to them. This method is easy, but it cannot be used to determine the effect of the incorrect or unrealistic values of the parameters on design and operation. Also, this method cannot be used to study all possible scenarios in discrete situations, especially in large-scale projects, and some combinations and states of uncertain variables may not be evaluated (Loucks et al. 1981).

13.3.3.2 Random Simulation Method (Sampling)

The other method used to investigate uncertainty is the sampling method, as in the case of Monte Carlo Simulation (MCS). This method is based on repetition, and thus, system behaviors are examined based on the effects of various factors changes (Bao and Mays 1990). The MCS method is a quantitative process for generating random numbers based on probabilistic distribution. In this method, firstly, uncertain variables that affect the performance of the network are identified. Then, considering these variables, and considering their natural variations, they consider a suitable probability distribution for each of them. In the next step, the distribution parameters for each variable are determined, and in each repetition, a random number series is generated for each of the variables is determined in the network. Finally, the network is simulated using these random numbers. This method is almost as sensitive as the method of analysis, with the use of this method, more combinations can be reviewed. In this method, because of the concept of probabilistic distributions for non-deterministic variables, a more appropriate estimation of uncertainty in the network is made. In spite of all of the above, for large issues such as a large number of computations and the time to run this method, the use of the MCS method is uncommon to investigate the uncertainty in these issues, but in a small case, it is possible to do a lot of simulations in a short time.

13.3.3.3 Fuzzy Theory

In a fuzzy set, since all the information about this set is described by its membership function. In fact, the membership function represents the fuzzy amount of a fuzzy set; in other words, the degree of membership of different members to a fuzzy set is shown using the numbers 0 and 1. In fuzzy logic optimization problems, in addition to the objective function, another function is defined that indicates this desirability function. The fuzzy theory usually applies to issues in which there are several non-homogeneous target functions. The use of fuzzy logic causes the problem to be out of rigidity, making decision making more logical and flexible, but it is not possible to easily select the best fuzzy function that this function should best capture the characteristics of their non-deterministic and supreme variables relative to each other. Give Also, in cases where the degree of influence of non-deterministic variables on the conditions of the problem is unknown, fuzzy logic cannot be used because in this method it is necessary to determine the extent and effectiveness of the indeterminate variables in the problem in determining its final outcome in order to determine the appropriate degree of desirability for them.

13.3.3.4 First-Order Analysis Method (FOAM)

The first-order analysis method (FOAM) also called the Delta method, is a relatively simple and useful method for investigating uncertainty. This method applies to many issues because of its simplicity. The FOAM uses uncertainty analysis in definitive modeling that includes non-deterministic parameters (parameters that are definitely not known). This method evaluates and characterizes the effect of uncertainty in the relations of a model, similar to the conditions used by non-deterministic parameters in the model (Mays and Tung 1992). In this method, first define the parameter we want to examine the uncertainty effect of the uncertain variables in it, as a function of these non-deterministic variables. Then its statistical parameters such as mean and variance are calculated. It is important to calculate these indices in the FOAM process. Finally, we calculate the coefficient of variation of the desired parameter, which allows us to examine the effect of the parameter in question on the uncertainty of the uncertain variables that affect it.

13.3.3.5 First-Order, Second-Moment Method (FOSM)

The design of water systems usually takes place in natural environments exposed to external stresses. The durability or resistance of a blue system means that the system has succeeded in its optimal performance without failing, under tensions and external loads. The resistance or durability of the systems, as well as the loads or stresses applied, are usually probable and, as a result, are uncertain because the resistance of the systems, as well as external stresses, are related to the natural world. By assessing the interaction between loading and resistance in a system, the degree of reliability of that system is evaluated. This assurance is that the resistance of the system is greater or equal to loading. In fact, the risk is the probability that the load will exceed the resistance of the system. Another method is to assess the reliability of using the FOSM method. According to this method, reliability is defined as a performance function for the system, and thus, reliability means that the function is greater than or equal to zero (Mays and Tung 1992).

13.3.3.6 First-Order Reliability Method (FORM)

The optimal design of water resource systems is a multi-objective and complex process that involves creating a balance between cost and system reliability. One of the ways to calculate system capacity reliability is provided by Xu and Goulter (1999). In this method, a probabilistic hydraulic model is performed based on the first-order reliability method (FORM), and in the case of non-deterministic variables, a reliability calculation is made. For example, in a water distribution network, reliability is expressed as the need for nodes at the minimum compressive height defined in the grid, provided that the needs and roughness of the pipes are non-deterministic, and therefore the reliability of the capacity is related to hydraulic failures. The uncertainty in the need for nodes and the roughness of the tubes causes uncertainty at the compressive height of the nodes. This relationship can be defined using a probabilistic hydraulic model. In this model, the values of non-deterministic parameters are replaced by their statistical information such as mean, variance, and probability distribution functions.

13.3.3.7 Chance-Constrained Models

Since the uncertainties in the variables affecting the design of water resources systems are very important, the researchers have tried to introduce these uncertainties into the system modeling by using various techniques and by extending the design optimization relationships for these systems. Therefore, non-deterministic variables in the modeling of water resources systems are considered as probabilistic variables with the known probabilistic distribution. In this method, we consider the basis for probabilistic relations in the problem and replace them with their definite state. The standard deviation of non-deterministic variables in these conditions is an intrinsic factor for examining the uncertainty in the model.

In the Chance-constrained models, first, they establish the definite relations equivalent to probabilistic hypotheses in the models for optimizing the design of the water resource system is established. Then the non-deterministic parameters into the modeling are entered. Finally, using explicit and uncertain relationships, these parameters are evaluated in the design and operation of the system. Since there are complex relationships in determining definite equivalents for probabilistic constraints (especially in non-deterministic conditions), and also because the uncertainty of probabilistic variables affects the level of reliability, this method cannot be used for large-scale problems. Also, in many cases, the probability variables of the CDF relationship are not known, and the problem is investigated for different probability distribution functions. For this reason, extensive statistical relationships are used, not mentioning the high amount of time needed.

13.3.3.8 Entropy Theory

A new method for examining and quantifying the surplus information of entropy theories of water systems, in which existing limited information is used. When information is obtained, uncertainties will be reduced, and the value of the information obtained will be equal to the amount of uncertainty. According to Shannon (1948), events that are more likely to occur give less information, and vice versa; therefore, an event with a low probability of occurrence gives more information. In fact, entropy theory is an indicator to lessen the lack of knowledge and knowledge about the characteristics of a system. Various types of entropy measurement indicators used to analyze information are as follows:

1. Boundary entropy, 2. entropy, 3. conditional entropy, and 4. entropy of information transmission.

In another division, each of the high entropies is divided into two-dimensional discrete entropy and continuous entropy.

In discrete entropy, the available information takes into account the interval of changes in the values of separation (disintegration) and the distribution tables of the variables, but in continuous entropy, it is assumed that the probability distribution of the variables follows the normal or logical distribution of variables. In a discrete state, after calculating the tables, the probability of occurrence in each state is calculated. Many quantitative and qualitative variables in water resource systems do not follow the probability distribution function of normal or logarithmic distributions. To correct this, the discrete entropy theory is applied.

In hydrological problems, this theory is usually used to evaluate the optimality of the information obtained from measurement at the quantitative or qualitative monitoring stations of water resources. If the cost of the monitoring is compared with the information it provides, the performance of the planned stations is determined economically. To do this, first, the problem is defined as an optimization problem to maximize the amount of information, so that the costs have the lowest value. Entropy theory does not depend on hypotheses such as the linearity or normality of non-deterministic variables, and Hence, it can be used in large areas with heterogeneous data as well as problems with multivariable properties in time and place; however, this method requires accurate and sufficient data in each region and should use a probabilistic distribution to determine the time variation of the data. In the case of normal and logarithmic multivariate distributions, the use of this method will provide good results.

13.3.4 Analysis of Efficiency and Reliability in Water Resources Systems

In each water resource system, determining the goals and options available is the basis on which water resource managers can develop appropriate quantitative models. To determine the benchmark for good performance, we can use very broad goals such as public health, individual and national security, economic development, happiness and well-being, and so on. The answer to the question of how to achieve a benchmark or objective usually leads the evaluation process to the benchmark or specific purpose of the system's performance and development tools for these criteria, which are management options available. A fundamental objective is a goal that the only possible answer to the question "why?" is: "Achieving this goal is what everybody wants."

There are several ways to summarize performance time-series data that may be derived from simulation analyzes. The original time series is placed in an unsatisfactory condition for a longer time than the inverse time series. However, its highest failure rate was less than the reversal time series. These properties can be controlled by reliability, reversibility, and vulnerability criteria. These three criteria are one of the most widely used performance analysis indicators for water resource systems, as presented by Loucks (1997).

13.3.4.1 Reliability

A more reliable system is not necessarily better than a less reliable system. Because of the reliability of how the rate of recovery and return to a satisfactory amount does not express the badness of an unsatisfactory amount. As shown in Fig. 13.3, the reliability of the original time series and the inverse time series is equal to 0.7.

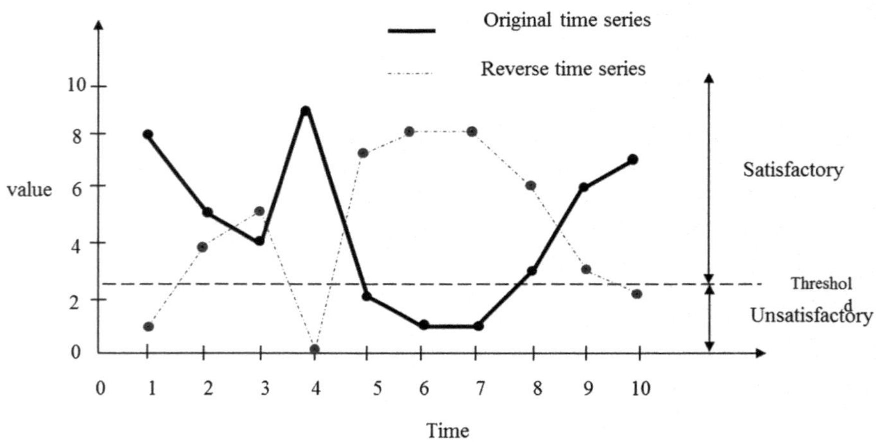

Fig. 13.3 Main time series and reverse system operation

13.3.4.2 Resiliency

The reversibility can be expressed as the probability that if a system is unsatisfactory, its next state will be satisfactory. This criterion is the probability of having an unsatisfactory amount in the t + 1 time series for an unsatisfactory amount in each t period. In fact, the reversibility is equal to the ratio of the number of times a satisfactory state is returned to after an unsatisfactory state to the total number of times the state is unsatisfactory.

According to Fig. 13.3, the reciprocity for the original time series is equal to 0.33 for the reverse-time series, which is 0.66.

13.3.4.3 Vulnerability

A vulnerability is a probability of the magnitude of the difference between the threshold value and the unsatisfactory amount of time series. In fact, the vulnerability is equal to the ratio of the highest positive deviation between the unsatisfactory states of the threshold boundary to the total positive deviations of the unsatisfactory states of the threshold boundary.

According to Fig. 13.3, the vulnerability for the original time series and the reverse-time series is equal to the following values:

$\frac{((2.5-2)+(2.5-1)+(2.5-1))}{3} = 1.17$	Original time series
$\frac{((2.5-1.2)+(2.5-0.2)+(2.5-2.2))}{3} = 1.3$	Reverse time series

13.4 Importance and Necessity of the Subject

Investigating uncertainty and other probabilistic and stochastic parameters is the first case in the discussion of risk and reliability in aqueous systems. Uncertainty is defined as the occurrence of quantitative or qualitative changes where the changes are not controlled by humans. If the necessary decisions are made in the design and operation of water supply systems due to the various types of uncertainty of the variables affecting them, the reliability of the water supply systems will increase.

In water resources systems, many parameters are used in models to determine hydraulic, economic, social impacts, etc. These parameters are uncertain. Also, the work environment of water resource planners is uncertain, and the information they use varies over time and place and is uncertain; therefore, it is necessary to consider different types of uncertainty in the design of water resources systems. As a result, considering the input parameters of the variables used in the models as well as the known uncertainties, it is possible to reduce the uncertainties of the outputs of the models so that they can be considered in management and planning. This leads to a more comprehensive view of the unpredictable conditions in water resources systems, and the decisions made are more flexible.

Given that many factors affecting the implementation of water resource systems are uncertain, so uncertainty exists at all stages, such as planning, development, exploitation, design, and so on. Measuring variables and parameters or processes in modeling may lead to uncertainty. Mathematical models use definitive methods to evaluate the performance of systems, while actual phenomena are usually uncertain. The discrepancy between the certainty of mathematical models and the uncertainty of real phenomena has a strong impact on the reliability of these models. Failure to pay attention to the uncertain parameters in the performance of the systems reduces the efficiency of the systems during operation, as well as the useful life of the systems, and their design, while their implementation and maintenance costs will increase.

Therefore, the study of the effect and intensity of uncertainty on the simulation of the performance of water resource systems and their consideration is very important. Otherwise, the reliability of the designs is reduced, and on the other hand, the profits considered by the planners and managers are not obtained, and the interest Investors will also suffer financial losses.

13.5 Method of Work and Process

Water resources issues are divided into three categories: modeling issues, simulation issues, and optimization problems, with random variables and uncertain parameters in each of them. Due to the importance and purpose of uncertainty analysis in water resource systems, various methods are used that their principles will vary in each of the above issues. Figure 13.4 shows the process of performing uncertainty analysis in a system.

Using the concept of probabilistic distributions for non-deterministic variables results in better estimation of uncertainty in the system. According to the amount of probability distribution parameters in each variable, it is possible to determine the uncertainty level of that variable and analyze its impacts on system performance.

13.6 Practical Example

Suppose we have a water allocation problem with three locations, according to Fig. 13.5. The profit from the assigned amount of water equal to N_i and the amount of water allocated in the period t is equal to M_{IT}. The amount of flow allocated in each period is equal to Q_{it}, which is equal to Q_t for each place of consumption (i).

We would like to use the Monte Carlo simulation to determine the probabilistic distribution of the profits of each location. Assume that the allocation policy for each location is by Figs. 13.6, 13.7, 13.8 and 13.9.

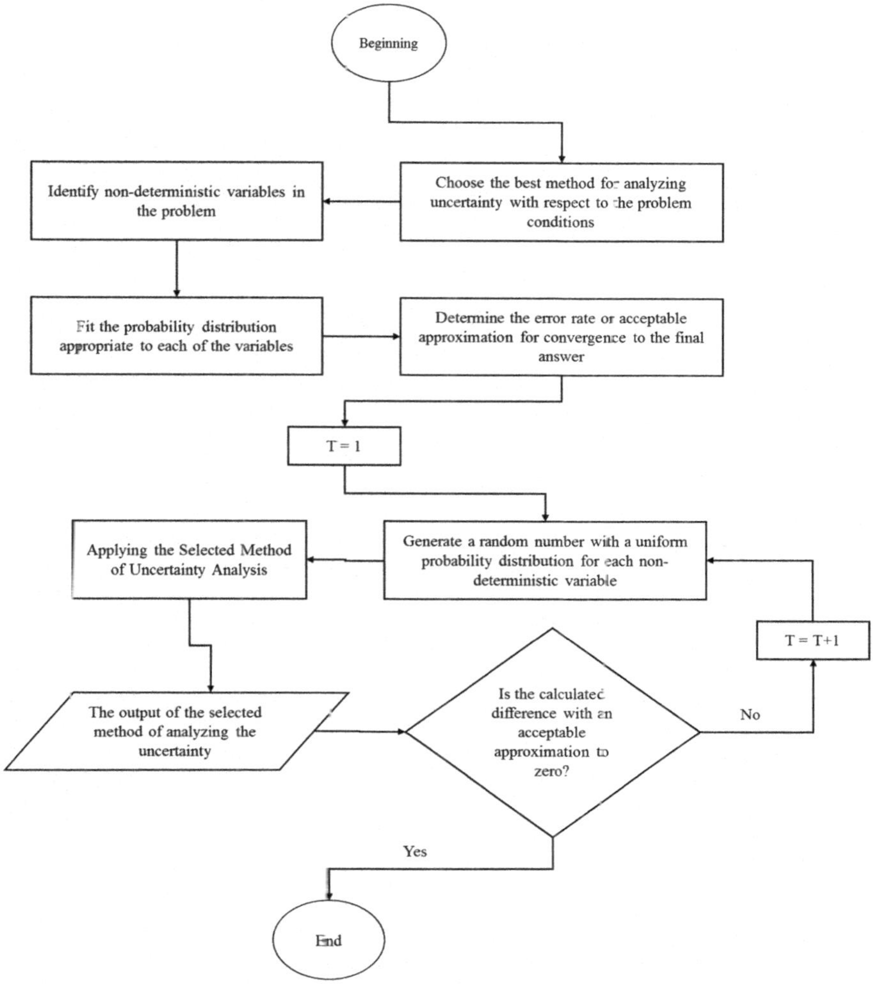

Fig. 13.4 The process of implementing the process of failure analysis

Each of the charts represents policies that prioritize each location. We now want to use the MCS method to obtain probabilistic distributions for allocations.

To do this, first, the available flows for each location (Q_t) should be plotted based on a probabilistic distribution for the period considered. Then, according to the obtained distribution, the flow amount is determined, and hence, the amount of allocations (M_{it}) is also calculated.

In the next step, concerning the allocation, it reduces its value from the flow of the river and, as a result, the next allocation is calculated according to the allocation policy outlined in Figs. 13.6, 13.7, 13.8 and 13.9.

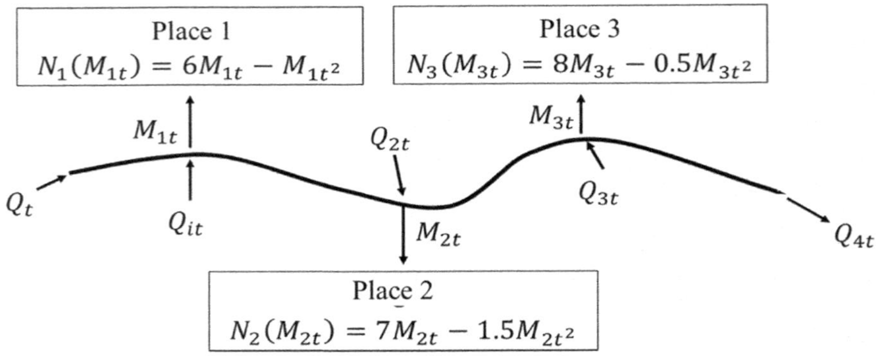

Fig. 13.5 Allocation of river flow at any time interval t

Fig. 13.6 Allocation policy for location 1

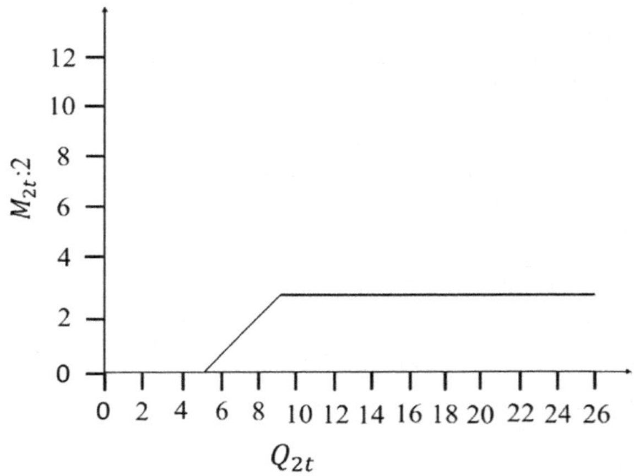

Fig. 13.7 Allocation policy for location 2

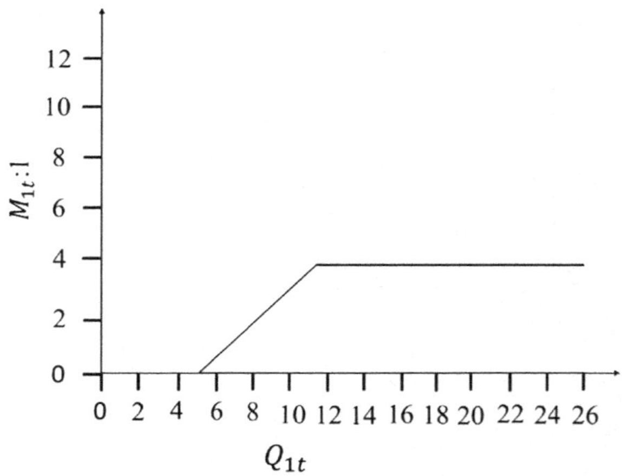

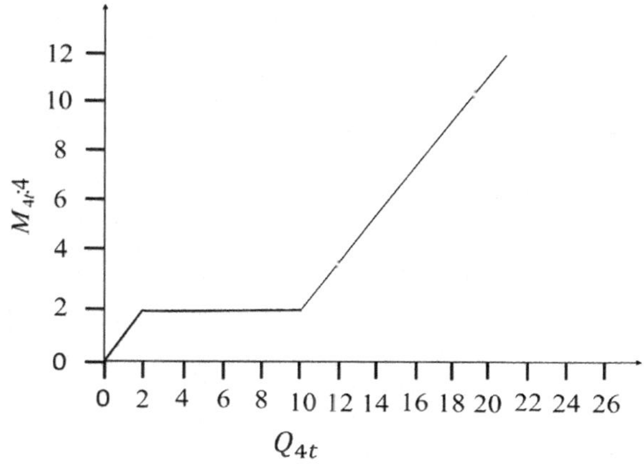

Fig. 13.8 Allocation policy for location 3

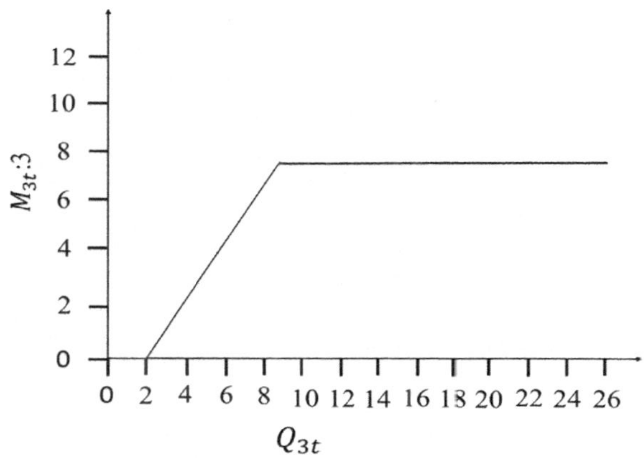

Fig. 13.9 Allocation policy for location 4

Finally, after repetitions, the probability distribution of allocations is determined for each location.

13.7 Summary

Water resource systems have dimensions with uncertain sources that cannot be predicted over time. Determining the sources of uncertainty in modeling and studying their random dimensions by probabilistic distributions is always part of the program of managers and policymakers of these systems. Failure to pay attention to the uncertainty of the systems increases risk and reduces reliability. Regardless of how much effort is being made to quantify and reduce the effect of uncertainties on model outputs, uncertainty will always remain. Managers and decision-makers who need to manage the risk of the systems and those who need to live with the risk of uncertainty require different reports and information on uncertainty. Elimination of these requirements should lead to much more informed decisions than when it comes down to neglecting uncertainty. This requires a cost that must be taken into account against the benefits of having uncertainty information.

Since the important part of the analysis of water resource systems is the simulation or modeling of the system, using the methods of this chapter to introduce variables uncertainty in the process of system analysis. Doing so, more realistic and more flexible solutions can be obtained from different issues in the study of water resource systems. One of the benefits of simulation methods is the involvement of engineering understanding and designer experience in making conceptual changes in design variables for trial and error testing. However, this method is very time consuming and may result in a non-optimal or non-economic response. In the process of optimizing the range of decision variables, the objective function, as well as the problem constraints, must be implicitly defined for the model. The optimization methods for decision type and making changes from one step to the next are more consistent and more consistent with simulation methods, which is why using mathematical logic during the solving process.

Considering the performance criteria of water resource systems in designing, operating, and analyzing these systems leads to better performing plans. The definition of performance criteria is based on the objectives of the plan under consideration. The selection and application of each performance criterion depend on the characteristics of the system. Due to the time and place variations of the variables and parameters affecting the utilization of water resource systems, the performance criteria of the system are also subject to variability and uncertainty, and the investigation of these scenarios can be useful in a more accurate analysis of water resource systems. If the decision is made taking into account the interactions between the economic, technical, and political objectives of each plan and its performance criteria over time or place, the utilization of more reliable water supply systems, longer system life and achieving the set goals for the system will be more convenient.

Acknowledgements The authors thank Iran's National Science Foundation (INSF) for the support of this research.

References

Bao Y, Mays LW (1990) Model for water distribution system reliability. J. Hydraulic Eng 116(9):1119–1137

Chen C, Huang GH, Li YP, Zhou Y (2013) A robust risk analysis method for water resources allocation under uncertainty. Stoch Env Res Risk Assess 27(3):713–723

Ezbakhe F, Pérez-Foguet A (2019) Embracing data uncertainty in water decision-making: an application to evaluate water supply and sewerage in Spain. Water Supply 19(3):778–788

Hashimoto T, Stedinger JR, Loucks DP (1982) Reliability, resiliency, and vulnerability criteria for water resources system performance evaluation. Water Resour Res 10(1):14–20

Huang YP, Li GH, Huang YF, Zhou HD (2009a) A multistage fuzzy-stochastic programming model for supporting sustainable water-resources allocation and management. Environ Model Softw 24:786–797
Huang GH, Li YP, Yang ZF, Nie SL (2008) IFMP: Interval-fuzzy multiprogramming for water resources management under uncertainty. Resour Conserv Recycling 52:800–812
Huang GH, Zhang YM, Zhang XD (2009b) Inexact De Novo programming for water resources systems planning. Eur J Oper Res 199:531–541
Lane BA, Sandoval-Solis S, Porse EC (2014) Environmental flows in a human-dominated system: integrated water management strategies for the Rio Grande/Bravo basin. River Res Appl. https://doi.org/10.1002/rra.2804
Liu X et al (2019) Optimization of planning structure in irrigated district considering water footprint under uncertainty. J Clean Prod 210:1270–1280
Loucks DP (1997) Quantifying trends in system sustainability. Hydrol Sci J 42(4):513–530
Loucks DP, Stedinger JR, Haith DA (1981) Water resource systems planning and analysis. Prentice-Hall
Mays LW, Tung YK (1992) Hydrosystems engineering and management. McGraw-Hill Book Co., Singapore
Moy WS, Cohon JL, Revelle CS (1986) A programming model for the analysis of reliability, resilience, and vulnerability of a water supply reservoir. J Water Resour Res 22(4):2135–2141
Safavi HR, Golmohammadi MH, Sandoval-Solis S (2016) Scenario analysis for integrated water resources planning and management under uncertainty in the Zayandehrud river basin. J Hydrol 539:625–639
Sarzaeim P, Bozorg-Haddad O, Fallah-Mehdipour E, Loáiciga HA (2017) Environmental water demand assessment under climate-change conditions. J Environ Monitor Assess 189(7). https://doi.org/10.1007/s10661-017-6067-3
Sarzaeim P, Bozorg-Haddad O, Zolghadr-Asli B, Fallah-Mehdipour E, Loáiciga HA (2018) Optimization of run-of-river hydropower plant design under climate change conditions. Water Resour Manag 32(12):3919–3934
Shannon CE (1948) A mathematical theory of communication. J. Bell System Tech 27:379–423
Stefano P, Giovanni MS, Paola Z (2004) A DSS for water resources management under uncertainty by scenario analysis. Environ Model Softw 20:1031–1042
Webby RB, Adamson PT, Boland PG, Howlett AV, Piantadosi J (2007) The Mekong applications of value at risk (VaR) and conditional value at risk (CVaR) simulation to the benefits, costs, and consequences of water resources development in a large river basin. Ecol Model 201:89–96
Xu C, Goulter IC (1999) Reliability-based optimal design of water distribution networks. J Water Resour Plann Manag (ASCE) 125(6):352–362
Zhuang XW et al (2018) Analyzing climate change impacts on water resources under uncertainty using an integrated simulation-optimization approach. J Hydrol 556:523–538

Correction to: Essential Tools for Water Resources Analysis, Planning, and Management

Omid Bozorg-Haddad

Correction to:
O. Bozorg-Haddad (ed.), *Essential Tools for Water Resources Analysis, Planning, and Management*, Springer Water, https://doi.org/10.1007/978-981-33-4295-8

The original version of the book was inadvertently published with incorrect affiliations for the author Barkha Chaplot in chapter 1 and for Prof. Hazi Mohammad Azamathulla in chapter 2. Correction to the Previously Published Version has been updated as in below:

In chapter 1

Author Barkha Chaplot's affiliation updated as in below:

Barkha Chaplot

Assistant Professor, Department of Geography, M.J.K. College Bettiah, Babasaheb Bhimrao Ambedkar Bihar University, Muzaffarpur, Bihar, India

e-mail: barkhachaplotjain@gmail.com

In Chapters 1 and 2

The author Omid Bozorg-Haddad's affiliation Designation "Distinguished Professor" inserted

In chapter 2

The author Hazi Mohammad Azamathulla's name and affiliation has been updated as in below:

Hazi Mohammad Azamathulla

Department of Civil and Environmental Engineering, University of the West Indies St. Augustine, St. Augustine, Trinidad and Tobago

e-mail: Hazi.Azamathulla@sta.uwi.edu

The updated versions of these chapters can be found at
https://doi.org/10.1007/978-981-33-4295-8_1
https://doi.org/10.1007/978-981-33-4295-8_2

O. Bozorg-Haddad (ed.), *Essential Tools for Water Resources Analysis, Planning, and Management*, Springer Water, https://doi.org/10.1007/978-981-33-4295-8_14

Printed by Printforce, the Netherlands